Springer Tracts in Modern Physics 86

Springer Tracts in Modern Physics

* Denotes a volume which contains a Classified Index starting from volume 36

B. H. Wiik   G. Wolf

Electron-Positron Interactions

With 238 Figures

Springer-Verlag Berlin Heidelberg GmbH 1979

Dr. Björn H. Wiik
Dr. Günter Wolf

DESY, Notkestrasse 85
D-2000 Hamburg 52, Fed. Rep. of Germany

Manuscripts for publication should be addressed to:

Gerhard Höhler
Institut für Theoretische Kernphysik der Universität Karlsruhe
Postfach 6380, D-7500 Karlsruhe 1, Fed. Rep. of Germany

*Proofs and all correspondence concerning papers in the process of publication
should be addressed to:*

Ernst A. Niekisch
Institut für Grenzflächenforschung und Vakuumphysik der Kernforschungsanlage
Jülich GmbH, Postfach 1913, D-5170 Jülich 1, Fed. Rep. of Germany

---

ISBN 978-3-662-15820-3   ISBN 978-3-540-34792-7 (eBook)
DOI 10.1007/978-3-540-34792-7

Library of Congress Cataloging in Publication Data. Wiik, Björn H. 1937-. Electron-positron interactions.
(Springer tracts in modern physics ; v. 86). Bibliography: p. Includes index. 1. Electron-positron interactions. 2. Hadrons. I. Wolf, Günter, 1937-joint author. II. Title. III. Series. QC1.S797 vol. 86 [QC794.8.H5]
530'.08s [539.7'54]   79-19120

2153/3130 — 5 4 3 2 1 0

Preface

The exploration of electron-positron interactions at high energies has become a powerful and fascinating tool of particle physics. There is a simple reason for this; the annihilation diagram, which yields an intermediate state with the quantum numbers of a photon, dominates the cross section at low energies and can be separated from higher-order diagrams at high energies. Furthermore the current is timelike and couples directly to the fundamental fermions, leptons and quarks, in a well defined way. These facts have led to a series of stunning discoveries which have established the existence of two new quarks and a new lepton.

In this volume we review the field of e^+e^- physics. We start out with a brief discussion of the properties of e^+e^- storage rings and some of the general problems associated with experiments at such machines. Tests of quantum electrodynamics, the classic domain of e^+e^- storage rings, are reviewed in Chap. 3. Chapters 4 - 11 are devoted to hadron production in e^+e^- annihilation at low and high energies. In particular the properties of the new hadrons containing a charmed quark are discussed in detail. The evidence for the new lepton τ and its properties are presented in Chap. 12. The next three chapters discuss the data on Υ and Υ' inclusive hadron production and jets. The last chapter presents the very first data from PETRA.

The volume is based on lectures given at various schools and conferences over the past few years. The emphasis is on the experimental results including discussions of the detectors.

We are grateful to Professor G. Höhler who stimulated this endeavour.

Hamburg, September 1979   B.H. Wiik G. Wolf

Contents

1. Introduction

Pioneering work on the feasibility of storing and colliding electrons was carried out at Stanford and Frascati from 1958 to 1962 /1/. The main physics motivation behind these first machines was the desire to test the validity of QED in a pure environment free from strong interaction effects. The first result of such tests - a measurement of the cross section for Møller scattering - was published /2/ in 1966 by a Princeton-Stanford collaboration. However, for reasons listed below the primary interest in e^+e^- collisions soon became hadron production via the one photon annihilation channel. Unlike the situation in hadron-hadron collisions one is here dealing with a well-defined system with the quantum numbers of a photon. Hadron production for c.m. energies up to 1 GeV was investigated with storage rings in Orsay /3/ and Novosibirsk /4/. It was found that the total cross section at these energies was dominated by the production of the known vector mesons ρ, ω, and ϕ. A wealth of information on various decay channels and on coupling constants resulted from this work. Since then the field of e^+e^- physics has developed at a breathtaking pace. It started at ADONE with the observation /5/ of a very large cross section for hadron production at c.m. energies above the resonance region. The cross section - extended to higher energies at CEA /6/ and SPEAR /7/ - had a surprisingly weak energy dependence. By analyzing final states, the SLAC-LBL collaboration found /8/ that the hadrons at high energies are produced preferentially back to back in two jets.

With the discovery in late 1974 of the J/ψ particle in proton-beryllium collisions at BNL /9/ and in e^+e^- annihilation at SPEAR /10/ a new chapter began. Within a few months further vector mesons /11/ of the same family were discovered and the total cross section was found to exhibit a step with a complex structure around 4 GeV. New related states observed in the radiative decays of the ψ' and the J/ψ were discovered at DORIS /12/ and SPEAR /13/.

The observations could be explained by introducing a new heavy quark Q and interpreting the new states as bound Q$\bar{\text{Q}}$ states /14/. The most natural choice for Q was to identify it with the charmed quark c, originally proposed to achieve symmetry between leptons and quarks and later to explain /15/ the absence of strangeness changing neutral currents.

Verification of the charm model required the observation of states with open charm, i.e., $c\bar{q}$ states resulting from combining an old quark q with the charmed quark. These states must decay weakly; they can be identified either by their semi-leptonic decays or by their narrowness. In 1976, measurements done at SPEAR provided the first evidence for the nonstrange charmed meson D with a mass of 1.87 GeV /16/. Simultaneously by investigating the energy region around 4 GeV, groups at DORIS /17/ observed events with hadrons and a single electron resulting from the semileptonic decay of a weakly decaying hadron of mass between 1.8 and 2.0 GeV. Weakly decaying new hadrons were also found in neutrino nuclear interactions at FNAL /18/. The analysis of inclusive η production in the 4 to 5 GeV region led DASP to the observation of the charged member F of the charmed meson family /19/.

In 1975 a group at SPEAR observed anomalous eμ events /20/ which they interpreted as resulting from pair production of a new heavy lepton. Subsequent experiments carried out for the last two years at DORIS and SPEAR on eμ and single lepton events gave conclusive evidence for the existence of a third charged lepton called τ with a mass of 1.78 GeV.

The recent upgrading of DORIS to 10 GeV total c.m. energy provided access to a new energy regime and allowed production of the τ in e^+e^- collisions previously discovered at FNAL in proton nucleus collisions /21/. The measurements by PLUTO /22/ and DASP2 /22/ revealed the τ as a narrow state, similar to J/ψ and ψ', and it seems almost certain by now that the τ is the onium of a fifth type of quark. In a heroic effort the DORIS machine group pushed the maximum energy beyond 10 GeV which permitted DASP2 /23/ and a collaboration from DESY-Hamburg-Heidelberg and Munich /23/ to observe the τ'. The comparison of the leptonic widths of τ and τ' showed that the fifth quark has charge 1/3 - and not 2/3.

Perhaps the most fascinating aspect of τ studies is the possibility for a thorough test of QCD. In QCD the direct hadronic decays of the τ (as well as those of the J/ψ) proceed via a three-gluon intermediate state. As a result the spatial configuration of final states from τ decay should be markedly different from the two-jet structure observed /8/ for nonresonant hadron production (SLAC-LBL). The data presented by the DORIS experiments show that the final states on and off the τ are indeed different in a manner consistent with the QCD predictions.

The present notes review the data on e^+e^- physics as they are available at the end of 1978 /24/.

2. Electron-Positron Storage Rings

Electron-positron annihilation in the GeV region has to be studied with storage rings where the full beam energy is available for particle production. For an accelerator with a stationary target only a fraction of the beam energy contributes to the c.m. energy $\sqrt{s}$, the remainder being lost in motion of the center of mass system.

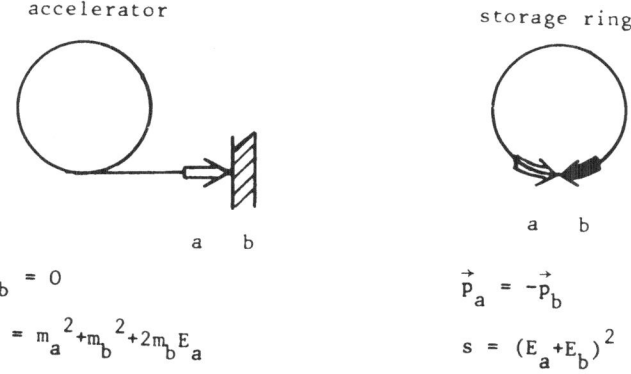

accelerator   storage ring

a b   a b

$\vec{P}_b = 0$ $\vec{P}_a = -\vec{P}_b$

$s = m_a^2 + m_b^2 + 2m_b E_a$ $s = (E_a + E_b)^2$

The c.m. energy grows with the square root of the beam energy; the growth rate is determined by the target mass and is minute for an electron target. It is instructive to compare storage ring and accelerator for e^+e^- and pp collisions.

	c.m. energy [GeV]	Beam energy [GeV] accelerator	Required storage ring energy [GeV]
pp	10	53	5
	100	5330	50
	1000	$5.3 \cdot 10^5$	500
e^+e^-	1	10^3	0.5
	10	10^5	5
	100	10^7	50

The price one pays for the favorable kinematics of a storage ring is low target density and therefore small counting rates.

2.1 Event Rate

a) Accelerator

For a fixed target of length L with nucleon density ρ, the rate N of events for a cross section σ is given by

nucleon density ρ

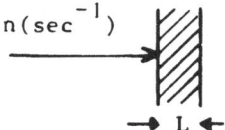

$$N(s^{-1}) = n \cdot \rho \cdot L \cdot \sigma. \tag{2.1}$$

Typically, $n = 10^{12}(s^{-1})$ and $\rho \cdot L = 10^{23} cm^{-2}$ yielding $N = 10^{35}(cm^{-2}s^{-1}) \cdot \sigma$ or 10^5 events/s for $\sigma = 1$ µb. Hence, accelerator experiments have typically a luminosity $\mathscr{L} \equiv N/\sigma$ of the order of $10^{35} cm^{-2}s^{-1}$.

b) Storage Ring

We consider head-on collisions between bunches of particles:

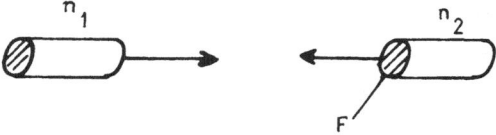

Defining n_1, n_2 number of particles per bunch
        F   beam cross section
        B   number of bunches per beam
        f   rotation frequency

the event rate for a cross section σ is given by

$$N = \frac{n_1 \cdot n_2 \cdot f}{F} \cdot B \cdot \sigma \equiv \mathscr{L} \cdot \sigma \ , \tag{2.2}$$

where $\mathscr{L}$ is the luminosity. A gaussian particle density distribution with rms radii σ_x, σ_y has a cross section of

$$F = 4\pi \sigma_x \sigma_y .$$

Expressing $\mathscr{L}$ in terms of the beam currents, $i = n \cdot e \cdot f \cdot B$ (e electron charge) one has

$$\mathscr{L} = \frac{1}{4\pi e^2} \ \frac{i_1 \cdot i_2}{\sigma_x \sigma_y \cdot f \cdot B} \tag{2.3}$$

For a typical set of $e^+ e^-$ storage ring parameters,

$$\begin{aligned}
B &= 1 \\
f &= 10^6 \text{s}^{-1} \\
i_i &= 50 \text{ mA} \cong 3.3 \cdot 10^{17} \text{ e/s} \\
\sigma_x &= 0.1 \text{ cm} \\
\sigma_y &= 0.01 \text{ cm}
\end{aligned}$$

the luminosity is $\mathscr{L} = 10^{31} \text{cm}^{-2} \text{s}^{-1}$ leading to N = 10 evts/s for $\sigma = 1$ µb. Hence, the luminosity of a storage ring is several orders of magnitude smaller than that of a typical accelerator experiment.

2.2 Luminosity

The maximum luminosity depends strongly on energy /25/. The main limiting factors are:

a) Synchrotron Radiation

The stored electrons emit photons. The radiated energy U_{syn} depends on the beam energy E and the bending radius ρ,

$$U_{syn}(\text{keV}) = 88 \ \frac{E^4 \ (\text{GeV})}{\rho (\text{m})} \ . \tag{2.4}$$

As an example, the energy loss in DORIS is 1 MeV per electron and turn at E = 3.5 GeV. The photon emission leads to an increase of the beam cross section,

$$\sigma_{x,y} \sim E. \tag{2.5}$$

b) Beam-Beam Interaction

It is clear that in order to maximize the luminosity and thereby the event rate one would like to store as much current as possible in a few bunches and to focus the beam to small transverse dimensions ($\sigma_x \sigma_y$ small) at the interaction point. Typical beam dimensions in the crossing point are σ_x = 0.1 cm (bend plane) and σ_y = 0.01 cm (normal to the bend plane). The stored current at low energies and a few bunches is in general limited by the beam-beam interaction. The particles in each bunch make betatron oscillations around the ideal orbit in vertical and horizontal directions. If the betatron frequency is a integral multiple (or submultiple) of the revolution frequency then the particles always traverse the magnets in the same orbit. In this case the same small imperfections in the magnetic field will be encountered on each revolution and its effect on the beam is amplified. In practice this will lead to a very rapid loss of the beam and one therefore selects a working point far away from such resonances. However, when the bunches cross, each bunch acts as an electromagnetic lens on the other - i.e., the crossing leads to a "smearing" of the working point by an amount

$$\Delta Q_x^- = \frac{r_e \, n_+ \, \beta_x^-}{\gamma \cdot 2\pi(\sigma_x + \sigma_y)\sigma_x} \quad \text{(equivalent for } \Delta Q_y\text{)} \tag{2.6}$$

Here r_e = the classical electron radius = 2.82 x 10^{-13}
      γ = E/m_e
      $\beta_{x,y}$ = the amplitude functions at the interaction point (~10 cm - 100 cm).

If this smearing is so large that any part of the bunch will have a betatron frequency close to an integer (or a submultiple) of the revolution frequency, the bunch will be lost. In practice stable operation can be achieved for $\Delta Q_{x,y} \leq 0.06$. This limits the number of particles per bunch; inserting (2.5) into (2.6), one obtains for the energy dependence of the maximum number of electrons per bunch

$$n \sim E^3 \tag{2.7}$$

and of the luminosity (with B fixed)

$$\mathscr{L} \sim \frac{n_+ \cdot n_-}{\sigma_x \cdot \sigma_y} \sim E^4. \tag{2.8}$$

c) Power Limitation

The total power radiated off by synchrotron radiation is given by

$$W = U_{syn}(n_+ + n_-) \, ef \cdot B. \tag{2.9}$$

It increases for a fixed B as $W \sim E^7$. Once the maximum power available is reached, the number of particles in the ring has to decrease as

$$n \sim E^{-4} \qquad (E > E_0), \tag{2.10}$$

which causes the luminosity to drop as

$$\mathscr{L} \sim E^{-10} \qquad (E > E_0). \tag{2.11}$$

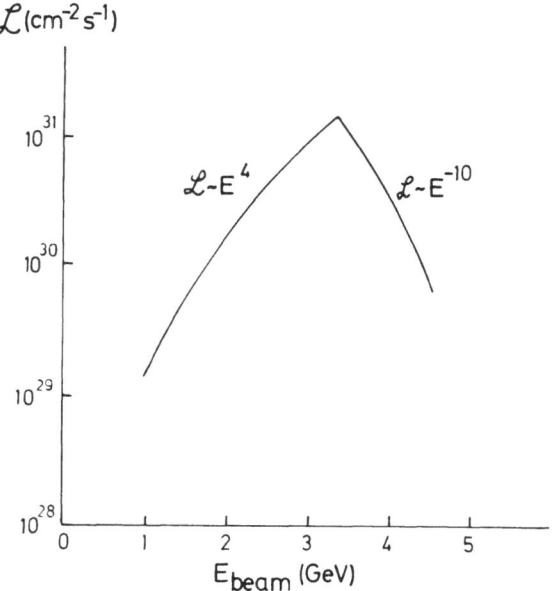

Fig. 2.1. A typical luminosity curve as a function of the beam energy

A typical luminosity curve is shown in Fig. 2.1. At energies below E_0 the luminosity can be increased beyond the B = 1 beam-beam limit by increasing the number of bunches. The maximum luminosity is obtained when at each energy, B is chosen such that the power limit is being reached. In this case $B \sim E^{-7}$ and $\mathscr{L} \sim E^{-3}$. A many-bunch operation requires a double ring structure since in a single ring the number of bunches is limited to half the number of interaction points in order to avoid bunch crossings outside these points.

2.3 Energy Spread

The energy spread σ_E of the beams is determined by synchrotron radiation. It depends on E and the ring radius ρ

$$\sigma_E \sim \frac{E^2}{\sqrt{\rho}} . \tag{2.12}$$

For DORIS and SPEAR

$$\sigma_E(\text{MeV}) \approx 0.9 \left(\frac{E}{2 \text{ GeV}}\right)^2 \tag{2.13}$$

2.4 Beam Polarization

The magnetic guide field together with the synchrotron radiation lead to a polarization of the beams with the positron (electron) spin parallel (antiparallel) to the

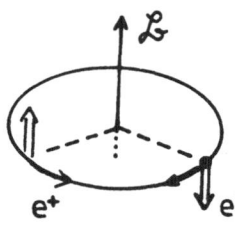

magnetic guide field /26/. The polarization arises because, e.g., in the case of an e^+ the transition from the $\vec{S}, \vec{B} = \downarrow\uparrow$ to the $\vec{S}, \vec{B} = \uparrow\uparrow$ state is energetic-ally favored over the inverse transition. Since the synchrotron radiation is strongly energy dependent, the buildup time for the polarization will depend strongly on E, as well. Defining ρ to be the bending radius in the magnets, and R the average radius of the ring, the degree of polarization as a function of time is given by

$$P(t) = P_0 \left(1 - e\right)^{-t/\tau}, \tag{2.14}$$

with

$$P_0 \quad = \frac{8\sqrt{3}}{15} = 0.92$$

and

$$\tau^{-1} \quad = \frac{5\sqrt{3}}{8} c \frac{r_e^2}{\alpha} \frac{\gamma^5}{\rho^2 R} , \tag{2.15}$$

where $\gamma = E/m_e$ and $\alpha = 1/137$. For DORIS and SPEAR one has

$$\tau \approx \frac{165 \ h}{E^5} \ , \quad E \ \text{in GeV} \tag{2.16}$$

$\tau = 5$ h for $E = 2$ GeV and $\tau = 15$ min for $E = 4$ GeV.

Since the average beam lifetime (defined here as the time over which the luminosity drops by a factor of e) is of the order of 3 - 5 hrs, studies with polarized beams at DORIS or SPEAR become practical for $E \gtrsim 3$ GeV.

The polarization is destroyed when the ring operates near a machine resonance. For this reason polarized beams will be obtained only for certain sets of machine parameters.

2.5 Examples of Existing Storage Rings

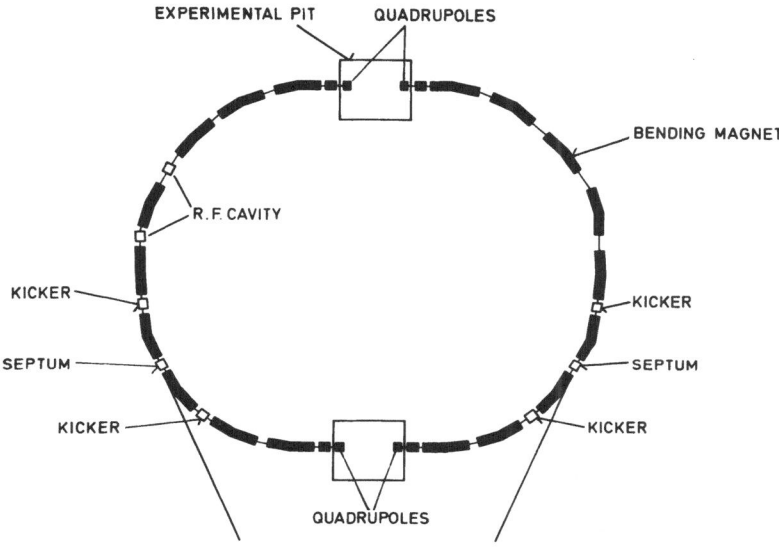

Fig. 2.2. Schematic layout of the storage ring SPEAR

Figure 2.2 shows the layout of the storage ring SPEAR which is a single ring structure with two interaction points.

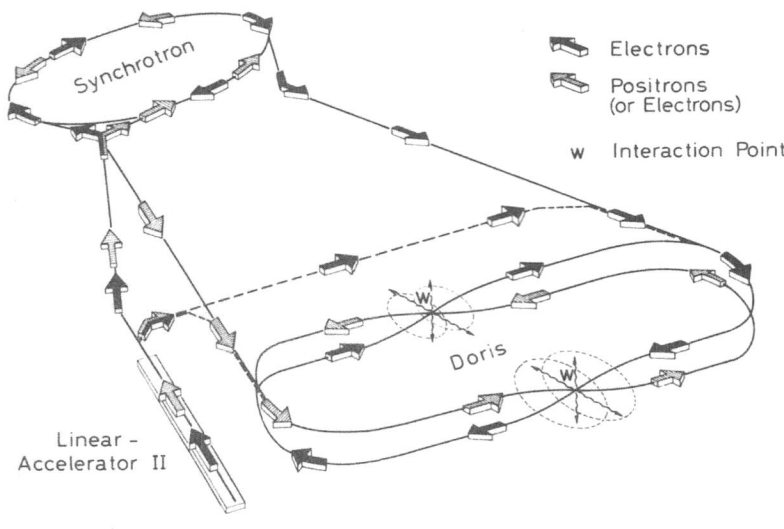

Fig. 2.3. The DESY-DORIS accelerator and storage ring complex

The accelerator and storage ring complex DESY-DORIS is shown in Figure 2.3. DORIS consists of two vertically separated rings which cross each other at two interaction points. The rings are filled in the following way: Electrons from a linear accelerator (LINAC) are injected into the synchrotron, accelerated to the desired energy, and transferred to DORIS: Positrons are produced at the halfway point of the LINAC by having the electron beam strike a target. The positrons with energies of a few MeV are collected by a magnetic horn, accelerated in the second half of the LINAC, and injected into the synchrotron. Each ring can be filled with up to 480 bunches leading to the following time structure of the beam current:

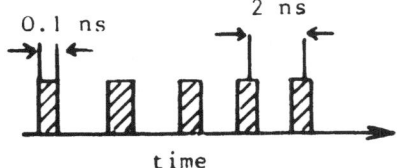

The double ring structure permits to store particle combinations other than (e^+, e^-), such as (e^-, e^-) or (e, p).

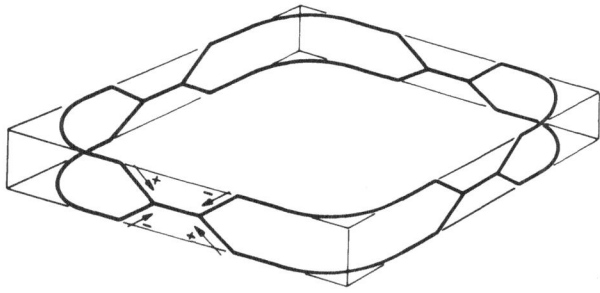

<u>Fig. 2.4.</u> Principle of operation of the new Orsay storage ring DCI

Figure 2.4 shows the ring arrangement for the new Orsay storage ring DCI. The effect of the beam-beam interaction is reduced by space charge compensation: In the interaction region an e^+ and e^- bunch travel together and collide with e^+ and e^- from the opposite direction. At present operations have started with one ring.

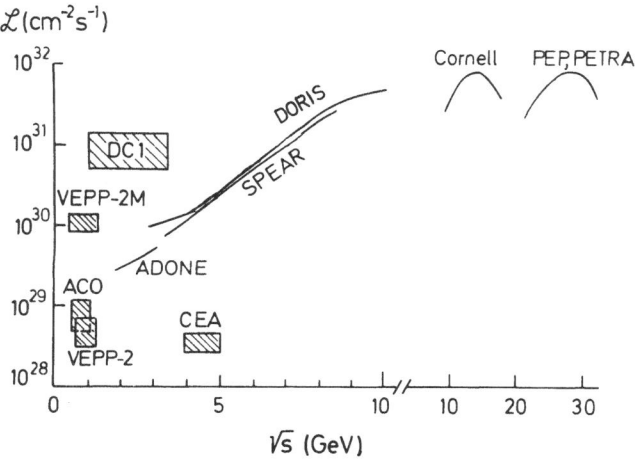

<u>Fig. 2.5.</u> Luminosity versus total c.m. energy for existing or planned e^+e^- storage rings

Table 2.1 and Figure 2.5 summarize the existing and planned ee storage rings.

Table 2.1. Existing and planned ee storage rings

Ring		Start of operation		Beam energy [GeV]
Ada	Frascati	1960	e^+e^-	0.25
Princeton-Stanford	Stanford	1962	e^-e^-	0.55
ACO	Orsay	1966	e^+e^-	0.2 - 0.55
VEPP-2	Novosibirsk	1966	e^+e^-	0.2 - 0.55
ADONE	Frascati	1969	e^+e^-	0.7 - 1.55
BYPASS	Cambridge (USA)	1971	e^+e^-	~1.5 - 3.5
SPEAR	Stanford	1972	e^+e^-	1.2 - 4.2
DORIS	Hamburg	1974	e^+e^-	~1 - 4.5
			e^-e^-,ep	
VEPP-2M	Novosibirsk	1975	e^+e^-	0.2 - 0.67
DCI	Orsay	1976	e^+e^-	0.5 - 1.7
VEPP-4	Novosibirsk	1978	e^+e^-	
PETRA	Hamburg	1978	e^+e^-	5 - 19
PEP	Stanford	1979	e^+e^-	5 - 19
CESR	Cornell	1979	e^+e^-	3 - 8
LEP	Europe	?	e^+e^-	22 - 130

2.6 Types of Storage Ring Detectors

The detectors which are commonly used at e^+e^- storage rings can be grouped into
three classes: nonmagnetic detectors, solenoid detectors, and magnetic spectrometers.

a) Nonmagnetic Detectors[1]

Figure 2.6 shows as an example the layout of the DESY-Heidelberg detector. It consists of a set of counter hodoscopes and drift chambers close to the beam pipe
followed by sodium-iodide and lead-glass shower counters. The inner detector is
surrounded by iron and drift chambers to detect muons. In order to improve the
spatial resolution for photons a vessel between the second and third cylindrical
drift chambers can be filled with mercury (thickness 2 rad lengths) for γ-ray conversion.

---

[1] See, e.g., the detector at ACO, the setups of the $\gamma\gamma$ and BB groups at ADONE, BOLD
at CEA, of the Stanford group at SPEAR, of the DESY-Heidelberg collaboration at
DORIS, at SPEAR, OLYA at VEPP-2M and the iron ball of the Colorado-Pennsylvania-
Wisconsin group.

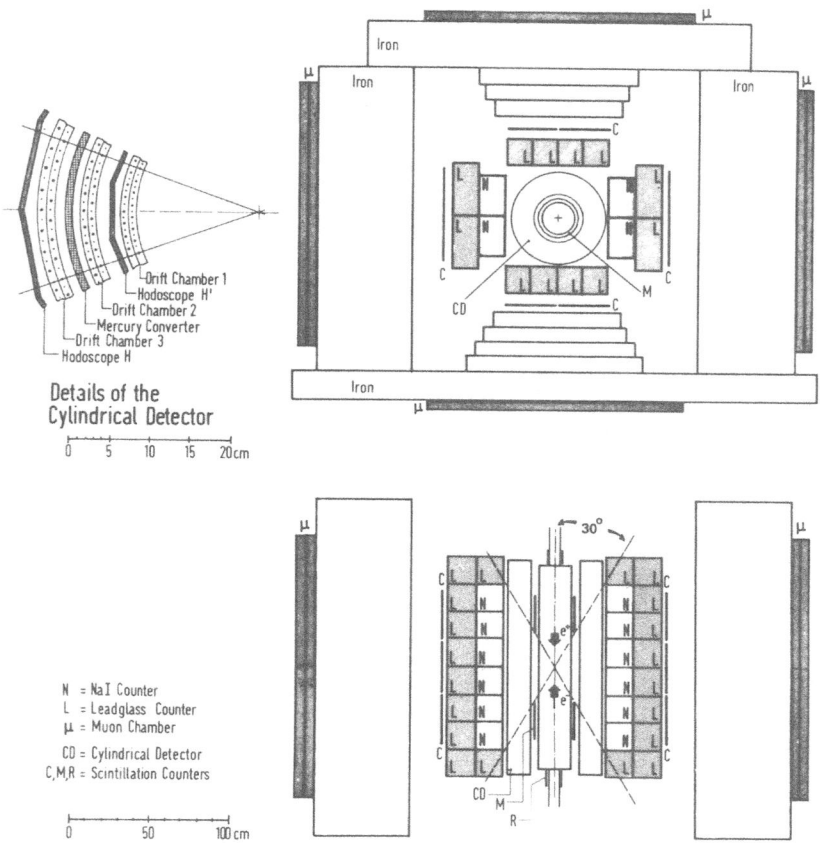

Details of the Cylindrical Detector

Drift Chamber 1
Hodoscope H'
Drift Chamber 2
Mercury Converter
Drift Chamber 3
Hodoscope H

0   5   10   15   20 cm

N = NaI Counter
L = Leadglass Counter
μ = Muon Chamber
CD = Cylindrical Detector
C,M,R = Scintillation Counters

0   50   100 cm

Fig. 2.6. The nonmagnetic detector of the DESY-Heidelberg collaboration

The nonmagnetic detector is capable of measuring directions of charged particles and photons. This allows a complete kinematic reconstruction of final states with four or less particles. It is in general possible to identify γ, e and μ and separate these particles from hadrons. Frequently, shower counters are used to identify electrons and to measure the energy of electrons and photons. The accuracy of the energy measurement depends on the type of shower counters used; typically

$$\sigma_E/E \approx 2\ \%\ E^{-1/4} \quad \text{for NaI}$$
$$6\ \%\ E^{-1/2} \quad \text{for lead glass}$$
$$15\ \%\ E^{-1/2} \quad \text{for lead-scintillator sandwich}$$

(E in GeV).

Magnetic detectors must not disturb the storage ring. That means that ∫Bdℓ taken over the interaction region has to be zero, either by compensating or by making B = 0 on the beam axis.

b) Solenoid Detectors

In Figure 2.7 a schematic view of the SLAC-LBL solenoid detector is given.[2] The solenoid is 3 m in diameter, 3 m long, and yields a homogeneous field of 4 kG parallel to the beam. The effect on the circulating beams is cancelled by two solenoidal compensation coils at either end of the detector. The solenoid is filled with trigger counters and cylindrical magnetostrictive chambers to detect charged particles and determine their momenta. Outside the coil shower counters detect photons and serve as electron identifiers. A detector of this type is suited for the analysis of few-body as well as multibody reactions. The momentum accuracy for charged particles is typically $\Delta p/p \approx 0.05 \cdot p$ (p in GeV/c). The solid angle covered is, e.g., 65 % of 4π for the SLAC-LBL setup, 86 % for the PLUTO detector at DORIS. Although this is fairly large, the lack of full coverage causes problems in the event analysis. Consider a final state with n particles produced isotropically. The probability for finding all n particles within the solid angle $\Delta\Omega$ is

$$P_n = \left(\frac{\Delta\Omega}{4\pi}\right)^n.$$

With $\Delta\Omega/4\pi$ = 65 % (86 %) the probability for detecting all particles is 8 % (40 %) for n = 6. Particle identification is possible for γ, e, and μ. In addition, by measuring the time of flight between the interaction point and the trigger counters lining the solenoid wall, the SLAC-LBL group is able to separate π/K/p up to ~0.7 GeV/c and K/p up to ~1.2 GeV/c.

---

[2] Other solenoid detectors are the Pluto detector at DORIS and MEA at ADONE.

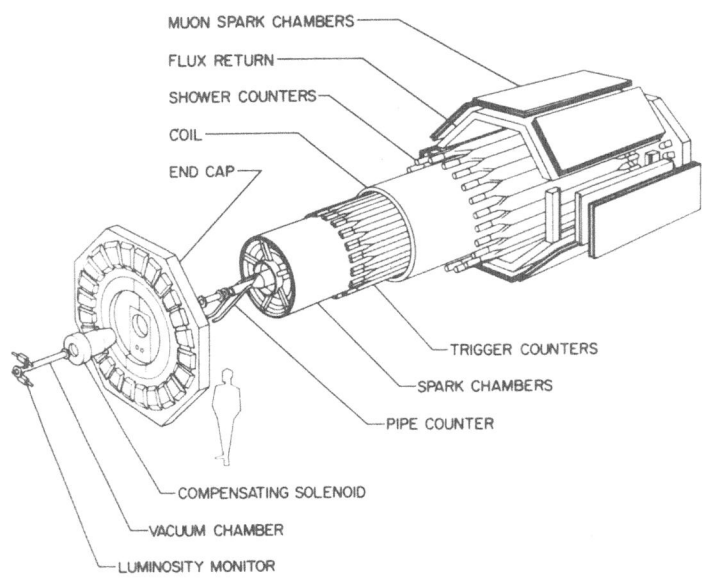

MUON SPARK CHAMBERS

FLUX RETURN

SHOWER COUNTERS

COIL

END CAP

TRIGGER COUNTERS

SPARK CHAMBERS

PIPE COUNTER

COMPENSATING SOLENOID

VACUUM CHAMBER

LUMINOSITY MONITOR

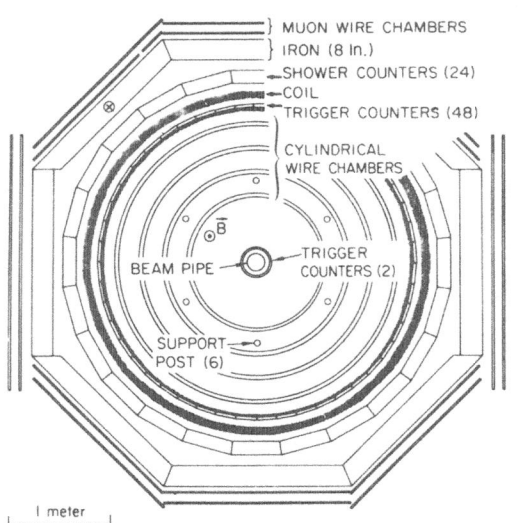

} MUON WIRE CHAMBERS

} IRON (8 In.)

SHOWER COUNTERS (24)

COIL

TRIGGER COUNTERS (48)

CYLINDRICAL
WIRE CHAMBERS

$\overline{B}$

BEAM PIPE

TRIGGER
COUNTERS (2)

SUPPORT
POST (6)

1 meter

Fig. 2.7. The solenoid detector of the SLAC-LBL collaboration

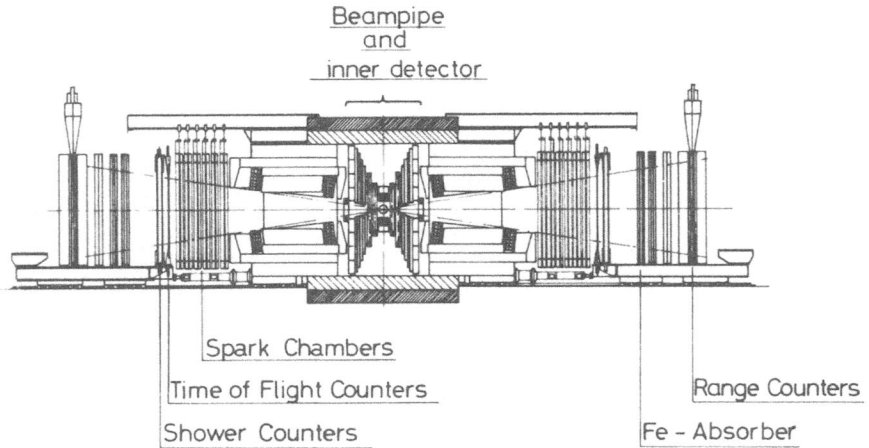

Beampipe
and
inner detector

Spark Chambers

Time of Flight Counters

Shower Counters

Range Counters

Fe – Absorber

(a)

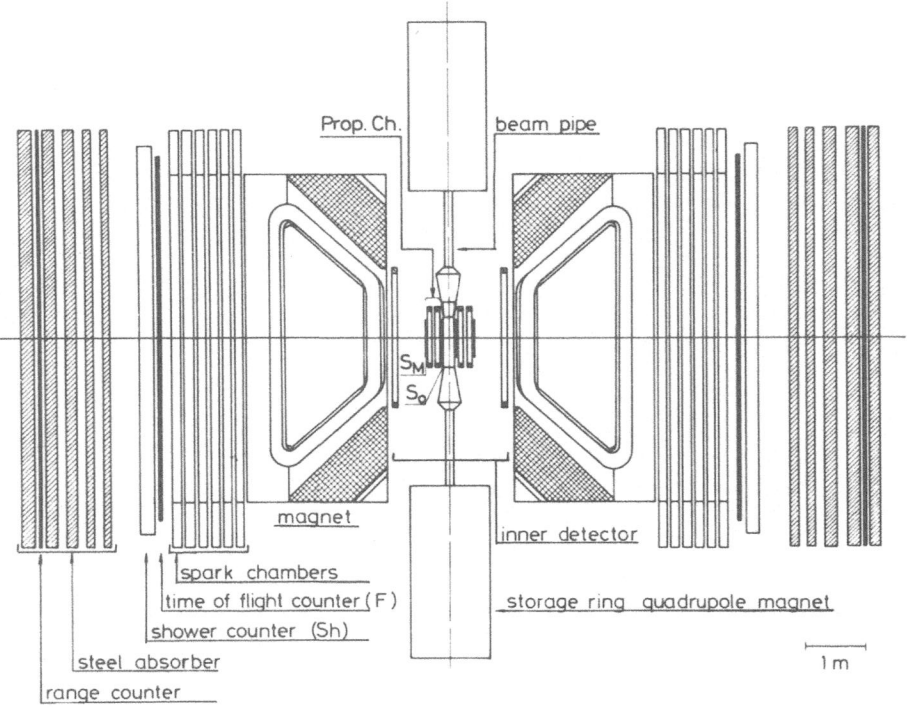

Prop. Ch.   beam pipe

S_M

S_0

magnet

inner detector

spark chambers

time of flight counter (F)

shower counter (Sh)

storage ring quadrupole magnet

steel absorber

range counter

1 m

(b)

Fig. 2.8. The double arm spectrometer DASP. (a) View from above, (b) view along the beam

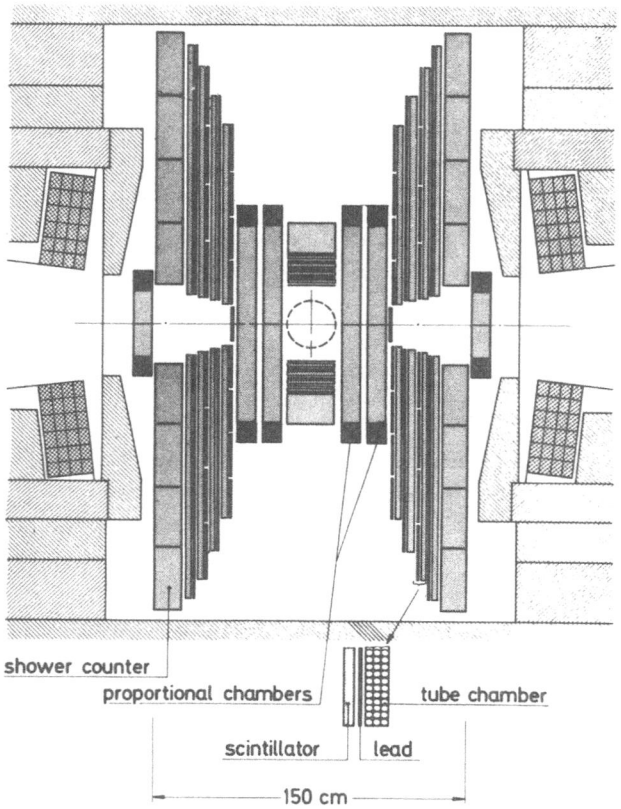

shower counter

proportional chambers

scintillator

tube chamber

lead

⊢————— 150 cm —————⊣

DASP — Inner Detector

Fig. 2.8. (c) The nonmagnetic detector located between the magnets as viewed along the beam

c) Magnetic Spectrometers

Figure 2.8 shows the double-arm spectrometer (DASP) of the Aachen-Hamburg-Tokyo collaboration used at DORIS.[3] It consists of a pair of identical magnetic spectrometers on either side of the interaction region and a nonmagnetic detector in

---

[3]A magnetic spectrometer is also used by the Maryland-Princeton-Pavia collaboration at SPEAR.

between. The magnets are of the H-type and have a maximum ∫Bdℓ of 18 kGm. The two magnets have opposite polarity leading to a vanishing field at the position of the beams. Particles passing through the magnet are detected by trigger counters close to the beam pipe, proportional and magnetostrictive chambers before and after the magnet, trigger- (time-of-flight) and shower counters. The shower counters are followed by 90 cm iron. Particles penetrating the iron are registered by a chamber after 40 cm of iron and a wall of scintillation (range) counters after 60 cm of iron.

Charged particles are identified as follows: muons by their ability to penetrate the iron; electrons by a large pulse height in one of the shower counters; pions, kaons, and protons are separated using the momentum and time-of-flight measurements. The time-of-flight resolution is 0.6 ns (FWHM), the path length ~5 m. This allows the separation of pions from kaons up to 1.5 GeV/c and π,K from protons up to 3 GeV/c. The geometrical solid angle accepted by the two spectrometers is 0.9 sr.

The nonmagnetic or inner detector located between the two magnets consists of proportional chambers, scintillation counters, proportional tube chambers, and shower counters (see Figure 2.8c). This part of the detector covers 70 % of 4π. It determines the directions of charged particles and photons to within ±2° and gives an energy measurement for photons and electrons. The efficiency for detecting photons is 50 % for E_γ = 50 MeV and above 90 % for E_γ > 100 MeV.

Detectors of the spectrometer type provide a precise momentum measurement for charged particles coupled with good particle identification in a limited part of the solid angle.

2.7 Experimental Background Problems

In this section we list some major background sources. For comparison we shall assume a luminosity of $\mathscr{L} = 10^{30} cm^{-2} s^{-1}$ and a cross section of 20 nb or 0.02 events/s for the annihilation process to be studied.

a) Beam-Gas Scattering

The gas pressure in the beam pipe is typically ~5·10^{-9} Torr (which is a factor of ~10^3 higher than in a pp ring due to the synchrotron radiation). The e^+,e^- beams interact with the gas nuclei mainly by quasi real γ-N scattering and produce background.

The beam-gas rate is given by

$$N_{beam-gas}^{(events/s)} = (\frac{I_+ + I_-}{e}) \rho \cdot L \ r \int \sigma_{\gamma N}(k) \frac{dk}{k} \qquad (2.17)$$

ρ   nucleon density

L   length of beam seen by the detector

$\sigma_{\gamma N}$ total cross section for γN scattering.

The factor r measures the relative flux of quasi real photons,

$$r \frac{dk}{k} = \frac{\alpha}{\pi} \ln \frac{0.1 \ GeV^2}{Q^2_{min}} \frac{dk}{k} \approx 7 \% \frac{dk}{k}.$$

For typical machine parameters one finds

$$N_{beam-gas} \approx 10^3 \ events/s,$$

which is roughly four or five orders of magnitude larger than the signal.

b) Cosmic Rays

The flux of cosmic ray particles incident on a horizontal surface is $2 \cdot 10^{-2} cm^{-2} s^{-1}$, or $2 \cdot 10^3/s$ for a detector covering 10 m^2 in area. These particles can in general trigger the detector and will produce unwanted background.

Both beam-gas background and cosmic ray events increase the event rate accepted by the data taking system by a factor of 10^2 - 10^3. They are, however, in general easily recognized at the analysis stage and then discarded.

c) Synchrotron Radiation

The number of photons with energy k > k_γ produced by synchrotron radiation per mA current, m path length, and s is roughly given by

$$N(k > k_\gamma) \approx 10^{16} \exp(-0.5 \frac{k_\gamma \rho}{E^3}) \tag{2.18}$$

ρ radius of bending magnet

In Figure 2.9 N(k > k$_\gamma$) is plotted for a current of 100 mA, electron path length 1 m, and bending radii of 12 m (DORIS bending radius in the ring) and 60 m (DORIS vertical bending radius before interaction point). The number of photons is seen to rise extremely rapidly from below 100 to 10^{18} photons/s. Photons above a certain minimum energy (~1 keV) will produce knock-on electrons in proportional chambers. Some shielding against synchrotron radiation is provided by the beam pipe. For the DASP experiment, which uses an aluminium pipe (wall thickness 2 mm) the critical γ energy and flux were found to be $k_\gamma \geq 30$ keV, $N_\gamma \sim 10^6 \gamma/s$. Therefore, synchrotron radiation produces a sharp limit for the maximum useful beam energy. This can be overcome by adding radiation shields in or around the beam pipe, which, on the other hand, are undesirable, e.g., for a clean photon detection.

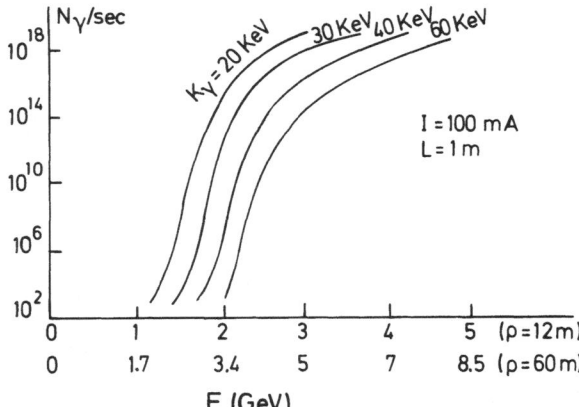

Fig. 2.9. The number of synchrotron photons radiated per second by electrons for a current of 100 mA, a path length of 1 m, and for bending radii of ρ = 12 m and 60 m, respectively

3. Purely Electromagnetic ee Interactions

3.1 General Remarks

We start with a brief discussion of the phenomenology of electromagnetic e^+e^- interactions. The electron will be assumed to have only electromagnetic inter-actions.

The lowest order processes are of order α^2, such as Bhabha scattering or μ-pair production.

The next higher order processes constitute radiative corrections to the first ones:

Although of order α^3, their contributions can be important. Since they depend strongly on the properties of the experimental setup (such as energy and angular resolution) the experimental results are usually presented with the contributions from radiative corrections removed.

A new class of processes is encountered in fourth order: the virtual photon clouds of the incident beams interact with each other. After integration over the photon

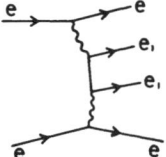

spectra the cross section is proportional to $\alpha^4 \ln^2(\frac{E}{m_e})$. Since $\ln \frac{E}{m_e} \sim 10$ for energies in the GeV region one power of α is essentially cancelled. Events of this type have been observed at Novosibirsk /27/ and Frascati /28/.

3.2 Tests of Quantum Electrodynamics

Purely electromagnetic processes can be calculated directly from QED /29/. A comparison between experiment and theory provides therefore a stringent test of the validity of QED. Such tests have been done for four different processes:

a) $e^- e^- \rightarrow e^- e^-$ (Møller Scattering)

Two diagrams with spacelike photons contribute:

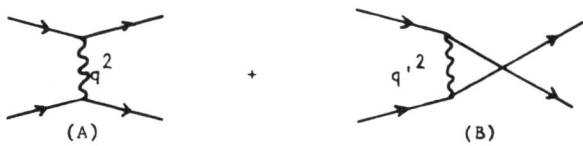

The QED differential cross section for producing an e^- at an angle Θ is given by

$$\frac{d\sigma}{d\Omega} = \frac{\alpha^2}{2s} \left\{ \frac{q'^4 + s^2}{q^4} + \frac{s^2}{q^2 q'^2} + \frac{q'^4 + s^2}{q'^4} \right\}$$

$$\quad\quad\quad\quad |A|^2 \quad\quad 2AB* \quad\quad |B|^2$$

with $q^2 = -s \cos^2 \Theta/2$ $q'^2 = -s \sin^2 \Theta/2$. $\quad\quad\quad\quad\quad$ (3.1)

The angular distribution is strongly peaked forward and backward.

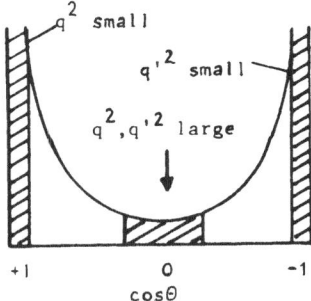

Deviations from the QED prediction could occur, e.g., due to strong interaction contributions at the vertices or in the photon propagator.

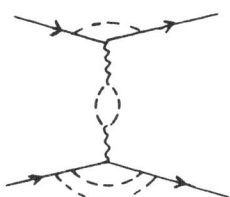

These contributions, which can only depend on the mass of the virtual photon, lead to the following modification of (3.1):

$$\frac{d\sigma}{d\Omega} = \frac{\alpha^2}{2s} \left\{ \frac{q'^4 + s^2}{q^4} \left| F(q^2) \right|^2 + \frac{s^2}{q^2 q'^2} \text{ Re } (F(q^2)F^*(q'^2)) \right.$$

$$\left. + \frac{q^2 + s^2}{q'^4} \left| F(q'^2) \right|^2 \right\} . \qquad (3.1')$$

It is customary to parametrize $F(q^2)$ by /30/

$$F(q^2) = 1 \mp \frac{q^2}{q^2 - \Gamma_{\pm}^2} , \qquad (3.2)$$

where $\Gamma_{\pm}$ is a cutoff parameter which characterizes the mass of the exchanged system, viz:

$$\frac{1}{q^2} \rightarrow \frac{1}{q^2} \mp \frac{1}{q^2 - \Gamma_{\pm}^2} .$$

Experimentally, the test on the validity of QED consists in a study of the shape of the angular distribution: the forward and backward peaks are determined at small four-momentum transfers for which QED is known to be correct. Deviations from QED will be most prominent in the central region ($\cos \Theta \approx 0$) corresponding to large values of q^2 and q'^2.

b) $e^+e^- \rightarrow e^+e^-$ (Bhabha Scattering)

Spacelike and timelike photon exchange contribute to Bhabha scattering.

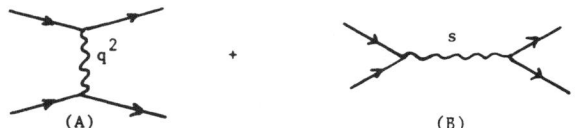

The differential cross section for finding an e^+ scattered at an angle Θ with respect to the e^+ beam is

$$\frac{d\sigma}{d\Omega} = \frac{\alpha^2}{2s} \left\{ \frac{q'^4 + s^2}{q^4} + \frac{2q'^4}{q^2 s} + \frac{q'^4 + q^4}{s^2} \right\} . \tag{3.3}$$

The angular distribution possesses a sharp forward peak due to the first diagram.

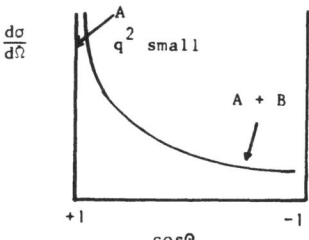

The modifications to the QED cross section are similar to those for Møller scattering. As before, the QED test consists in a measurement of the shape of the angular distribution. Bhabha scattering tests the photon propagator in the spacelike and timelike region.

c) $e^+e^- \rightarrow \mu^+\mu^-$ (μ Pair Production)

This is the simplest of all QED reactions. It proceeds via timelike photon exchange.

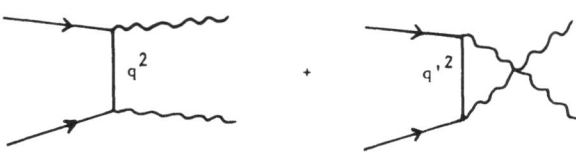

The differential cross section reads

$$\frac{d\sigma}{d\Omega} = \frac{\alpha^2}{4s} \beta_\mu \left\{ (1 + \cos^2 \Theta) + (1 - \beta_\mu^2) \sin^2 \Theta \right\} , \tag{3.4}$$

where $\beta_\mu = p_\mu/E_\mu$. For $p_\mu \approx E_\mu$

$$\frac{d\sigma}{d\Omega} = \frac{\alpha^2}{4s} (1 + \cos^2 \Theta). \tag{3.5}$$

The integrated cross section is given by

$$\sigma_{\mu\mu} \approx \frac{4\pi}{3} \frac{\alpha^2}{s} = \frac{\pi}{3} \frac{\alpha^2}{E^2} = \frac{21.9 \text{ nb}}{E^2} \qquad \text{(E in GeV)} \tag{3.6}$$

Possible deviations from QED will depend on s and can only be detected by measuring the absolute magnitude of the cross section. This can be done, e.g., by comparing μ-pair production to small angle Bhabha scattering.

d) $e^+e^- \rightarrow \gamma\gamma$ (Two-Photon Annihilation)

Two-photon annihilation proceeds via electron exchange.

The differential cross section is given by

$$\frac{d\sigma}{d\Omega} = \frac{\alpha^2}{2s} \left\{ \frac{q'^2}{q^2} + \frac{q^2}{q'^2} \right\} \tag{3.7}$$

where $q^2 = - s \cos^2\theta/2$, $q'^2 = - s \sin^2\theta/2$. The angular distribution is strongly peaked towards forward and backward angles.

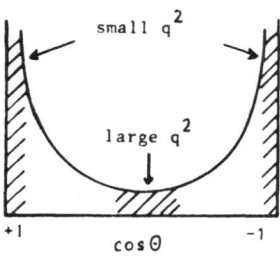

Deviations from the QED prediction will show up as a change in the shape of the angular distribution. A measurement of two-photon annihilation will not test the electron propagator as one might think at first. As was shown by KROLL /31/ the effect of corrections to the electron propagator cancel each other due to charge conservation. The electron propagator can be tested, however, in processes with closed electron loop diagrams (see below). The measurement of two-photon annihilation provides instead a test on the contribution from the "sea gull" term /31, 32/.

The parametrization of $F(q^2)$ used here is

$$F(q^2) = 1 \pm \frac{q^4}{\Lambda_\pm^4} .$$ (3.8)

The four QED processes discussed above have the same s dependence, $\sigma \sim s^{-1}$. In Figure 3.1 the differential cross sections are compared for a beam energy of 1 GeV. The largest cross section by far is for Bhabha scattering at small angles. For this reason and since QED is known to work at small momentum transfers, Bhabha scattering at small angles is generally used to measure the luminosity.

A comparison between the QED predictions and the measured cross sections for Bhabha scattering, μ pair production, and two-photon annihilation is shown in Figs. 3.2-4. The experimental results are seen to agree with theory. The cutoff parameters deduced from these measurements are summarized in Table 3.1. Note that the sensitivity on deviations from QED increases with energy. An accuracy of the cross section $\Delta\sigma$ results in a limit on the cutoff parameter proportional to

$$\Gamma_\pm \sim s^{1/2}(\tfrac{\Delta\sigma}{\sigma})^{-1/2} \qquad \Lambda_\pm \sim s^{1/2}(\tfrac{\Delta\sigma}{\sigma})^{-1/4} .$$ (3.9)

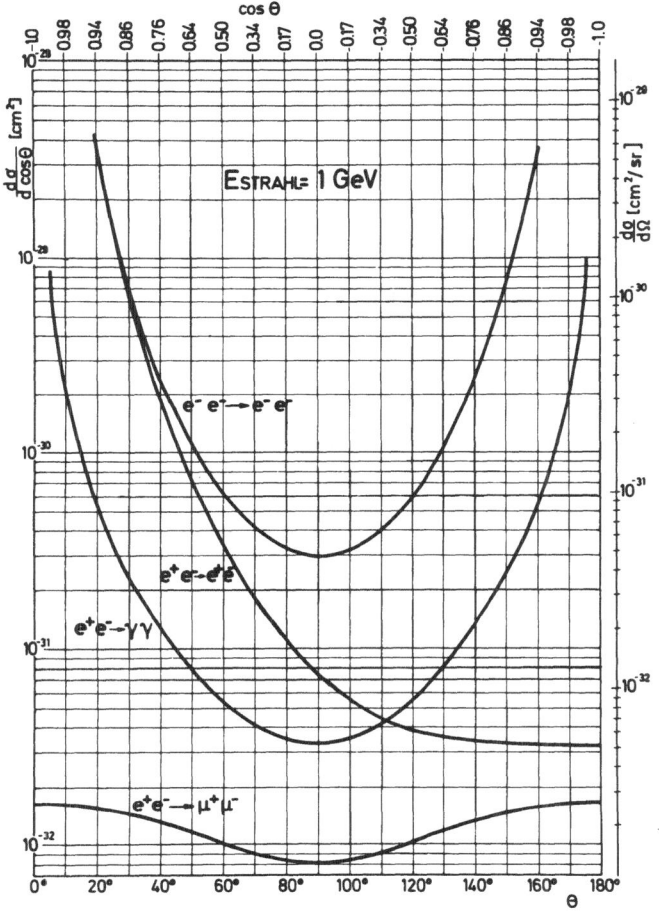

<u>Fig. 3.1.</u> Differential cross sections for $e^-e^- \rightarrow e^-e^-$, $e^+e^- \rightarrow e^+e^-$, $e^+e^- \rightarrow \gamma\gamma$, and $e^+e^- \rightarrow \mu^+\mu^-$ for a beam energy of 1 GeV

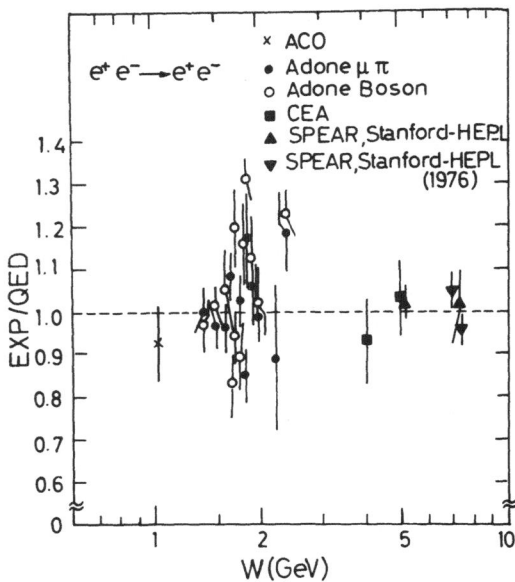

Fig. 3.2. $e^+e^- \to e^+e^-$: Ratio of the measured cross section to the QED prediction /32/

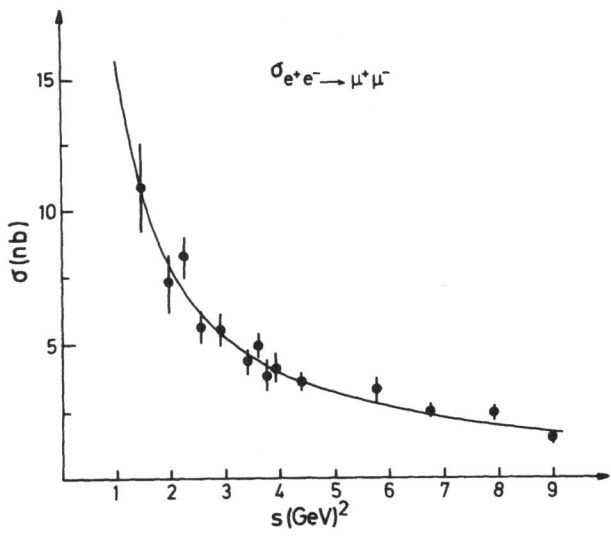

Fig. 3.3. $e^+e^- \to \mu^+\mu^-$: Ratio of the measured cross section to the QED prediction /33/

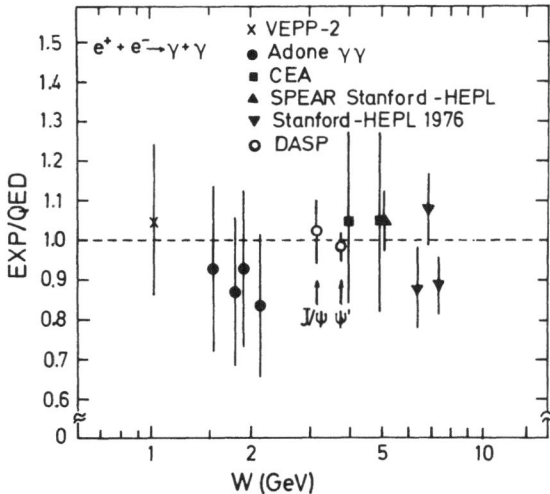

Fig. 3.4. $e^+e^- \to \gamma\gamma$: Ratio of the measured cross section to the QED prediction /12,32/

Table 3.1. Test of QED

Reaction	Experiment	Cutoff parameter
$e^-e^- \quad e^-e^-$	Princeton-Stanford /33/	> 2.5 GeV
$e^+e^- \quad e^+e^-$	SLAC-LBL /34/	Λ_+ > 15 GeV, Λ_- > 19 GeV
	Stanford-Pennsylvania /35/	Λ_+ > 16 GeV, Λ_- > 22.6 GeV
$e^+e^- \to \mu^+\mu^-$	Bologna /36/	Γ > 10 GeV
.	Stanford-Pennsylvania /35/	Λ_+ > 30 GeV, Λ_- > 30.5 GeV
$e^+e^- \to \gamma\gamma$	Stanford-Pennsylvania /35/	Λ_+ > 6.6 GeV, Λ_- > 7.9 GeV

In conclusion, according to e^+e^- experiments, which test mainly the photon propagator, QED is valid down to distances of the order of 10^{-15} cm.

The electron propagator has been tested by a DESY-Lund collaboration /37/ in Delbrück scattering and in photon splitting represented by the following diagrams:

No evidence for a breakdown of QED was observed.

4. Phenomenology of Hadron Production

4.1 General Remarks

The lowest order e^+e^- scattering processes leading to hadron production are:

One-photon annihilation

$\sim \alpha^2$

Radiative corrections to one-photon annihilation

$\sim \alpha^3$

Two-photon scattering

$\sim \alpha^4 \ln^2(\frac{E}{m_e})$

As in the purely electromagnetic case the two-photon scattering contribution is effectively of order α^3 after integration over the photon spectra. Furthermore, while the cross section for one-photon annihilation probably decreases as s^{-1} with energy (see below) the two-photon contribution increases $\sim \ln s$ and eventually will win over the one-photon contribution. The importance of the two-photon process was first recognized by LOW /38/ and by KESSLER and COWORKERS /38/. Evidence for this mechanism was found at Frascati, where three events of the type $e^+e^- \rightarrow e^+e^-\pi^+\pi^-$, $e^+e^-\pi^+\pi^-\pi^0$ /39/ were observed. The kinematics of the two-photon processes favor the emission of hadrons along the beam direction. At presently available energies ($s \leq 70$ GeV^2) and because present detectors do not cover angles close to the beams, two-photon contributions can be neglected. From now on we shall only consider the one-photon channel.

4.2 Properties of the One-Photon Channel

The hadron system produced by one-photon annihilation has the quantum numbers of the photon, $J^{PC} = 1^{--}$. For this reason the angular momentum L of the incident e^+ and e^- is limited to 0 and 2. Since $L = R \cdot E$, the radius of interaction will be of order $\frac{1}{E}$

leading to a total cross section for hadron production of

$$\sigma^{tot} \approx \alpha^2 \pi R^2 = \frac{\alpha^2 \pi}{E^2} \approx \frac{60 \text{ nb}}{E^2} \qquad \text{E in GeV.} \qquad (4.1)$$

From this simple-minded exercise we expect σ^{tot} to decrease with energy as s^{-1}.

Because $J^P = 1^-$ the most general angular distribution with respect to the beam direction for a particle h produced via $e^+e^- \to hX$ is of the form

$$\frac{d\sigma}{d\Omega} = a + b \cos^2\theta, \qquad (4.2)$$

which is radically different from typical angular distributions in hadron-hadron collisions.

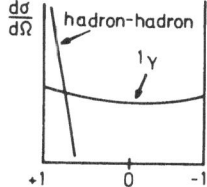

4.3 σ^{tot} in the Quark-Parton Model

The observation of scaling in deep inelastic electron-nucleon scattering led to the hypothesis that the photon-hadron interaction is basically a photon-quark interaction /40/. As a consequence we expect $e^+e^- \to$ hadrons to proceed via the formation of a quark-antiquark pair. We assume the quarks have spin 1/2 and are pointlike. Then the cross section for producing a free $q\bar{q}$ pair is the same as for producing a $\mu^+\mu^-$ pair [see (3.6)] except that the quark charge Q_i replaces the muon charge 1:

$$\sigma(e^+e^- \to q\bar{q}) = Q_i^2 \, \sigma_{\mu\mu} = Q_i^2 \frac{4\pi\alpha^2}{3s} \qquad (\beta_q = 1 \text{ is assumed}). \qquad (4.3)$$

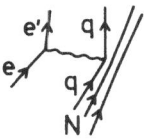

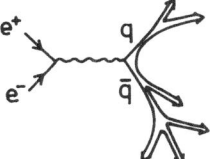

Assuming further that the produced $q\bar{q}$ pair turns into hadrons with probability one, the total hadron cross section is found by summing over all possible $q\bar{q}$ pairs

$$\sigma^{tot} = \sum_i Q_i^2 \, \sigma_{\mu\mu} \, . \qquad (4.4)$$

As we see from (4.4) the quark model predicts the total hadron cross section to decrease with energy as $\sigma^{tot} \sim s^{-1}$ and its magnitude to be of the order of the μ pair production. The value of Q_i^2 depends on the specific quark scheme:

quark model	Q_i^2
u, d, s	$\frac{4}{9} + \frac{1}{9} + \frac{1}{9} = \frac{2}{3}$
(u, d, s) X color	$3 \cdot \frac{2}{3} = 2$
(u, d, s, c) X color	$3 \cdot \frac{10}{9} = \frac{10}{3}$
Han-Nambu (u, d, s)	4
Han-Nambu (u, d, s, c)	6

Because of the expected behavior (4.4) it is customary to define the ratio $R \equiv \sigma^{tot}$ ($e^+e^- \to$ hadrons)$/\sigma_{\mu\mu}$.

The discussion above concerned the high energy behavior of σ^{tot}. At low energies we expect to produce nonstrange vector mesons which have the same quantum numbers as the photon.

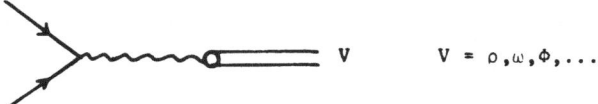

$$V \qquad V = \rho, \omega, \phi, \ldots$$

Let us try to guess the cross section from the quark model. In the quark model V is a $q\bar{q}$ pair. Therefore $e^+e^- \to V$ can be described by the following quark diagram:

for which we know the cross section

$$\sigma_{q\bar{q}} = Q_i^2 \, \sigma_{\mu\mu} \, .$$

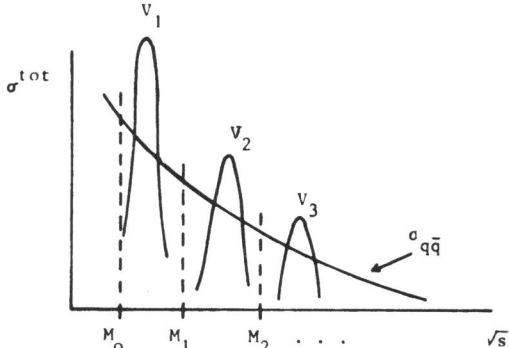

At low energies the cross section for $q\bar{q}$ production (which is an unphysical process since quarks do not become free) has to equal <u>on the average</u> the cross section for vector meson production (which is the physical process) /41/

$$\int \sigma_V dM = \int \sigma_{q\bar{q}} dM, \tag{4.5}$$

where $M = \sqrt{s}$. Defining σ_{peak} as the cross section at the mass M_V of the vector meson and Γ the vector meson width one finds

$$\sigma_{peak} \cdot \Gamma = \sigma_{q\bar{q}}(M_V) \int \frac{\sigma_{q\bar{q}}(M)}{\sigma_{q\bar{q}}(M_V)} dM$$

$$\sigma_{peak} = \sigma_{q\bar{q}}(M_V) \frac{M_V^2}{\Gamma} \int \frac{dM}{M^2}$$

$$= \frac{4\pi\alpha^2}{3} (\Sigma Q_i)^2 \frac{1}{\Gamma} \int \frac{dM}{M^2} \tag{4.6}$$

In order to fix the integration limits we shall assume a Veneziano-like spectrum for each physical $q\bar{q}$ system. Considering only the first vector meson of each series, M_0 will be the physical threshold for a given $q\bar{q}$ system. The upper integration limit M_1 we take halfway between the first and the second vector meson. This prescription gives the numerical results shown in Table 4.1.

Finally, the full cross section contribution is given by a Breit-Wigner distribution /29/.

$$\sigma_{e^+e^- \to V} = \sigma_{peak} \frac{M_V^2}{s} \frac{\Gamma^2/4}{(M_V - \sqrt{s})^2 + \Gamma^2/4} . \tag{4.7}$$

Table 4.1. Cross sections for $e^+e^- \to V$ calculated from (4.6)

V	$(\Sigma Q_1)^2$	M_0	M_1 [GeV]	σ_{peak} [nb]
$\rho^0 = \frac{1}{\sqrt{2}}(u\bar{u} - d\bar{d})$	$\frac{1}{2}$	$2m_\pi$	1.2	10^3
$\omega = \frac{1}{\sqrt{2}}(u\bar{u} + d\bar{d})$	$\frac{1}{18}$	$3m_\pi$	1.2	10^3
$\Phi = s\bar{s}$	$\frac{1}{9}$	$2m_k$	1.6	10^3
$J/\psi = c\bar{c}$	$\frac{4}{9}$	$m_{J/\psi}$	$m_{\psi'}$	$1.6 \cdot 10^5$

Figure 4.1 shows the expected low and high energy behavior of σ^{tot}.

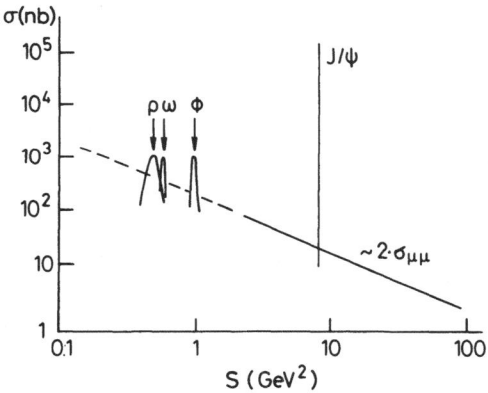

Fig. 4.1. The energy behavior of the total cross section for $e^+e^- \to$ hadrons expected from the quark model calculation (see text)

For completeness we calculate also the V ee partial width Γ_{ee} and the photon vector meson coupling e/γ_V. The vector meson cross section expressed in terms of Γ_{ee} reads

$$\sigma_{e^+e^- \to V} = \frac{3\pi}{s} \frac{\Gamma_{ee}\Gamma}{(M_V - \sqrt{s})^2 + \Gamma^2/4} \tag{4.8}$$

$$\Gamma_{ee} = \frac{M_V^2 \Gamma}{12\pi} \sigma_{peak}$$

$$= \frac{\alpha^2 (\Sigma Q_i)^2}{9} M_V^2 \int \frac{dM}{M^2} \ . \tag{4.9}$$

The partial width Γ_{ee} is related to the photon-vector meson coupling e/γ_V by

$$\Gamma_{ee} = \frac{\alpha^2}{\gamma_V^2/4\pi} \frac{M_V}{12} \tag{4.10}$$

and therefore

$$\frac{\gamma_V^2}{4\pi} = \left\{ \frac{4}{3} (\Sigma Q_i)^2 M_V \int \frac{dM}{M^2} \right\}^{-1} . \tag{4.11}$$

For higher vector mesons (V_n, $n \gg 1$) with a mass squared spacing of ΔM^2 (≈ 1 GeV^2) (4.10, 11) reduce to

$$\Gamma_{ee} \cong \frac{\alpha^2}{18} (\Sigma Q_i)^2 \frac{\Delta M^2}{M_V} \tag{4.10'}$$

and

$$\gamma_V^2/4\pi \cong \left\{ \frac{2}{3} (\Sigma Q_i)^2 \frac{\Delta M^2}{M_V^2} \right\}^{-1} . \tag{4.10''}$$

The strength of the γ - V coupling decreases within each vector meson series with increasing mass, $\gamma_V^{-2} \sim M_V^{-2}$.

5. The Total Cross Section

The total cross section for e^+e^- annihilation into hadrons, σ_{tot}, is computed from the number of events N observed, the integrated luminosity L, and the acceptance A for hadronic events. Furthermore a correction factor f has to be applied to account for radiative effects in the initial state. The result is

$$\sigma_{tot} = \frac{N}{L} \cdot \frac{f}{A} .$$

(5.1)

In general the accuracy of the measured σ_{tot} values is limited by systematical errors and not by statistics. The largest uncertainties are caused by the incomplete coverage of the solid angle. As an example, the so-called 4π detectors of the SLAC-LBL and PLUTO groups for charged particle triggering cover only 65 % and 86 % of 4π, respectively. Furthermore, the detection for neutral particles is even less complete. In the SLAC-LBL case the efficiency for detecting hadron events is ~35 % below 3 GeV rising to ~65 % above 7 GeV. Extrapolation to the full solid angle can be done by means of a Monte Carlo program that includes assumptions on the multiplicities of charged and neutral particles, the dynamics of the production process (e.g., jet formation), etc. The assumptions can be checked by comparing the Monte Carlo data with the measured results. The systematic uncertainty of A determined in this way is typically 5 - 15 %. The luminosity is determined from small angle (few degrees) and/or large angle Bhabha scattering, $e^+e^- \rightarrow e^+e^-$. The systematic errors mainly due to acceptance and radiative corrections are on the order of a few percent.

The radiative correction factor f primarily accounts for processes where the incoming electron (positron) has emitted a photon: as a result the total c.m. energy available for hadron production is reduced and the c.m. system is moving in the laboratory frame leading to a change in acceptance. In order to apply radiative corrections, a good knowledge of the total cross section and the behavior of the final states at lower energies is required. The uncertainty of this correction, in a region where σ_{tot} is smooth, is typically a few percent but may be considerably larger if σ_{tot} has a structure.

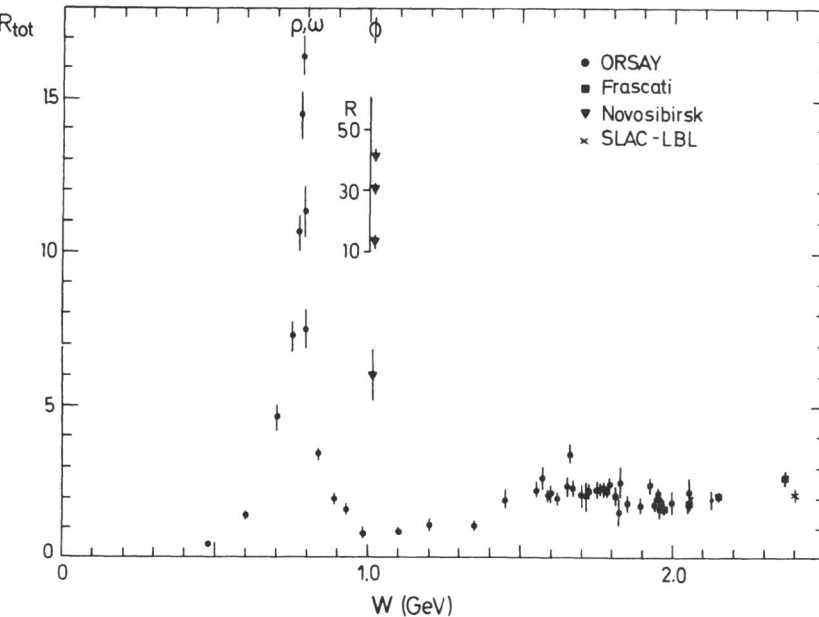

Fig. 5.1. The ratio R of the total hadron cross section to the μ pair cross section as measured by /42 - 45/

Figure 5.1 shows the ratio $R = \sigma_{tot}/\sigma_{\mu\mu}$ at low energies up to 2.4 GeV /42-45/.[1] The error bars include only statistical uncertainties. To these an overall systematic uncertainty of 10 - 15 % has to be added.

The energy region up to 1.1 GeV is dominated by the excitation of the vector mesons ρ, ω and Φ. Between 1.1 and 1.4 GeV R seems to be constant with a value near one. Between 1.4 and 1.5 GeV R rises from one to two and stays then almost constant up to 2.4 GeV.

Figure 5.2 shows the R measurements up to the highest energy reached so far, 17 GeV /42-49/. Note that the contribution from τ production has been removed in all experiments. The most spectacular structures seen are due to the excitation of J/ψ, ψ' and Τ, Τ'. Ignoring these and the structures seen at 4.0 to 4.5 GeV, qualitatively the R measurements above 1.5 GeV can be characterized by two steps: between 1.5 and 3.8 GeV R is approximately constant with a value of 2 to 2.5.

[1] The three data points between 1.1 and 1.4 GeV were obtained by adding the cross section data for the final states $\pi^+\pi^-$ (Novosibirsk), $\pi^+\pi^-\pi^0$ (Orsay), $\pi^+\pi^+\pi^-\pi^-$ and $\pi^+\pi^-\pi^0\pi^0$ (Frascati, Novosibirsk, Orsay)

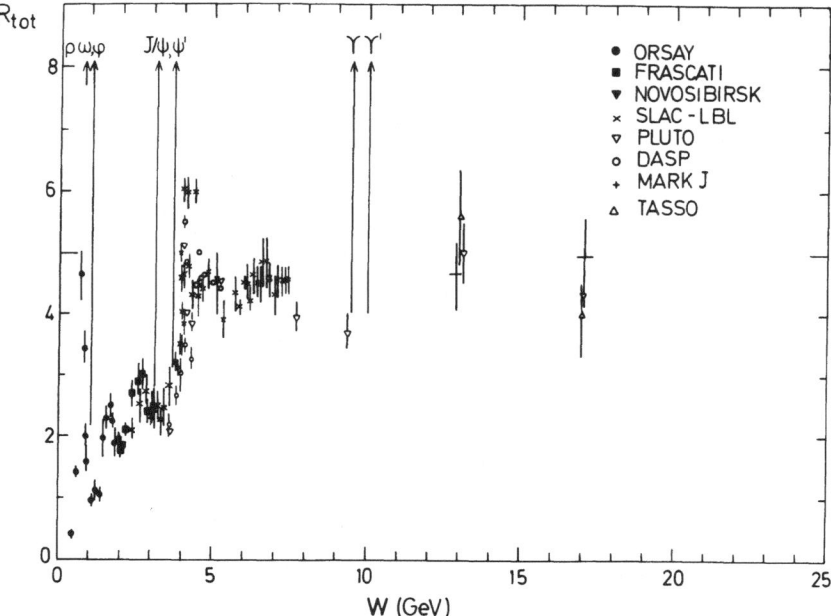

Fig. 5.2. The ratio of the total hadron cross section to the μ-pair cross section as measured by /42-49/

Near charm threshold (at 4 GeV) R rises sharply to reach a new level of 4 to 4.5, which persists up to 17 GeV. We expect a small step above the τ, τ' resonances. The data are insufficient to allow any conclusion on this point.

The behavior of R is in striking semiquantitative agreement with the simple quark model (see Figs. 5.3 and 5.4) which predicts

$$R(u,d) \quad\quad = 5/3 \quad\quad \text{below } s\bar{s} \text{ threshold}$$
$$R(u,d,s) \quad\quad = 2 \quad\quad \text{above } s\bar{s} \text{ and below } c\bar{c} \text{ threshold}$$
$$R(u,d,s,c) \quad = 3\ 1/3 \quad \text{above } c\bar{c} \text{ and below } b\bar{b} \text{ threshold and}$$
$$R(u,d,s,c,b) = 3\ 2/3 \quad \text{above } b\bar{b} \text{ threshold.}$$

A sixth quark with charge 2/3 would raise R to 5.

The deviations of the measured R values from the quark model predictions which are on the order of 0.5 to 1.5 units and which are most prominent just above a quark threshold may be explainable in terms of instanton effects and (perturbative) gluon corrections. The effect of the gluon corrections is to increase R by a factor $(1 + \alpha_s(s)/\pi)$:

$$R = \sum_i Q_i^2 \left(1 + \alpha_s(s)/\pi\right), \tag{5.2}$$

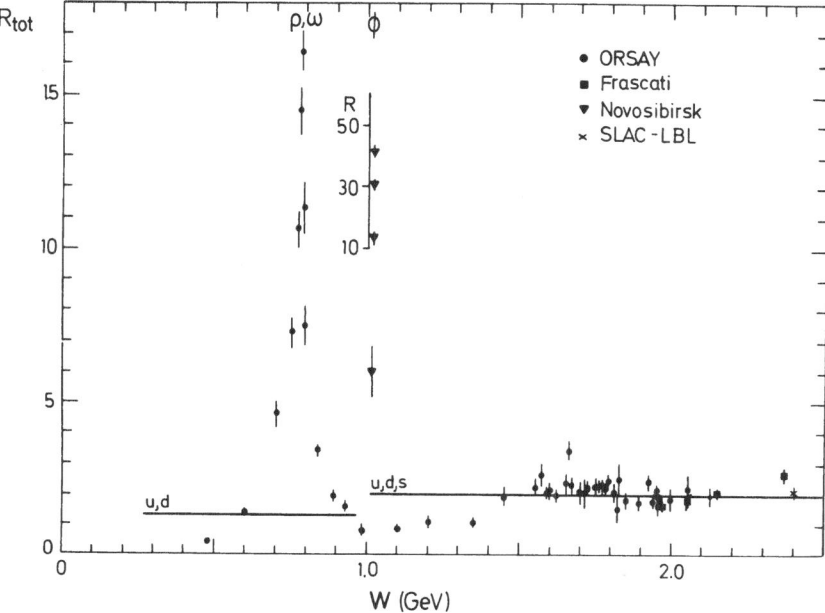

Fig. 5.3. The ratio R and the prediction of the simple quark model

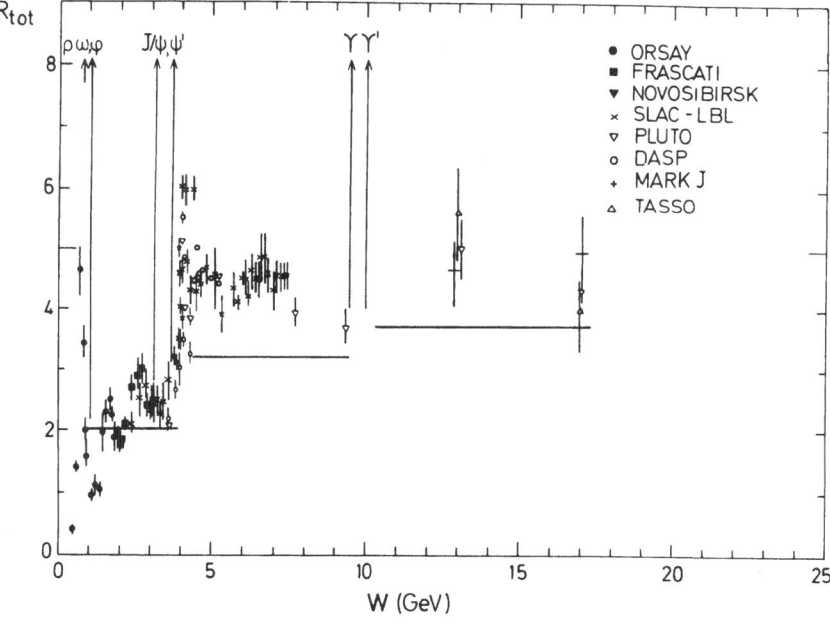

Fig. 5.4. The ratio R and the prediction of the simple quark model

where $\alpha_s = g_s^2/4\pi$ measures the gluon-quark coupling strength, defined in analogy to the fine structure constant, $\alpha = e^2/4\pi$. The gluon-quark coupling strength is energy dependent in asymptotically free theories (running coupling constant)

$$\alpha_s(s) = \frac{12\pi}{(33 - 2N_f) \ln(s/\Lambda^2)}, \qquad (5.3)$$

where N_f is the number of quark flavors excited (e.g., $N_f = 4$ for u,d,s,c) and Λ is a constant. Neutrino experiments yield for Λ a value around 0.5 GeV^{-1}. The energy

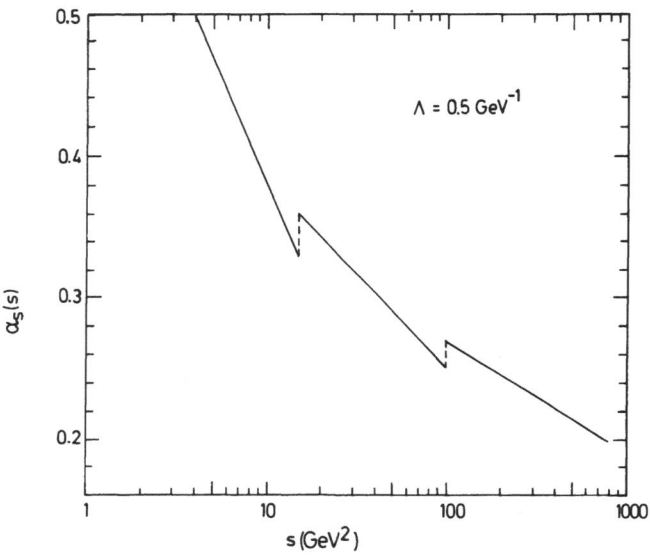

Fig. 5.5. The running coupling constant $\alpha_s(s)$

dependence of $\alpha_s(s)$ with $\Lambda = 0.5$ GeV is shown in Fig. 5.5. We see that the corrections predicted by (5.2) for an energy of 5 GeV is of the order of 10 % or 0.4 units in R, which is about half of the observed one. The agreement is expected to become better as energy increases. Indeed, at 9.4 GeV the R value predicted by (5.2) agrees well with experiment

R(PLUTO) = 3.7 ± 0.3 plus 15 % systematic error,
R(u,d,s,c) · (1 + α_s/π) = 3.6.

6. e^+e^- Annihilation at Low Energies

In this section we will consider the low-energy region and the characteristics of some exclusive final states.

6.1 Excitation of ρ, ω, Φ

The cross-section measurements for ρ, ω, Φ determine Γ_{ee} and the γ - V coupling, γ_V. As an example Figure 6.1 shows the cross section for $e^+e^- \to K_S^0 K_L^0$ near the Φ /43/.

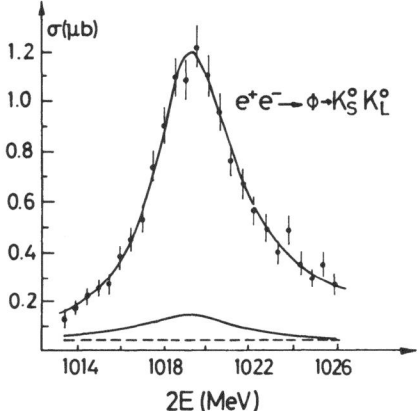

Fig. 6.1. The cross section for $e^+e^- \to K_S^0 K_L^0$ in the region of the Φ /43/

While in annihilation γ_V is measured for a photon mass squared $s = M_V^2$, experiments on photo- and electroproduction of vector mesons yield via VDM γ_V at $s = 0$ and $s = -Q^2 < 0$.

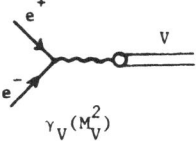
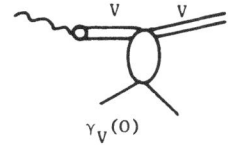
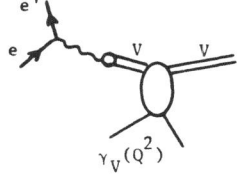

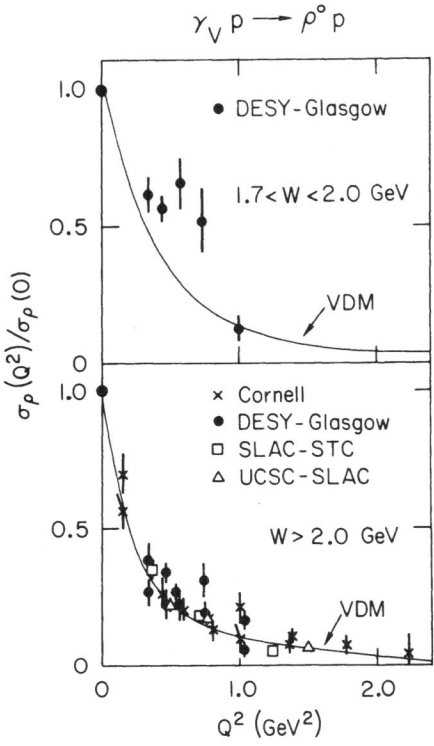

The cross section for rho production by virtual photons, $\gamma_V p \rightarrow \rho^0 p$, as a function of the mass squared, Q^2, of the virtual photon; W is the total c.m. energy of the $\rho^0 p$ energy /51/

In Figure 6.2 the measurements of rho electroproduction are compiled in terms of $\sigma_\rho(Q^2)/\sigma_\rho(0)$ /50, 51/. The curve was calculated assuming $\gamma_V(Q^2) = \gamma_V(0)$. With these measurements we can study the question whether γ_V depends on the photon mass. Table 6.1 summarizes the experimental results on Γ_{ee} and γ_V at $s = M_V^2$, 0, -1 GeV^2.

Table 6.1. Experimental results on the Vee partial width Γ_{ee} and on the γ - V coupling

| | Γ_{ee} [keV] | $\gamma_V^2/4\pi$ | | |
		$s = M_V^2$	$s = 0$	$s = -1$ GeV^2
ρ	6.4 ± 0.8 /52/	0.54 ± 0.07 /52/	0.65 ± 0.1 /53/	0.6 ± 0.3 /51/
ω	0.76 ± 0.17 /52/	4.7 ± 1.1 /52/	6 ± 1.5 /53/	
ϕ	1.1 ± 0.15 /43/	4.3 ± 0.5 /43/	5.1 ± 0.6 /53/	

One concludes from Table 6.1 that the data are consistent with no dependence on the photon mass. The assumption that γ_V is independent of s is basic to the vector dominance model.

6.2 Evidence for Vector States Between 1 and 2 GeV

Recent experiments performed at DESY /54/, Frascati /44, 55/, and Orsay /42, 56/ suggest that the 1 to 2 GeV region is a bonanza for new vector mesons. Table 6.2 lists the states for which evidence has been claimed; the majority of them needs confirmation by further measurements.

Table 6.2. Summary of the evidence for vector mesons below 2.2 GeV

State	Mass [GeV]	Width [MeV]	Γ_{ee} [keV]	I^G	Remarks
ρ	773 ± 3	152 ± 3	6.4 ± 0.8	1^+	
ω	783	10.0	0.76 ± 0.17	0^-	
ϕ	1020	4.1	1.3 ± 0.15	0^-	
1100	$1097 \, ^{+16}_{-19}$	$31 \, ^{+24}_{-20}$			$\gamma p \to p e^+ e^-$ /54/
ρ'(1250)	~1260	~110		1^+	$\gamma p \to p \pi^+ \pi^- \pi^0 \pi^0$ /57/, $p e^+ e^-$ /54/ $e^+ e^- \to \pi^+ \pi^-$ (see Sect. 6.3)
1550	1533 ± 21	202 ± 70	2.7	1^+	$\gamma p \to p 4\pi$ /58/, $p e^+ e^-$ /54/ $e^+ e^- \to 4\pi$ /42, 56/
1660	$1652 \pm 7 \pm 10$	42 ± 17	0.18	0^-	$e^+ e^- \to 3\pi$, 5π /42, 56/
1680	1690 ± 14	180 ± 87	1.7	1^+	$e^+ e^- \to \pi^0 \pi^0 X$ /42, 56/
1770	$1772 \pm 6 \pm 10$	49 ± 25		0^+	$e^+ e^- \to 3\pi$, 5π /42, 44, 55, 56/
2130		~30			$e^+ e^- \to K\pi X$ /44, 55/

The 1100 state was observed by a DESY-Frascati group at DESY studying photoproduction of $e^+ e^-$ pairs /54/. The experiment is sensitive to small vector meson contribution by searching for an interference with the Bethe-Heitler amplitudes. This is illustrated by the following diagrams.

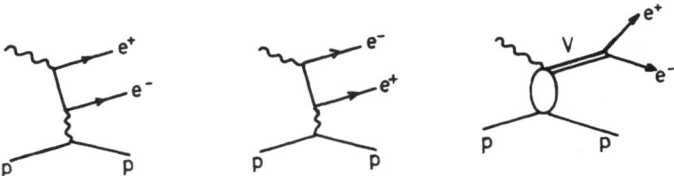

For a diffractively produced vector meson the Bethe-Heitler amplitudes interfere with the dispersive part of the vector meson propagator. The interference leads to a charge asymmetry since the e^+e^- system produced by the Bethe-Heitler process has C = + while in the vector meson case it has C = -.

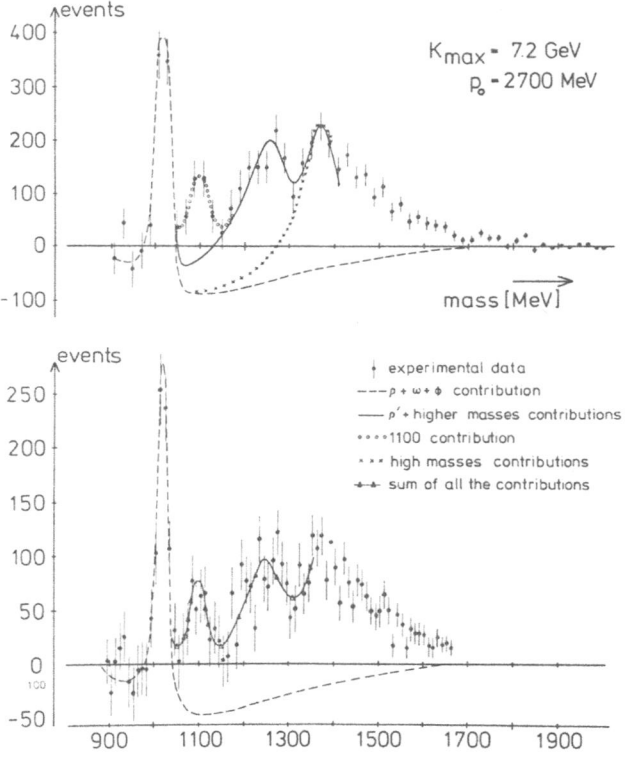

Fig. 6.3. The interference spectrum at 13°, 15°, and 16° plotted in 20 MeV (top) and 10 MeV (bottom) intervals. Plotted is the integral $\int (N_+ - N_-) \frac{x}{|x|} dx$, where $N_\pm$ denote the events with a positron (electron) in the right arm of the spectrometer, and x is the asymmetry parameter, $x = P_R \Theta_R - P_L \Theta_L$, where $P_{L(R)}$ and $\Theta_{L(R)}$ are the momenta and projected angles in the horizontal plane of the lepton going through the left (right) spectrometer arm /54/

Figure 6.3 shows the interference term which is proportional to the amplitudes for Bethe-Heitler and vector meson production,

$$A \sim \sqrt{\sigma_{BH}} \cdot \sqrt{\sigma_V}.$$

The dominant structure is due to the $\Phi(1020)$. The enhancement around 1100 MeV represents the evidence for the 1100 structure. If the interference spectrum plotted in Figure 6.3 is described by a sum of contributions from ϕ, ω, Φ and from vector mesons above 1.2 GeV the authors find the 1100 peak to be 7 s.d. off this curve. This result, however, depends on the details of the vector meson parameters. A conservative estimate from an eyeball fit yields 3 - 4 s.d. The mass and width of the 1100 structure are given in Table 6.2. The product of forward production cross section times branching ratio for the e^+e^- decay is measured to be

$$\left. \frac{d\sigma}{dt} \right|_{t=0} \cdot B_{e^+e^-} = 5 \cdot 10^{-5} \text{ μb GeV}^{-2}.$$

This number is roughly one to two orders of magnitude smaller than the corresponding values measured for ρ, ω and ϕ production: $4.3 \cdot 10^{-3}$ μb GeV^{-2}, $0.9 \cdot 10^{-3}$ μb GeV^{-2}, and $1.1 \cdot 10^{-3}$ μb GeV^{-2}, respectively.

The $\rho'(1250)$ will be dealt with in a separate section.

Some evidence for a narrow state at 1500 MeV state had been found earlier by groups measuring at the Frascati storage ring. More data taken in this energy region did not substantiate this claim.

The upper limit on the integrated resonance cross section for a 1 MeV wide resonance is 0.6 nbGeV (90 % C.L.) which corresponds to 6 % of the J/ψ cross section.

An experimental group working at the Orsay storage ring DCI observed four structures between 1550 and 1800 MeV (see Table 6.2 and Figs. 6.4a-c). It is clear from Figure 6.4 that more data are needed to confirm their existence. The 1550 and 1680 states showed up in even pion final states, the 1660 and 1770 structures in odd pion states (assuming that the charged particles observed were all pions). The first two states have therefore positive G parity and isospin T = 1 while the latter two have G = -, T = 0. It is interesting to note from Table 6.2 that as in the case of ρ, ω, and ϕ the isospin zero states have much smaller widths than those with T = 1. The 1550 state is probably identical to the $4\pi^{\pm}$ structure seen in photoproduction near the same mass value /58/. Evidence for its existence was also reported from other e^+e^- experiments /54, 59/.

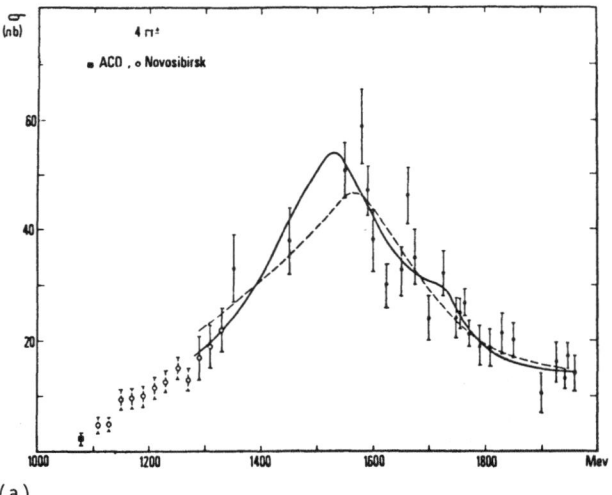

(a)

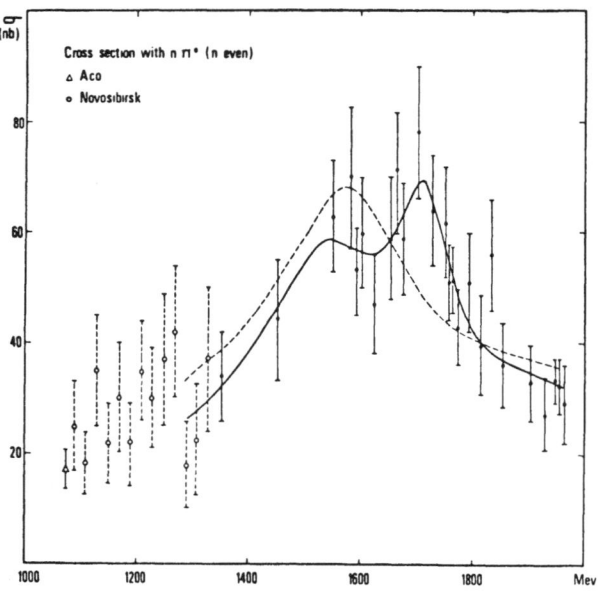

(b)   (----: Single Breit-Wigner fit)

Fig. 6.4. (a) Cross section for $e^+e^- \to 4\pi^\pm$. (b) Cross section for the sum of $e^+e^- \to 2\pi^\pm 2\pi^0$, $4\pi^\pm 2\pi^0$, and $2\pi^\pm 4\pi^0$ /56/

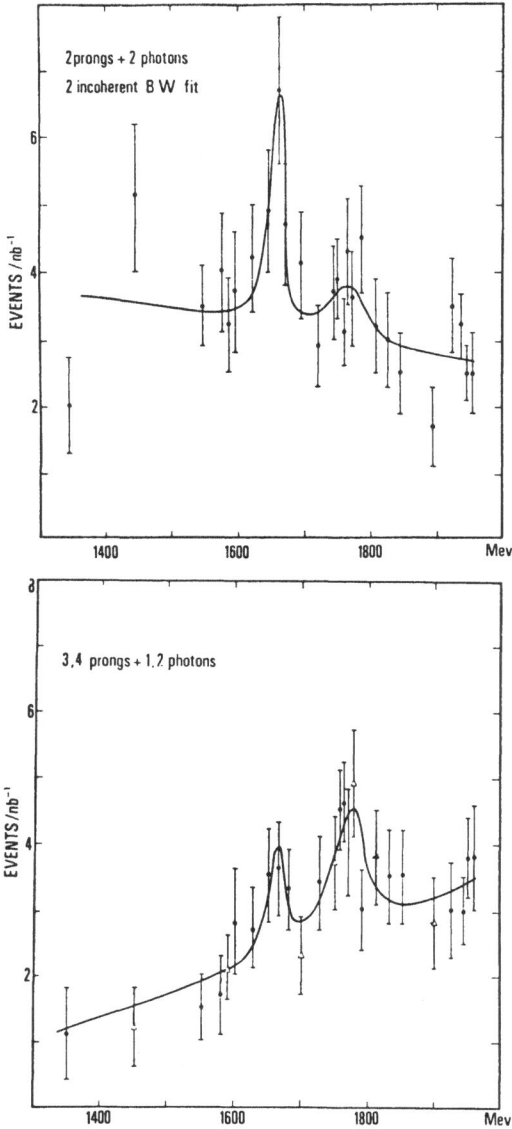

Fig. 6.4. (c) Event yield for $e^+e^- \rightarrow$ states with one or two photons observed /56/

The 2130 MeV state was observed by the MEA group /55/ as a narrow ($\Gamma \sim$ 30 MeV) enhancement in the cross section for $K^{\pm}\pi^{\mp}$ production where the $K\pi$ system is in the mass region of the $K^*(890)$ (see Figure 6.5). Actually, pions and kaons were not identified in this experiment; rather, the charged particles were assumed to be π or K.

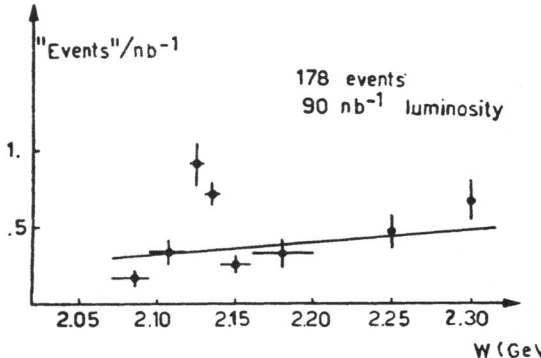

Fig. 6.5. $(\pi K)^0$ yield vs W /55/

6.3 The Pion Form Factor

Two-pion production, $e^+e^- \to \pi^+\pi^-$, measures the pion form factor, F_π, in the time-like region. The differential cross section is given by

$$\frac{d\sigma}{d\Omega} = \frac{\alpha^2}{8s} \beta_\pi^3 \sin^2\theta \; |F_\pi(s)|^2 \qquad (6.1)$$

with $\beta_\pi = P_\pi/E_\pi$. Near the rho, F_π has a Breit-Wigner behavior,

$$F_\pi(s) = \frac{m_\rho^2}{m_\rho^2 - s - im_\rho\Gamma} \cdot F_o. \qquad (6.2)$$

From the normalization condition,

$$F(0) = 1,$$

it follows that $F_o = 1$ if we assume the Breit-Wigner equation (5.2) to be valid down to $s = 0$. Finite width corrections and the contribution from the ω via its two-pion decay modify this simple expression:

$$F_\pi(s) = F_{GS}(s) + g e^{i\phi_{\rho\omega}} T_\omega(s). \qquad (6.3)$$

$F_{GS}(s)$ represents the rho in the form given by GOUNARIS and SAKURAI /60/

$$F_{GS}(s) = \frac{m_\rho^2(1 + d\,\Gamma_\rho/m_\rho)}{(m_\rho^2 - s) + f(s) - (im_\rho/\Gamma_\rho)(K/K)^3(m_\rho/\sqrt{s})}, \qquad (6.3a)$$

where K, K_ρ are the pion momenta for the energies $\sqrt{s}$ and m_ρ, the quantity $d = 0,48$, and

$$f(s) = \Gamma_\rho \, m_\rho^2 / K_\rho^3 \, \{K^2 \, (h(s) - h(m_\rho^2)) - (s - m_\rho^2) \, K_\rho^2 h'(m_\rho^2)\}$$

with

$$h(s) = \frac{2}{\pi} \frac{K}{\sqrt{s}} \ln \, (\sqrt{s} + 2K)/(2m_\pi)$$

$$h'(s) = \frac{d(h(s))}{ds}$$

The second term in (6.3a) describes the ω contribution

$$g \;\; = \frac{6\Gamma_\omega}{\alpha m_\omega} \; B_{\omega \to \pi\pi}^{1/2} \; B_{\omega \to ee}^{1/2} \; \left(\frac{m_\omega^2}{m_\omega^2 - 4m_\pi^2} \right)^{3/4}$$

$$\Gamma_\omega \;\; = \frac{m_\omega^2}{m_\omega^2 - s - im_\omega \, \Gamma_\omega}$$

and $\phi_{\rho\omega}$ measures the relative phase between ρ and ω.

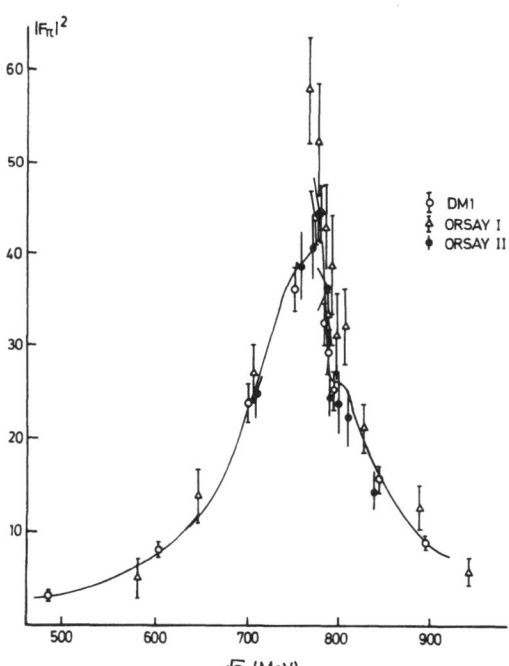

Fig. 6.6. The square of the pion form factor as a function of the total c.m. energy

Figure 6.6 shows F_π between 0,48 and 0.9 GeV as measured at ORSAY /42/. The dominant feature is the excitation of the rho. The recent data exhibit also the $\rho-\omega$ interference. The authors fitted (6.3) to their data with the following result [F_π(o) was constrained to unity]: m_ρ = 780.3 ± 3.5 MeV, Γ_ρ = 139.1 ± 3.9 MeV, d = 0.518 ± 0.025, $\sqrt{B_{\omega\pi\pi}}$ = 0.169 ± 0.042, $\phi_{\rho\omega}$ = 112.9° ± 12.6°. The inclusion of higher mass terms (e.g., a ρ' at 1200 MeV) changes the fit parameters only slightly. The fit provides also a value for the mean square radius of the pion which can be compared to results obtained by other methods:

$\langle r^2 \rangle$ = 0.458 ± 0.014 f^2 e^+e^- annihilation, timelike photon /42/

    = 0.31 ± 0.04 f^2 inverse electroproduction, spacelike photons /61/

    = 0.55 ± 0.17 f^2 π^+ electroproduction, spacelike photons /62/.

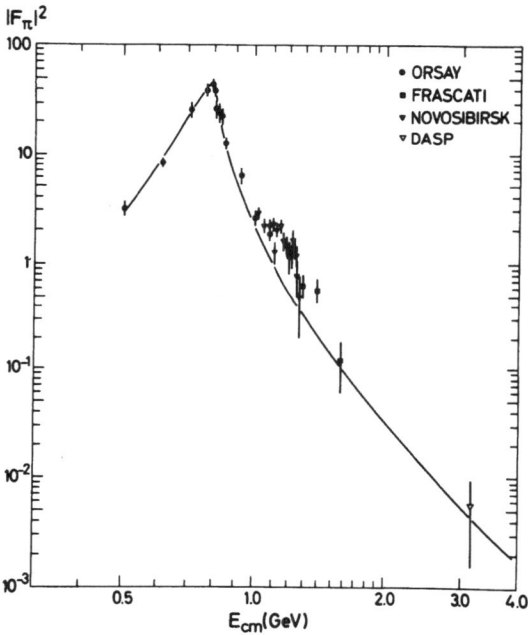

Fig. 6.7. The square of the pion form factor as a function of the total c.m. energy

Figure 6.7 shows the measurements on F_π in the low- and high-energy regions. We first remark on the F_π values measured at 3.1 and 3.7 GeV. The high-energy behavior of F_π, $F_\pi \sim s^{-1}$, leads to a rapid decrease of $\pi^+\pi^-$ production ($\sigma_{\pi^+\pi^-} \sim s^{-3}$) and

makes a precise measurement of F_π at high energies rather difficult. However, at
3.1 and 3.7 GeV the one-photon channel is enhanced by the resonant production of
J/ψ and ψ'. Since the J/ψ and ψ' do not decay directly into $\pi^+\pi^-$ (see Sect. 7) the
$\pi\pi$ branching ratio measured for J/ψ and ψ' /63-65/ determines the pion form factor

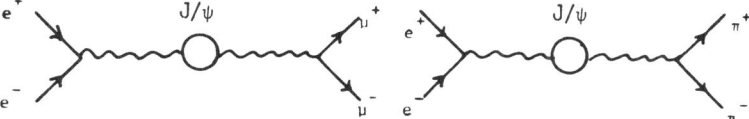

With $\beta = 1$ one has

$$|F_\pi|^2 = 4 \; \frac{\Gamma_{\pi^+\pi^-}}{\Gamma_{\mu\mu}},$$

where $\Gamma_{\pi\pi}$ and $\Gamma_{\mu\mu}$ are the partial widths for decay into $\pi^+\pi^-$ and $\mu^+\mu^-$, respectively.

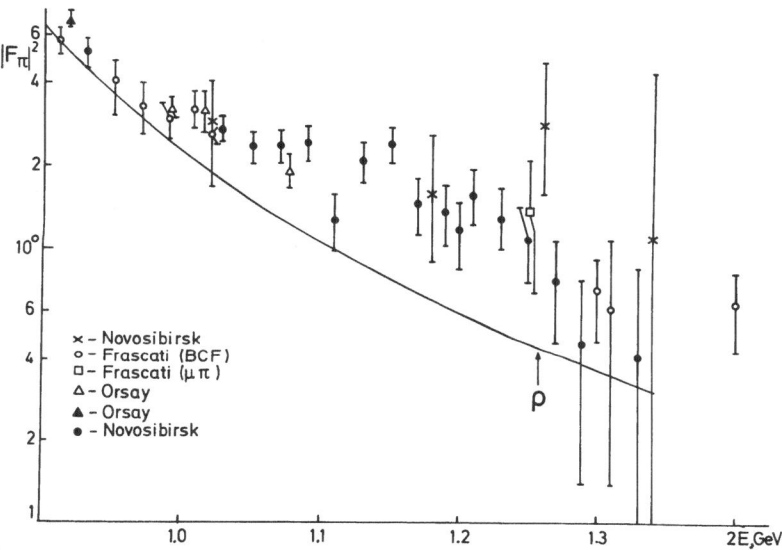

Fig. 6.8. The square of the pion form factor as a function of the total c.m. energy /43/

The rho pole (curve in Figure 6.7) describes the data rather well except for a
bump centered around 1.2 GeV. This region is shown on an expanded scale in Figure
6.8, which includes the new data points from NOVOSIBIRSK /43/. The Novosibirsk

52

group has fitted the data with a ρ pole plus a ρ' at 1.25 GeV with a width of 150 MeV

$$F_\pi(s) = \left\{ \frac{m_\rho^2(1 + d\frac{\Gamma_\rho}{m_\rho})}{m_\rho^2 - s - im_\rho \Gamma_\rho} + \frac{f_{\rho'\pi\pi}}{\gamma_{\rho'}} \frac{\gamma_\rho}{f_{\rho\pi\pi}} \frac{m'^2}{m'^2 - s + im'\Gamma'} \right\} , \qquad (6.4)$$

where $f_{\rho\pi\pi}$ and $f_{\rho'\pi\pi}$ measure the vector meson coupling to the $\pi\pi$ system, e.g., $\Gamma_\rho = \frac{2}{3}\left(f_{\rho\pi\pi}^2/4\pi\right) K_\rho^3/m_\rho^2$. The data are consistent with this ansatz. The fit gave for the ρ' contribution to $\pi^+\pi^-$ production

$$\sigma_{e^+e^- \to \rho' \to \pi^+\pi^-} (\sqrt{s} = m_{\rho'}) = 9 \pm 3 \text{ nb}. \qquad (6.5)$$

Furthermore, the products of the coupling constants $f_{\rho'\pi\pi}/\gamma_{\rho'}$, and $f_{\rho\pi\pi}/\gamma_\rho$ were found to have opposite signs. This result depends of course on the assumptions that $m_\rho' = 1.25$ GeV and that contributions other than those from ρ and ρ' can be neglected. From the result (6.5) we obtain

$$\frac{f_{\rho'\pi\pi}}{\gamma_{\rho'}} \cdot \frac{\gamma_\rho}{f_{\rho\pi\pi}} = -(0.10 \pm 0.02), \qquad (6.6)$$

which is in surprising agreement with the value of -0.11 found from the analysis of F_π in the spacelike region /51/. The data on $F_\pi(Q^2)$ for $Q^2 < 0$ come mainly from π

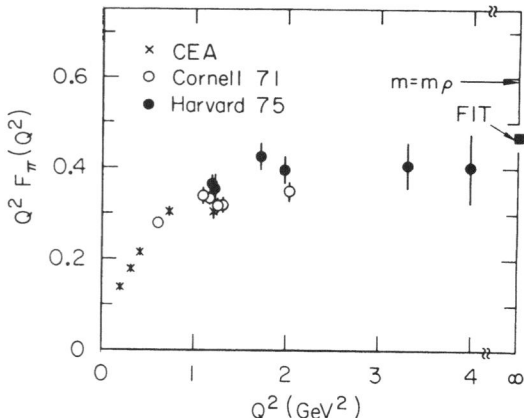

Fig. 6.9. The square of the pion form factor for negative mass squared of the virtual photon /51, 66/

electroproduction (eN → eN'$\pi^{\pm}$) /66/. Figure 6.9 shows the Q^2 dependence of F_π multiplied by Q^2. After the turn-on at low Q^2 the quantity $Q^2 F_\pi(Q^2)$ reaches a plateau, $Q^2 F_\pi(Q^2) = 0.47 \pm 0.01$ GeV^2 at $Q^2 = \infty$ /51/. This result together with the assumption that only ρ and ρ' contribute lead to the branching ratio quoted above.

The quark model relation (4.10) predicts for the ratio of $|\gamma_{\rho'}/\gamma_\rho|$ a value around 2.3 which, together with the result given in (6.6), can be used to calculate the branching ratio for the ρ' into two pions. One finds $\Gamma_{\rho'\pi\pi}/\Gamma_{\rho'} \simeq 0.06$, a value which does not contradict the $\pi\pi$ phase shift results /67/.

Finally, note that there is also consistency with the photoproduction data. In $\gamma p \to p\pi^+\pi^- X$ a bump is seen in the $\pi^+\pi^- X$ mass spectrum (where $X \approx \pi^0\pi^0$) around 1.25 GeV; the bump is diffractively produced and consistent with what would be expected for a ρ' /57, 68/. Association of this bump with the $\rho'(1250)$ gave $(\gamma_\rho/\gamma_{\rho'})^2 \approx$ $\sigma(\gamma p \to \rho'p)/\sigma(\gamma p \to \gamma p) \approx 0.2$.

Is there supporting evidence for a $\rho'(1250)$ from other annihilation channels? In order to study this question we shall look at four-pion final states.

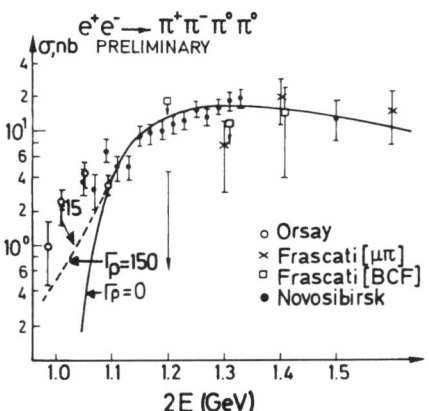

Fig. 6.10. The cross section for $e^+e^- \to \pi^+\pi^+\pi^-\pi^-$ /43, 69/

6.4 Four-Pion Production

Figures 6.10 and 6.11 summarize the cross-section measurements on these channels,

$$e^+e^- \to \pi^+\pi^+\pi^-\pi^- \tag{6.7}$$

$$\to \pi^+\pi^-\pi^0\pi^0 \tag{6.8}$$

up to $\sqrt{s} = 1.5$ GeV /43/. Both cross sections rise rapidly above threshold. Four-charged pion production is expected to be dominated by the $\rho''(1560)$ decaying mainly into $\rho^0\epsilon$ (ϵ is a shorthand for a scalar-isoscalar $\pi\pi$ system). The ρ'' contribution to reaction (6.8) follows from isospin invariance

$$\Gamma_{\rho'' \to \pi^+\pi^-\pi^0\pi^0} = \frac{1}{2}\Gamma_{\rho'' \to \pi^+\pi^+\pi^-\pi^-} \ .$$

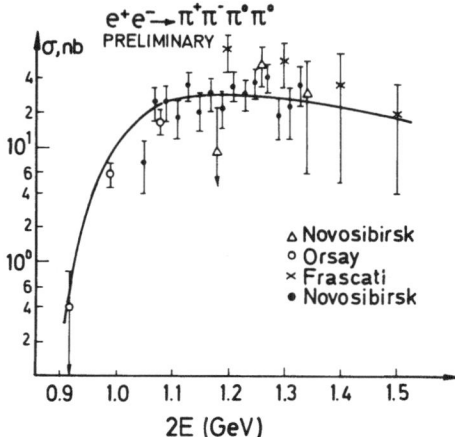

Fig. 6.11. The cross section for $e^+e^- \to \pi^+\pi^-\pi^0\pi^0$ /43/

A strong nonresonant contribution to (5.8) is expected from $\omega\pi^0$ production /70/

In Figure 6.12 the $\pi^+\pi^-\pi^0\pi^0$ cross section is compared to $\frac{1}{2}\sigma_{4\pi}\pm$ and to the sum of $\frac{1}{2}\sigma_{4\pi}\pm$ and the nonresonant $\omega\pi^0$ contribution, $\sigma_{\omega\pi}$. $\sigma_{\pi^+\pi^-\pi^0\pi^0}$ and $\sigma_{4\pi}\pm$ have different energy behavior near threshold: the rise of the former is much faster than of the latter. There is no compelling evidence for the production of a $\rho'(1250)$. However, if the $\rho'(1250)$ is produced and if it decays via $\omega\pi$, the nonresonant and resonant $\omega\pi$ amplitudes have to be treated together.

The high energy behavior of $\sigma_{4\pi}\pm$ is shown /71/ in Figure 6.13. For $\sqrt{s} \gtrsim 2$ GeV $\sigma_{4\pi}\pm \sim s^{-2.8}$. Such an s-dependence would be expected, e.g., for ρ dominance,

$$\sigma_{4\pi}\pm \sim \frac{1}{s}\left(\frac{m_\rho^2}{m_\rho^2 - s}\right)^2 \ . \tag{6.9}$$

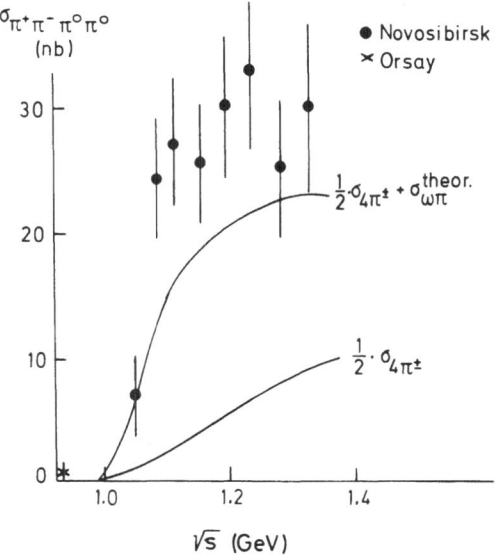

Fig. 6.12. The cross section for $e^+e^- \to \pi^+\pi^-\pi^0\pi^0$ /43/ compared to the sum of one-half of the measured cross section for $e^+e^- \to \pi^+\pi^+\pi^-\pi^-$ (from Figure 6.10) and the theoretical cross section for $e^+e^- \to \pi^0\omega$ /70/

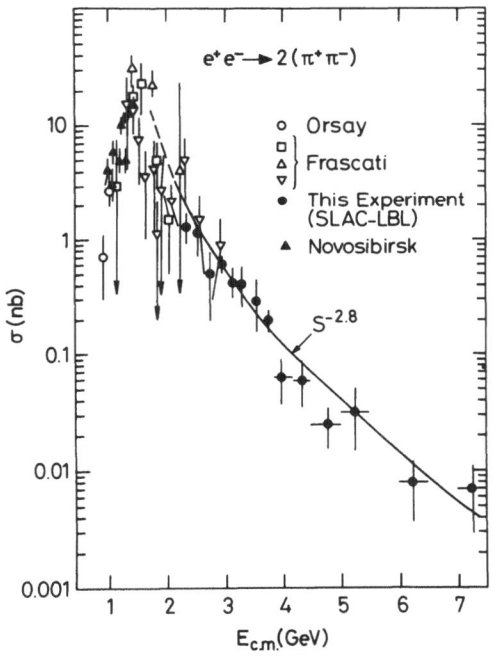

Fig. 6.13. The cross section for $e^+e^- \to 2\pi^+2\pi^-$ /71/

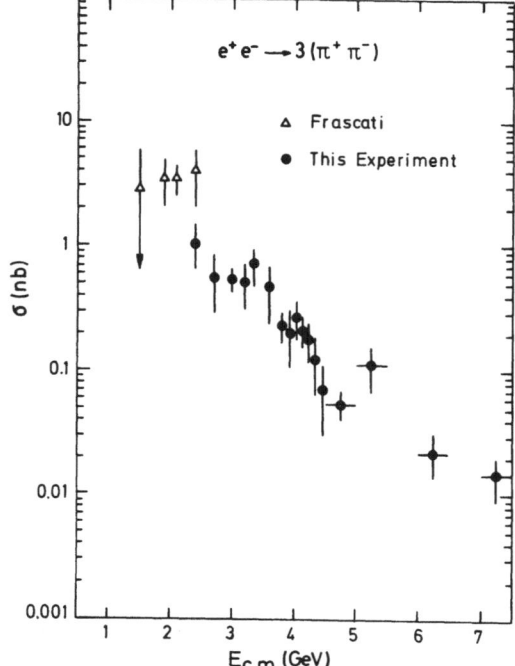

Fig. 6.14. The cross section for $e^+e^- \to 3\pi^+3\pi^-$ /71/

6.5 Six-Pion Production

The cross section for $e^+e^- \to 6\pi^\pm$ is given /71/ in Figure 6.14. It has approximately the same high-energy behavior as 4π production.

7. The New Particles J/ψ and ψ'

The first member of the new family of particles was discovered at Brookhaven as the
J particle in the e^+e^- spectrum of p-Be collisions (see Figure 7.1) /9/, pBe $\to$ e^+e^-X

BNL: Aubert et al.
28 GeV/c pBe $\to e^+e^-$X

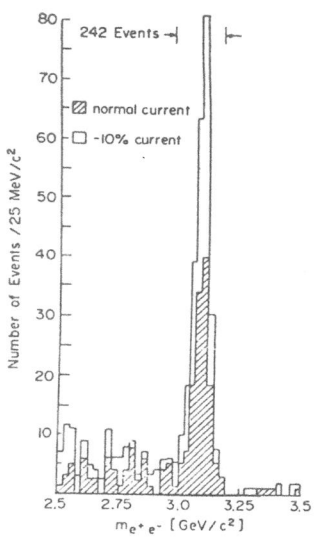

Fig. 7.1. The e^+e^- effective mass spectrum
from the reaction pBe $\to$ e^+e^-X /9/

and at SPEAR as the ψ particle in e^+e^- annihilation (see Figure 7.2) /10/

$e^+e^- \to$ hadrons
$\to e^+e^-$
$\to \mu^+\mu^-$.

The most exciting property of the J/ψ(3100) was its small decay width ($\Gamma < 2$ MeV).
Within a few days the J/ψ was also observed at Frascati, and DORIS and SPEAR had
found the next higher state ψ'(3700), shortly afterwards confirmed at DORIS. Since

58

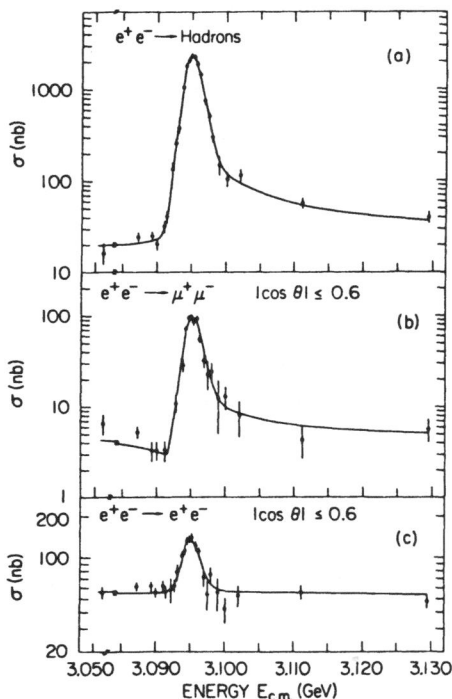

Fig. 7.2. Energy dependence of the cross sections $e^+e^- \to$ hadrons, $e^+e^- \to \mu^+\mu^-$ and $e^+e^- \to e^+e^-$ in the vicinity of the J/ψ /10/

then, the quantum numbers and decay properties of the J/ψ and ψ' have been so thoroughly investigated that the decay channels of J/ψ and ψ' are now better known than, e.g., for ω or Φ.

7.1 J/ψ Production Cross Section and Width

In e^+e^- annihilation the J/ψ shows up as a strong, narrow signal in hadron and μ-pair production, and in Bhabha scattering (see Figure 7.2) /10/. In the peak the enhancement factor is roughly 100 for σ^{tot}, 10 for $\sigma_{\mu\mu}$ and 2 for σ_{ee}. If we assume for the moment $J^P = 1^-$ then the cross section for

$$e^+e^- \to J/\psi \to f$$

can be written as

$$\sigma = \frac{3\pi}{s} \frac{\Gamma_{ee}\,\Gamma_f}{(M_0 - \sqrt{s})^2 + \Gamma^2/4} . \tag{7.1}$$

The observed width of ~2 MeV is of the order of the energy spread of the beams (see Sect. 2). For this reason, the width Γ cannot be read off directly from the resonance curve. On the other hand, the cross section integrated over the resonance contribution is independent of machine parameters /72/

$$\int_{\text{resonance}} \sigma^{\text{tot}} \, dM = \frac{6\pi^2}{M_o^2} \frac{\Gamma_{ee} \, \Gamma_h}{\Gamma}$$

$$\int_{\text{resonance}} \sigma_{\mu\mu} \, dM = \frac{6\pi^2}{M_o^2} \frac{\Gamma_{ee} \, \Gamma_{\mu\mu}}{\Gamma} \quad \text{and} \tag{7.2}$$

$$\int_{\text{resonance}} \sigma_{ee} \, dM = \frac{6\pi^2}{M_o^2} \frac{\Gamma_{ee}^2}{\Gamma} \, ,$$

where Γ_h is the hadronic and Γ_{ee} and $\Gamma_{\mu\mu}$ the leptonic decay widths. By determining separately $e^+e^- \rightarrow$ h, ee, and $\mu\mu$ and assuming that $\Gamma = \Gamma_h + \Gamma_{ee} + \Gamma_{\mu\mu}$ it is possible to solve for the widths of all three channels. Since Γ_{ee} and $\Gamma_{\mu\mu} << \Gamma_h$ the total hadron cross section determines essentially Γ_{ee}, and Γ_{ee} together with σ_{ee} yield Γ_h.

Radiative Corrections

The initial electron and positron lose energy through internal bremsstrahlung.

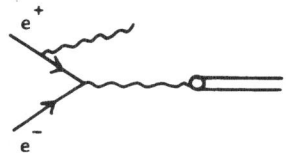

Roughly speaking, due to the deacceleration in the collision the $e^+(e^-)$ acts on the $e^-(e^+)$ as a target of thickness x (in radiation lengths) /73/

$$x \simeq \frac{2\alpha}{\pi} \{\ln \frac{\sqrt{s}}{m_e} - 1\}, \tag{7.3}$$

which for $\sqrt{s}$ = 3.1 GeV gives x $\approx$ 3.5 %. If the sum of the beam energies $E_+ + E_- = M_o$ before the collision, and the radiated photon energy K_γ is large compared to the resonance width, $K_\gamma >> \Gamma$, the e^+e^- pair will not contribute to the production of the resonance, but will be counted in the luminosity. This leads to an apparent reduction of the cross section. On the other hand, if $E_+ + E_- > M_o$ before the collision, by emitting a photon of energy $K_\gamma \approx E_+ + E_- - M_o$ the e^+e^- pair can produce the resonance. This will lead to an apparent enhancement of the cross section above the resonance mass (radiative tail).

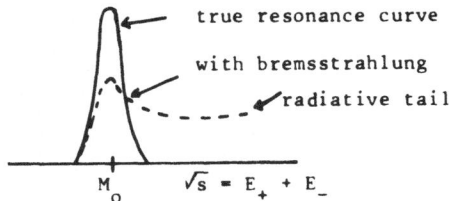

true resonance curve

with bremsstrahlung

radiative tail

M_o $\sqrt{s} = E_+ + E_-$

In order to get a feeling for the size of the effect we compute the probability for the emission of a photon with $K_\gamma > 2\Gamma$

$$P(K_\gamma > 2\Gamma) \approx 2X \int_{K=2}^{E} dK/K = 2X \ln E/2\Gamma. \tag{7.4}$$

For E = 1.55 GeV and Γ = 70 keV the probability is ~65 %. A more precise calculation leads to P $\approx$ 40 % /72, 74/; i.e., the observed cross sections are 40 % too small. The radiatively corrected cross sections and resonance parameters are given in Table 7.1.

Table 7.1. Resonance parameters of the J/ψ

	DORIS /75-77/	FRASCATI /78/	SLAC-LBL /79/
Mass [MeV]	3097 ± 1	3103 ± 6	3095 ± 4
$\int\sigma_h$ dM [μb MeV]	8.0 ± 1.2	9.6 ± 1.7	10.4 ± 1.5
$\int\sigma_{ee}$ dM [nb MeV]		790 ± 200	790
$\int\sigma_{\mu\mu}$ dM [nb MeV]			
Γ_{tot} [keV]	58 ± 13	67 ± 25	69 ± 15
Γ_{ee} [keV]		4.6 ± 0.8	4.8 ± 0.6
$\Gamma_{\mu\mu}$ [keV]	4.4 ± 0.6	4.6 ± 1.0	4.8 ± 0.6

The J/ψ width of $\Gamma \cong$ 70 keV is roughly a factor of 10^4 smaller than the width expected for a conventional meson of 3.1 GeV mass. This is illustrated by Figure 7.3, a plot of width versus mass for nonstrange mesons.

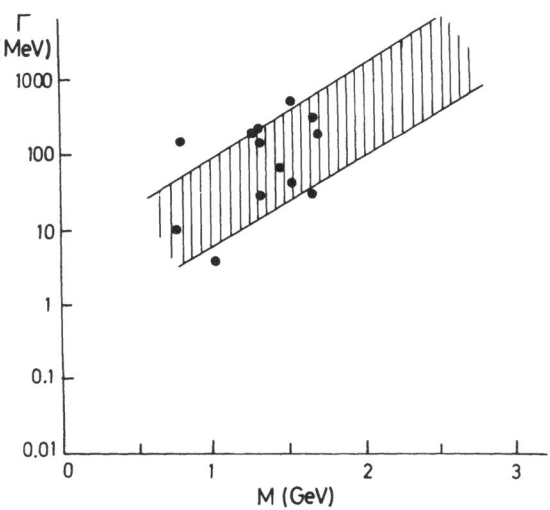

Fig. 7.3. A plot of width versus mass for nonstrange mesons. Data taken from the tables of the Particle Data Group

7.2 Spin and Parity of the J/ψ

Spin and parity of the J/ψ can be determined by studying the interference between its μ-pair decay and μ-pair production via QED. Table 7.2 lists the differential

cross sections expected for different J^P assignments /80/. A spin zero particle has helicity zero and will therefore not interfere with the QED contribution. For $J^P = 1^-$ the cross section $\sigma_{\mu\mu}(s)$ shows an interference effect which is destructive below the resonance and constructive above (provided J/ψ couples to ee and μμ with the same sign) while the angular distribution will be forward-backward symmetrical.

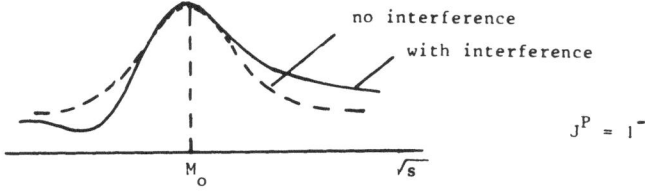

For an axial vector the opposite is true; there will be no interference effect in $\sigma_{\mu\mu}(s)$ but a forward-backward asymmetry which changes sign at the resonance mass will be observed. A vector boson with vector <u>and</u> axial vector coupling will lead

Table 7.2. μ pair production in the neighborhood of a resonance; g/e, g_V/e, g_A/e are the scalar, vector, and axial vector couplings of the resonance R to e^+e^- and $\mu^+\mu^-$; μ-e universality is assumed /80/

J^P_R	Differential cross section $\dfrac{d\sigma}{d\Omega} =$	Interference in $\sigma_{\mu\mu}(s)$	F/B asymmetry
$0^{\pm}$	$\dfrac{2\alpha^2}{3s}\left\{ \underbrace{\dfrac{3}{8}(1+\cos^2\Theta)}_{\lvert QED\rvert^2} + \underbrace{\dfrac{g^4}{e^4}\dfrac{s^2}{(M_0^2-s)^2+M_0^2\Gamma^2}}_{\lvert R\rvert^2} \right\}$	no	no
1^{-}	$\dfrac{\alpha^2}{4s}(1+\cos^2\Theta)\left\{ \underbrace{1}_{\lvert QED\rvert^2}\; \underbrace{-\,2\dfrac{g_V^2}{e^2}\dfrac{(M_0^2-s)s}{(M_0^2-s)^2+M_0^2\Gamma^2}}_{2QED\ R^*} + \underbrace{\dfrac{g_V^4}{e^4}\dfrac{s^2}{(M_0^2-s)^2+M_0^2\Gamma^2}}_{\lvert R\rvert^2} \right\}$	yes	no
1^{+}	$\dfrac{\alpha^2}{4s}\left\{ (1+\cos^2\Theta)\left(1+\dfrac{g_A^4}{e^4}\dfrac{s^2}{(M_0^2-s)^2+M_0^2\Gamma^2}\right) - 2\cos\Theta\,\dfrac{g_A^2}{e^2}\dfrac{(M_0^2-s)s}{(M_0^2-s)^2+M_0^2\Gamma^2} \right\}$	no	yes
$1^{+}1^{-}$	$\dfrac{\alpha^2}{4s}\left\{ (1+\cos^2\Theta)\left(1 - 2\dfrac{g_V^2}{e^2}\dfrac{(M_0^2-s)s}{(M_0^2-s)^2+M_0^2\Gamma^2} + \dfrac{(g_V^2+g_A^2)^2}{e^4}\dfrac{s^2}{(M_0^2-s)^2+M_0^2\Gamma^2}\right)\right.$ $\left. - 2\cos\Theta\left(\dfrac{g_A^2}{e^2}\dfrac{(M_0^2-s)s}{(M_0^2-s)^2+M_0^2\Gamma^2} + \dfrac{2g_A^2 g_V^2}{e^4}\dfrac{s^2}{(M_0^2-s)^2+M_0^2\Gamma^2}\right) \right\}$	yes	yes

to both types of interference effects. For spin two or three the interference in $\sigma_{\mu\mu}(s)$ will be constructive both below and above the resonance.

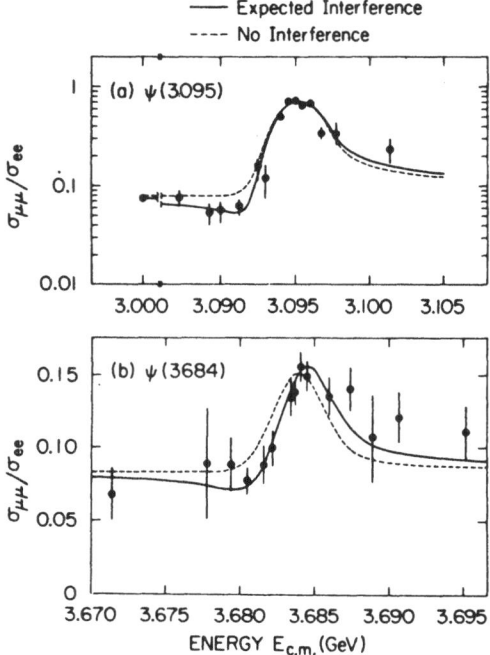

Fig. 7.4. Ratio of lepton-pair cross sections, $\sigma_{\mu\mu}/\sigma_{ee}$, within the detector acceptance, in the vicinity of the J/ψ and ψ', respectively /79/

Experimental Results

Figure 7.4 shows $\sigma_{\mu\mu}$ near the J/ψ mass. One observes an interference effect which is destructive for $\sqrt{s} < M_{J/\psi}$ and constructive above. This excludes $0^{\pm}$, 1^{+}, 2^{+}, 3^{+},... as possible J^P values and makes 2^{-}, 3^{-} unlikely assuming μ-e universality. The angular distribution averaged over the J/ψ is of the form (see Figure 7.5a)

$$\frac{d\sigma}{d\Omega} \sim 1 + \cos^2\theta.$$

No higher $\cos\theta$ terms are observed which makes J^P = 2^{-}, 3^{-},... unlikely. Furthermore, the forward-backward ratio is zero over the full resonance mass region (see Figure 7.5b). Therefore, the J/ψ is not a vector boson that has both vector and axial vector couplings. The data are consistent with J^P = 1^{-}. The negative parity implies negative charge conjugation

$$(\mu^+\mu^-) = CP(\mu^+\mu^-) = -C(\mu^+\mu^-) = - \text{sgn}(C) (\mu^+\mu^-).$$

Hence the J/ψ is a vector meson with J^{PC} = 1^{--}.

64

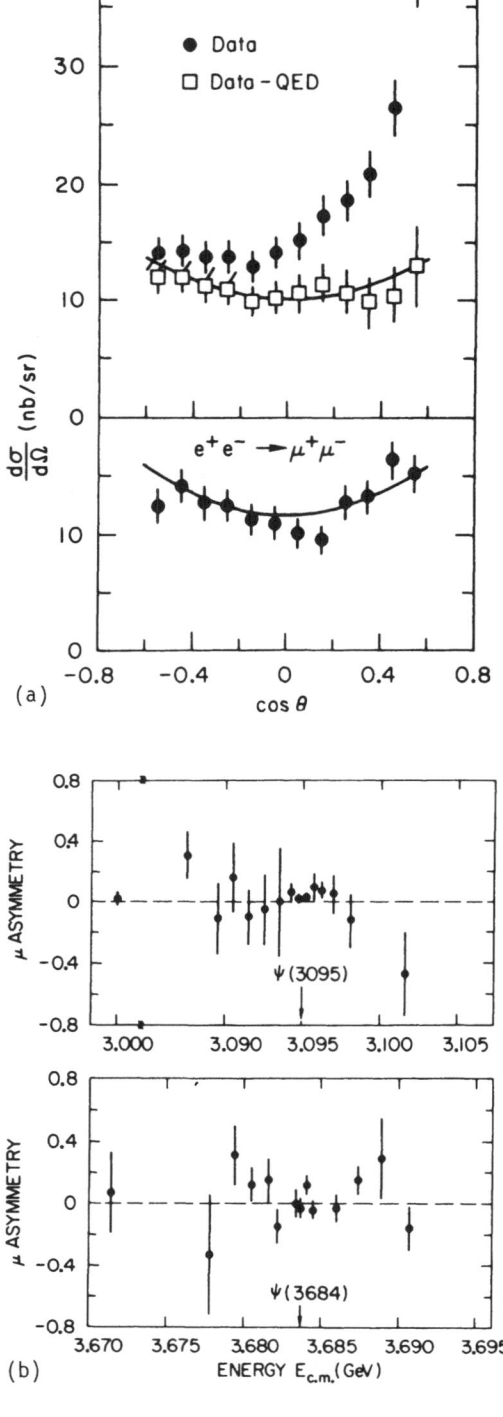

Fig. 7.5. (a) Differential cross sections for J/ψ production and decay into leptons /79/. (b) Forward-backward asymmetry for $\mu^+\mu^-$ production in the vicinity of the J/ψ and ψ' /79/

7.3 G Parity and Isospin of J/ψ

The G parity of a neutral, nonstrange meson can be read off from its decay into
pions: a G parity even (odd) meson decays into an even (odd) number of pions. The
DASP /64, 81/ and SLAC-LBL /82/ measurements yielded the following branching ratios:

$$B = \Gamma_f/\Gamma$$

$J/\psi \rightarrow \pi^+\pi^-$	$(1.0 \pm 0.5) \cdot 10^{-4}$	/64, 81/
$\rightarrow \pi^+\pi^-\pi^0$	$(1.6 \pm 0.6) \cdot 10^{-2}$	/82/
$\rightarrow 2\pi^+2\pi^-$	$(0.4 \pm 0.1) \cdot 10^{-2}$	/82/
$\rightarrow 2\pi^+2\pi^-\pi^0$	$(4.3 \pm 0.5) \cdot 10^{-2}$	/77, 82/
$\rightarrow 3\pi^+3\pi^-$	$(0.4 \pm 0.2) \cdot 10^{-2}$	/82/
$\rightarrow 3\pi^+3\pi^-\pi^0$	$(2.9 \pm 0.7) \cdot 10^{-2}$	/82/
$\rightarrow 4\pi^+4\pi^-\pi^0$	$(0.9 \pm 0.3) \cdot 10^{-2}$	/82/

Decay modes with an odd number of pions are clearly preferred. However, decays
into an even number of pions are also present. This is demonstrated in Figures 7.6

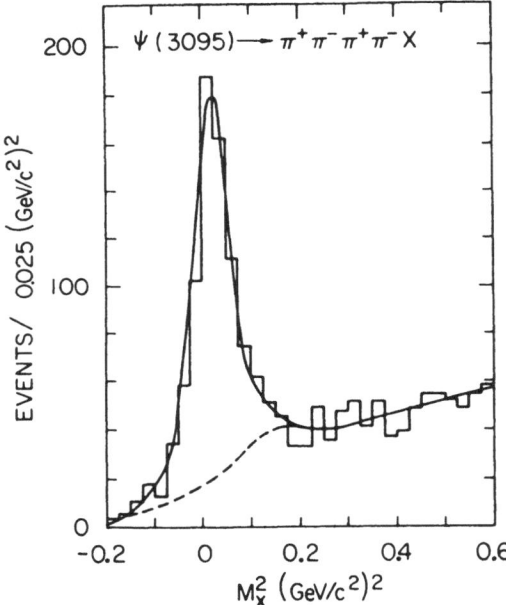

Fig. 7.6. Missing mass squared distributions for events with four-charged hadrons
of total 0 and missing momentum < 0.2 GeV/c from J/ψ decay /82/

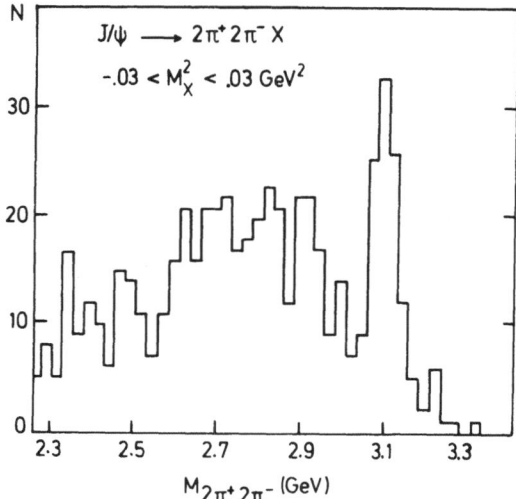

Fig. 7.7. The 4π mass distribution for events of the type $J/\psi \to 2\pi^+ 2\pi^- X$ /82/

and 7.7 for the four-charged pion final state. The missing mass distribution for $J/\psi \to \pi^+\pi^+\pi^-\pi^- X$ peaks strongly near $M_X^2 = m_\pi^2$ but a small contribution from $M_X \approx 0$ cannot be excluded (Figure 7.6). The 4π mass distribution for events in the peak shows a clear signal at the J/ψ mass (Figure 7.7) which gives evidence for the transition $J/\psi \to \pi^+\pi^+\pi^-\pi^-$.

The presence of decays into states with both odd <u>and</u> even number of pions indicates that G is not a good quantum number. However, some breaking of G parity is expected from the one-photon decay. In the J/ψ mass region three different processes can contribute: direct decay, decay via one photon, and nonresonant production.

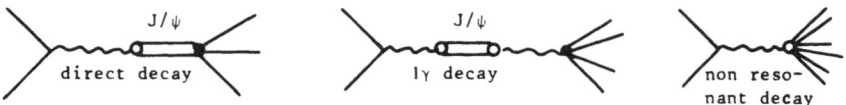

The nonresonant contribution is small and will be neglected. The one-photon contribution can be calculated from $\Gamma_{\mu\mu}$ and the ratio of $\sigma^{tot}/\sigma_{\mu\mu}$ off the resonance:

$$\left| \rangle\!\!-\!\!\langle \right|^2 : \left| \rangle\!\!-\!\!\langle_\mu^\mu \right|^2 = \left| \rangle\!\!-\!\!\langle \right|^2 : \left| \rangle\!\!-\!\!\langle_\mu^\mu \right|^2$$

$$\frac{\Gamma_{\gamma 1 \to hadrons}}{\Gamma_{\mu\mu}} = \left(\frac{\sigma^{tot}}{\sigma_{\mu\mu}} \right)_{off\ resonance} = R. \tag{7.5}$$

Table 7.3 summarizes the partial widths for leptonic, one-photon, and direct decays (Γ_d) of the J/ψ. In calculating Γ_d it was assumed that the direct and the one-photon decays do not interfere; Γ_d is seen to exceed Γ_γ by roughly a factor of four to five.

Table 7.3. Partial widths and branching ratios of the J/ψ final state

	Γ_f [keV]	Γ_f/Γ
all	69 $\pm$ 15	1.0
leptonic (e^+e^- + $\mu^+\mu^-$)	9.6 $\pm$ 1.2	0.138 $\pm$ 0.018
hadrons through 1γ	12 $\pm$ 2	0.17 $\pm$ 0.03
direct	47.4 $\pm$ 14	0.69 $\pm$ 0.04

The G parity and the isospin of the J/ψ can only be determined from its direct (strong) decay. We will assume that the J/ψ has a definite G parity and that the direct decay is G-parity conserving. If G(J/ψ) is odd (even) then the even (odd) pion final states must be due to one-photon decay and their contribution relative to μ-pair production must be the same on and off the resonance.

MULTIPION FINAL STATE

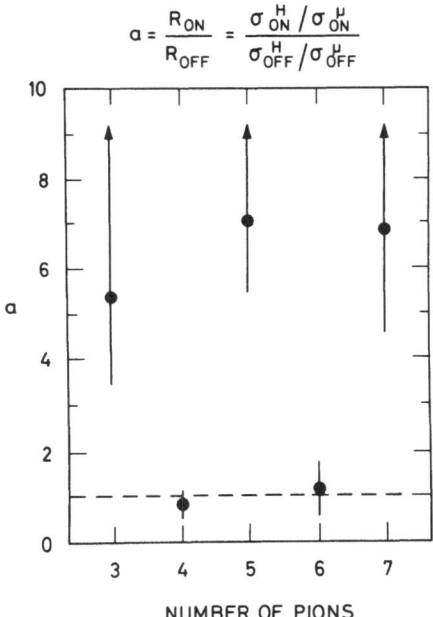

$$a = \frac{R_{ON}}{R_{OFF}} = \frac{\sigma^H_{ON}/\sigma^\mu_{ON}}{\sigma^H_{OFF}/\sigma^\mu_{OFF}}$$

a

NUMBER OF PIONS

Fig. 7.8. Comparison of the ratio of multipion to muon-pair cross sections at the J/ψ (ON) and at 3.0 GeV (OFF) /82/

Figure 7.8 shows the ratio α,

$$\alpha = \frac{\sigma^H_{on}/\sigma^\mu_{on}}{\sigma^H_{off}/\sigma^\mu_{off}}$$

as a function of the number of pions N_π in the final states H. α is consistent with unity for N_π = even and much larger than one for N_π = odd. Thus the even-pion final states can be attributed to one-photon decay and the data are consistent with $G(J/\psi) = -1$.

Isospin: Since the J/ψ decays directly into three pions $T \leq 3$. The relation $C = (-1)^T \cdot G$ together with $G = C = -1$ restricts then T to the values 0 and 2. A decision between these two values for example can be made by looking for the decay into $p\bar{p}$. Figure 7.9 shows the $p\bar{p}$ mass distribution for collinear $p\bar{p}$ pairs from the process

$$J/\psi \rightarrow p\bar{p}X.$$

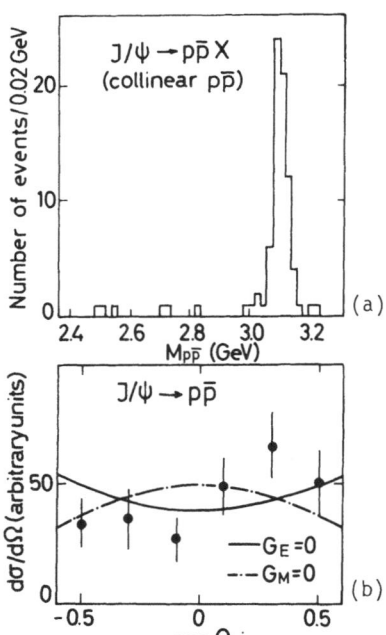

Fig. 7.9. (a) Effective $p\bar{p}$ mass distribution for collinear $p,\bar{p}$ pairs from $J/\psi \rightarrow p\bar{p}X$. (b) Polar angle distribution for $J/\psi \rightarrow p\bar{p}$; Θ is the angle between the proton and the positron beam /81/

A clear signal for the decay $J/\psi \rightarrow p\bar{p}$ is observed /64, 81/. The decay distribution is consistent with a $1 + \cos^2\Theta$ behavior as expected if the magnetic form factor dominates. The branching ratio is measured to be

$$\Gamma_{p\bar{p}}/\Gamma_{tot} = (2.5 \pm 0.4) \cdot 10^{-3} \qquad \text{DASP /64/}$$
$$(3.7 \pm 1.5) \cdot 10^{-3} \qquad \text{PLUTO /77/}$$
$$(1.98 \pm 0.15) \cdot 10^{-3} \qquad \text{SLAC-LBL /83, 84/}$$

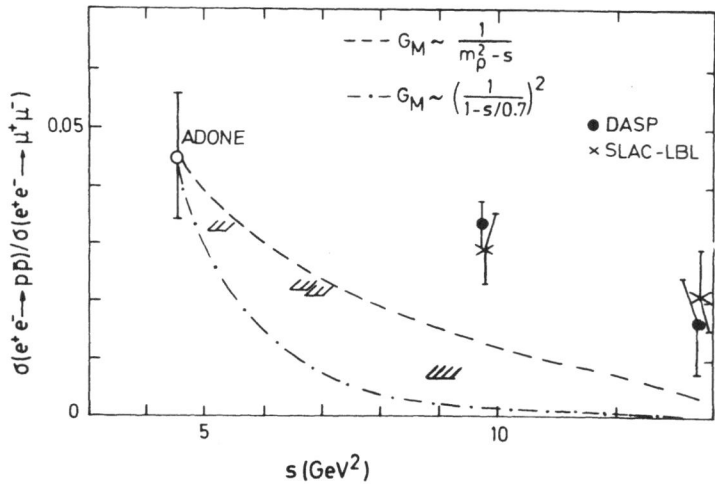

Fig. 7.10. The ratio $\sigma(e^+e^- \rightarrow p\bar{p})/\sigma(e^+e^- \rightarrow \mu^+\mu^-)$ as measured in e^+e^- collisions /64, 77, 81, 83, 85, 86/ and upper limits (///) from the inverse reactions $p\bar{p} \rightarrow e^+e^-$, $\mu^+\mu^-$ /87/. The curves show a rho-pole and a dipole behavior for G_M normalized at $s = 4.4$ GeV^2

The question whether the $p\bar{p}$ decay is a direct decay can be answered with Figure 7.10 where the ratio of $p\bar{p}$ to $\mu^+\mu^-$ production is plotted as a function of energy. This ratio is clearly smaller below the J/ψ than at the resonance; the $p\bar{p}$ decay therefore is a direct one. This excludes T = 2; therefore $T^G = 0^-$.

The J/ψ has the quantum numbers of a normal isoscalar vector meson. But its decay into normal hadrons is strongly suppressed. Is it really a strongly interacting particle? Experiments studying J/ψ production by photons off nuclei answer in the affirmative; the t distribution shows a coherent part from which a total J/ψ nucleon cross section of ~1 mb at $E_\gamma \approx 100$ GeV can be deduced using VDM /88/.

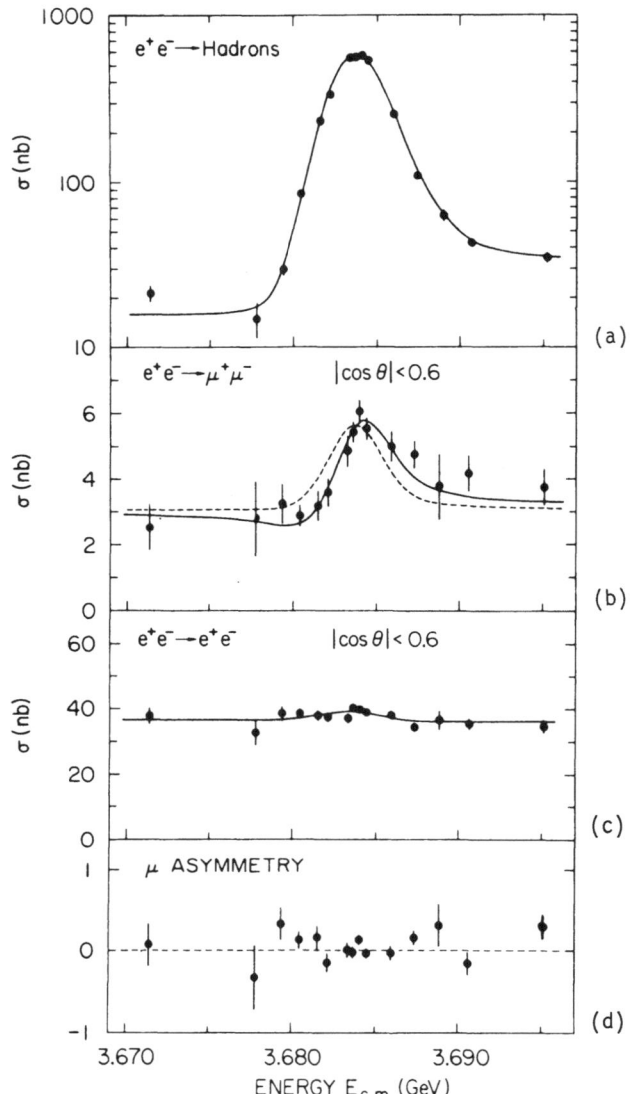

Fig. 7.11. Energy dependence of the cross sections $e^+e^- \to$ hadrons, $e^+e^- \to \mu^+\mu^-$, and $e^+e^- \to e^+e^-$ in the vicinity of the ψ' /89/

7.4 ψ' Production and Decay Properties

Figure 7.11 gives the energy dependence of σ^{tot}, $\sigma_{\mu\mu}$ and $\sigma_{e^+e^-}$ near 3.7 GeV. A sharp ($\Gamma < 4$ MeV) signal called the ψ' is seen in the hadronic production at a mass of 3.68 GeV. An enhancement is also seen in $\sigma_{\mu\mu}$, though much weaker than for the J/ψ.

No signal is observed in the Bhabha cross section. The mass and width parameters of the ψ' are given in Table 7.4.

Table 7.4. Resonance parameters of the ψ'

	DORIS /76/	SLAC-LBL /89/
Mass [MeV]	3686 ± 6	3684 ± 5
$\int \sigma_h$ dm [μb MeV]	3.42 ± 0.7	3.7 ± 0.6
Γ_{tot} [keV]	198 ± 57	228 ± 56
Γ_{ee} [keV] assuming $\Gamma_{ee} = \Gamma_{\mu\mu}$	2.0 ± 0.3	2.1 ± 0.3

By the same line of argument as used for the J/ψ , the ψ' was shown /89/ to have $J^P = 1^-$ and C = -1. As an example Fig. 7.11b shows the interference between the ψ' decay and the QED contribution to μ-pair production. Isospin and G parity were determined through the cascade decay of the ψ',

$$\psi' \rightarrow J/\psi + X \qquad\qquad (7.6)$$

which accounts for 57 ± 8 % of the ψ' decays /89/. The cascade can be seen in

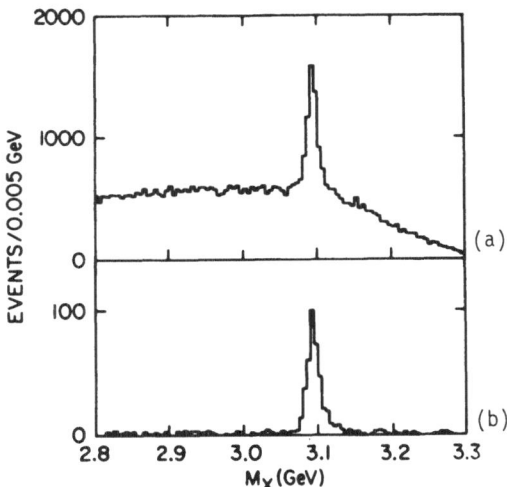

Fig. 7.12. (a) Missing mass of the system recoiling from an observed $\pi^+\pi^-$ pair. (b) Same as (a) for those events in which the observed charged particles satisfy within errors, conservation of total momentum and energy /89/

72

several ways. Figure 7.12 shows the missing mass spectrum for $\psi' \to \pi^+\pi^- X$, where the spike at $M_X = 3.1$ GeV is due to $X = J/\psi$. Figure 7.13 shows a strong J/ψ signal in the $\mu^+\mu^-$ effective mass spectrum from $\psi' \to J/\psi + X$.

$$\vert_{\to \mu^+\mu^-}$$

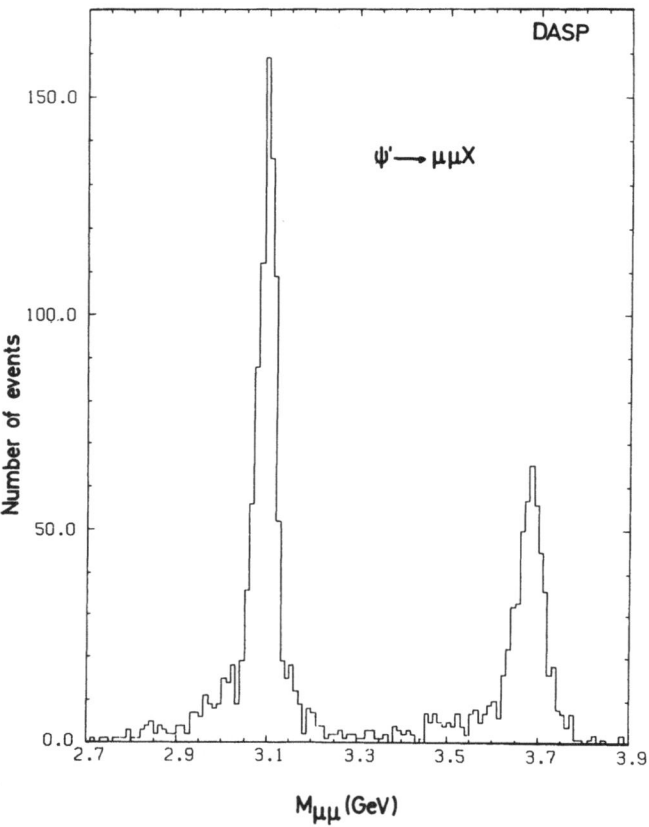

Fig. 7.13. Distribution of the $\mu^+\mu^-$ effective mass measured at the ψ' /12/

Table 7.5 lists the measured branching ratios for several cascade decays and the theoretical predictions for different isospin and G parity assignments assuming isospin conservation. Only $T^G = 0^-$ agrees with the data. Hence the ψ' has the same quantum numbers as the J/ψ.

Table 7.5. Branching ratios for cascade decays of the ψ'. $B_f = \dfrac{\Gamma_{\psi' \to J/\psi f}}{\Gamma_{\psi' \to J/\psi \text{ all}}}$

	Measured B		Relative B expected for T^G =		
Final state f	DASP /12/	SLAC-LBL /89/	1^-	1^+	2^-
$\pi^+\pi^-$	0.63 ± 0.10	0.56 ± 0.10	2	1	1
$\pi^0\pi^0$	0.32 ± 0.10	0.30 ± 0.05	1	0	2
η	0.065 ± 0.026	0.075 ± 0.014	allowed	0	0

7.5 SU$_3$ Assignment of J/ψ and ψ'

The hadronic two-body decay modes of J/ψ and ψ' provide a measure of the SU$_3$ properties of these resonances provided the decay conserves SU$_3$.

a) J/ψ,ψ' → $\pi\pi$,$K\bar{K}$

We first consider the decays into $\pi\pi$ and $K\bar{K}$ (and use J/ψ to denote also the ψ'). G parity forbids the decay J/$\psi \to \pi^+\pi^-$. Therefore, if the J/ψ is an SU$_3$ singlet the decay into K^+K^- and K^0K^0 is forbidden as well since these states are reached by a rotation in the T_3,Y plane.

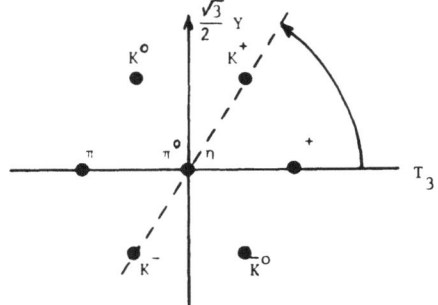

If J/ψ is an SU$_3$ octet the decays into K^+K^- and K^0K^0 are allowed and should have equal strengths, $\Gamma_{K^+K^-} = \Gamma_{K^0K^0}$. Table 7.6 lists the corresponding branching ratios as measured by the DASP /64, 81/ and SLAC-LBL /65, 83/ collaboration. Judging from the branching ratio for decay into p$\bar{\text{p}}$, the $\pi^+\pi^-$ and KK decay channels appear to be strongly suppressed. This favors the singlet assignment for J/ψ and ψ'.

Table 7.6. J/ψ and ψ' branching ratios B = Γ_f/Γ for two-body decays and the relative sizes for SU_3 singlet and octet assignments /90/

Decay mode	DASP /64, 81/	SLAC-LBL /65, 83, 84/	$\underset{\sim}{1}$	$\underset{\sim}{8}$
J/ψ → $\pi^+\pi^-$	$1.0 \pm 0.5 \cdot 10^{-4}$	$< 2 \cdot 10^{-4}$	0	0
K^+K^-	$2.2 \pm 0.9 \cdot 10^{-4}$	$< 1.2 \cdot 10^{-4}$	0	1
$K^0\overline{K^0}$		$< 0.8 \cdot 10^{-4}$	0	1
$p\bar{p}$	$2.5 \pm 0.4 \cdot 10^{-3}$	$1.98 \pm 0.15 \cdot 10^{-3}$		
ψ' → $\pi^+\pi^-$	$7.7 \pm 5.3 \cdot 10^{-5}$	$< 5 \cdot 10^{-5}$	0	0
K^+K^-	$9.6 \pm 7.0 \cdot 10^{-5}$	$< 5 \cdot 10^{-5}$	0	1
$K^0\overline{K^0}$			0	1
$p\bar{p}$	$1.4 \pm 0.8 \cdot 10^{-4}$	$2.3 \pm 0.7 \cdot 10^{-4}$		

b) J/ψ → πρ, KK*(892)

The decays into πρ and KK*(892) permit another test of the SU_3 assignment. Both decays are allowed by G parity and SU_3. If the J/ψ has singlet and octet components, the decay amplitudes will be of the form

$$A(\pi^\pm\rho) = A_1 - 2A_8$$
$$A(K^\pm K^*) = A_1 + A_8. \tag{7.7}$$

Different approaches were followed by SLAC-LBL /65, 82, 83/, DASP /81/, and DESY-Heidelberg /91/ to identify these decay modes. The SLAC-LBL group and DESY-Heidelberg analyzed the exclusive final states $\pi^+\pi^-\pi^0$ and $K\bar{K}$ while in the DASP experiment ρ and K* were observed as missing mass peaks in the $\pi^\pm$ and $K^\pm$ inclusive particle spectra (see Figure 7.14). The resulting branching ratios are listed in Table 7.7.

After phase space correction (which increases B_{KK*} by 1/0.85) one finds

$$\left|\frac{A_8}{A_1}\right| \cos\delta \quad \begin{array}{l} = -0.10 \pm 0.06 \quad \text{(DASP)} \\ = -0.12 \pm 0.06 \quad \text{(SLAC-LBL)} \end{array}$$

where δ is the relative phase between A_1 and A_8. The octet contribution is small and compatible with zero.

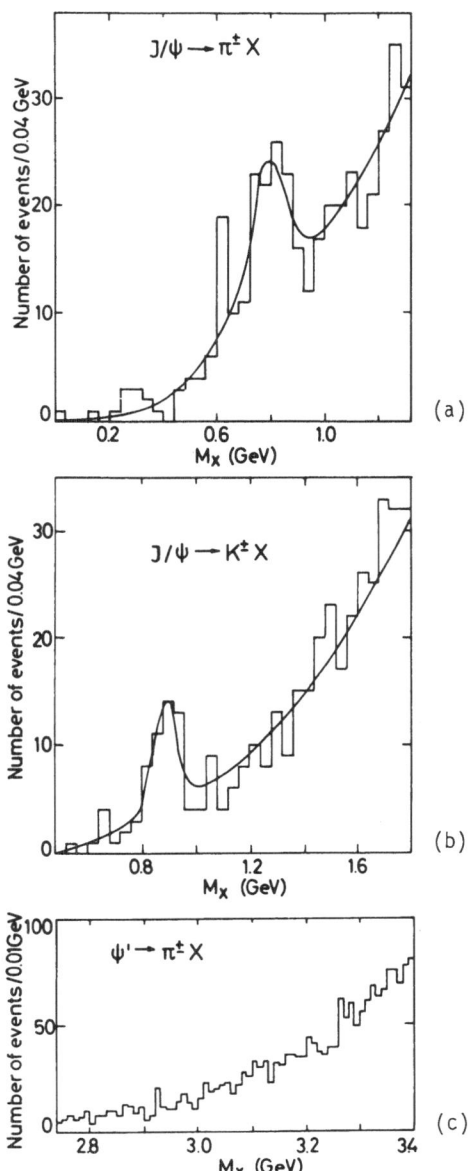

Fig. 7.14. Missing mass spectra observed in the decay: (a) $J/\psi \to \pi^{\pm}X$, (b) $J/\psi \to K^{\pm}X$, (c) $\psi' \to \pi^{\pm}X$ /81/

Table 7.7. Branching ratios for $J/\psi \to \pi\rho$ and KK^*

Decay mode	DASP /64, 81/	D-HD /91/	SLAC-LBL /82, 83/
$J/\psi \to \pi^{\pm}\rho$	$0.81 \pm 0.17 \cdot 10^{-2}$	0.62 ± 0.2	$0.83 \pm 0.2 \cdot 10^{-2}$
$\pi\rho$	$1.1 \pm 0.3 \cdot 10^{-2}$	1.0 ± 0.2	$1.24 \pm 0.3 \cdot 10^{-2}$
$K^{\pm}K^*$ (892)	$0.38 \pm 0.09 \cdot 10^{-2}$		$0.32 \pm 0.06 \cdot 10^{-2}$
$K^0 K^{*0}$(892)	$0.33 \pm 0.14 \cdot 10^{-2}$		$0.27 \pm 0.06 \cdot 10^{-2}$

If we assume J/ψ to be a singlet the contribution A_γ from one-photon decay to the KK^* channels is given by

$$A(K^0 K^{0*}) = A_1 - A_\gamma$$
$$A(K^+ K^{*-}) = A_1 + A_\gamma \tag{7.8}$$

The SLAC-LBL data yield $\left|\dfrac{A_\gamma}{A_1}\right| \cos\delta = 0.04 \pm 0.14$ the DASP data yield 0.02 ± 0.08; i.e., the one-photon decay is small.

c) $J/\psi \to MM'$

Consider the general case of J/ψ decaying into a pair of mesons $M,\bar{M}'$. Assume $M,\bar{M}'$ to be nonstrange mesons and the decay $J/\psi \to M\bar{M}'$ to be forbidden by G parity or isospin (which is the case, e.g., if $M' = M$); M and M' do not have to belong to the same SU_3 multiplet. Suppose M_s, $\bar{M}_s$ are strange mesons that can be reached from $M,\bar{M}'$ by rotation within the SU(3) multiplet. Then

$$J/\psi \to M_s \bar{M}'_s \quad \text{is} \quad \begin{Bmatrix} \text{forbidden} \\ \text{allowed} \end{Bmatrix} \quad \text{for} \quad \begin{Bmatrix} \text{a singlet} \\ \text{an octet} \end{Bmatrix}$$

Table 7.8. Branching ratios for J/ψ

Decay mode	Branching ratio		$\underset{\sim}{1}$	$\underset{\sim}{8}$
$\pi^{\pm} A_2$	$< 4.3 \cdot 10^{-3}$	/65, 83/	0	allowed
$K^{\pm}K^*(1420)$	$< 3.3 \cdot 10^{-3}$	/65, 83/	0	allowed
$K^{*0}(892) \overline{K^{*0}(892)}$	$< 5 \cdot 10^{-3}$	/65, 83/	0	allowed
$K^{*0}(892) \overline{K^{*0}(1420)}$	$6.7 \quad 2.6 \cdot 10^{-3}$	/65, 83/	allowed	allowed
$K^*(1420) \overline{K^*(1420)}$	$< 2.9 \cdot 10^{-3}$	/65, 83/	0	allowed

In particular, if $M_S = M_S'$ and the J/ψ is a singlet, the decay J/$\psi \to M_S\bar{M}_S'$ is forbidden. Table 7.8 lists the branching ratios for various other decay channels. Again, the data are consistent with an SU_3 singlet assignment.

d) J/ψ,ψ' → B$\bar{B}'$

Apart from the decay into p$\bar{p}$ the information on two-body baryonic decay channels is scanty. Table 7.9 lists the measured branching ratios and the values corrected for phase space effects relative to the p$\bar{p}$ decay. Within the rather large errors the phase space corrected branching ratios for p$\bar{p}$, $\Lambda\bar{\Lambda}$ and $\Xi^-\bar{\Xi}^-$ are the same as expected for an SU_3 singlet assignment, provided there is no interference with the one-photon decay /39/.

Table 7.9. Baryonic decays of J/ψ and ψ'

Decay mode	Branching ratio		Corrected for phase space rel. to p$\bar{p}$
J/ψ → p$\bar{p}$	$2.4 \pm 0.2 \cdot 10^{-3}$	/64, 65, 77, 83, 84/	$2.2 \pm 0.3 \cdot 10^{-3}$
→ $\Lambda\bar{\Lambda}$	$1.6 \pm 0.8 \cdot 10^{-3}$	/89, 92/	$2.4 \pm 1.2 \cdot 10^{-3}$
→ $\Lambda\bar{\Sigma}$	$< 0.4 \cdot 10^{-3}$	/89, 92/	
→ $\Xi^-\bar{\Xi}^-$	$\sim 0.4 \cdot 10^{-3}$	/89, 92/	$\sim 1.4 \cdot 10^{-3}$
ψ' → p$\bar{p}$	$1.9 \pm 0.5 \cdot 10^{-4}$	/64, 83/	$2.3 \pm 0.7 \cdot 10^{-4}$
→ $\Lambda\bar{\Lambda}$	$< 4 \cdot 10^{-4}$	/92/	
→ $\Xi^-\bar{\Xi}^-$	$\sim 2 \cdot 10^{-4}$	/92/	$\sim 4 \cdot 10^{-4}$

7.6 Inclusive Particle Spectra from J/ψ and ψ'

In this section we discuss the charged π, K, and $\bar{p}$ spectra from J/ψ and ψ' decay as measured by the DASP group /75/. In gathering these data a pure inclusive trigger was employed. The trigger efficiency was thus independent of the final state. Pions, kaons, and protons (antiprotons) were identified by time-of-flight and momentum measurements. Since the proton yield was contaminated by beam-gas scatters, only the antiprotons were used in the analysis.

The excitation curves (i.e., the particle cross sections integrated over momentum as a function of the total energy $\sqrt{s}$) were used to integrate cross sections over

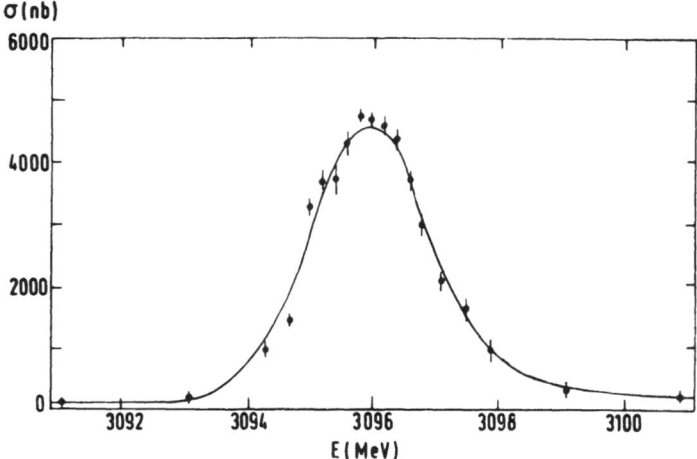

Fig. 7.15. Cross section for inclusive charged hadron production, $e^+e^- \to h^{\pm}X$ at the J/ψ as a function of the total c.m. energy integrated over particle momenta between 0.5 and 1.3 GeV/c /75/

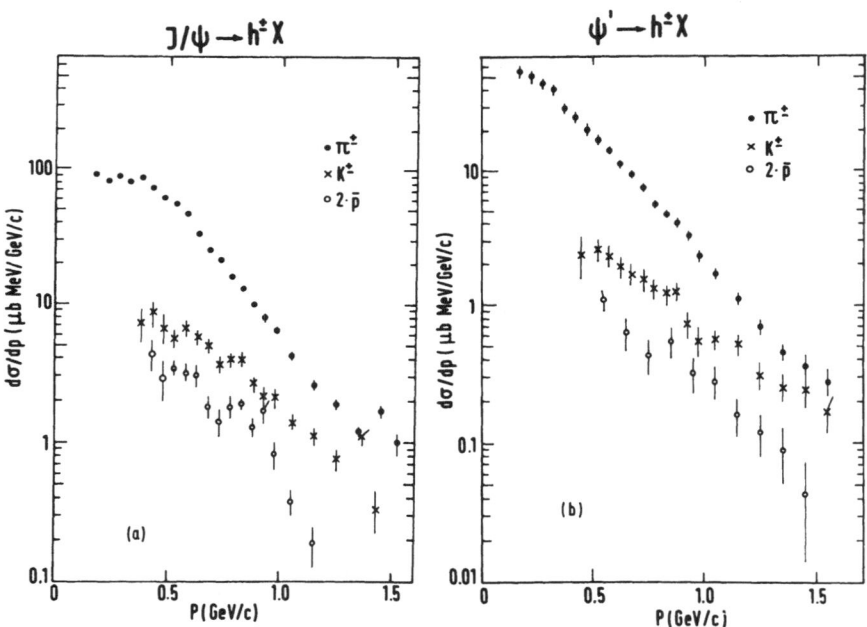

Fig. 7.16. Differential cross sections, dσ/dp, for $\pi^{\pm}$, $K^{\pm}$, and $\bar{p}$ for the J/ψ and ψ' integrated over the resonance as described in the text. The $\bar{p}$ yield has been multiplied by a factor of 2 /75/

the resonance. As an example Fig. 7.15 shows the cross section for production of charged hadrons with momenta between 0.5 and 1.3 GeV/c in the region of the J/ψ. The resonance contribution was integrated over the total c.m. energy and corrections were made for radiative effects in the initial state. The cross sections below represent radiatively corrected energy integrals over the resonance contribution, e.g., dσ/dp stands for

$$\int_R \frac{d\sigma(E)}{dp} \, dE.$$

Figure 7.16. shows the momentum spectra for $\pi^\pm$ (i.e., the sum of π^+ and π^-), $K^\pm$ and $2 \cdot \bar{p}$ from J/ψ and ψ' decay. Both resonances show qualitatively the same behavior. The pion and kaon yields decrease exponentially with momentum, but with different slopes. At 0.5 GeV/c the K yield is a factor of ~10 below the pion yield but approaches the pion yield with increasing momentum. The $2 \cdot \bar{p}$ yield is a factor of 2 below the K cross section. For the J/ψ the $\pi^\pm \rho$ and $K^\pm K^*$ decay channels show up as enhancements near the kinematical limit.

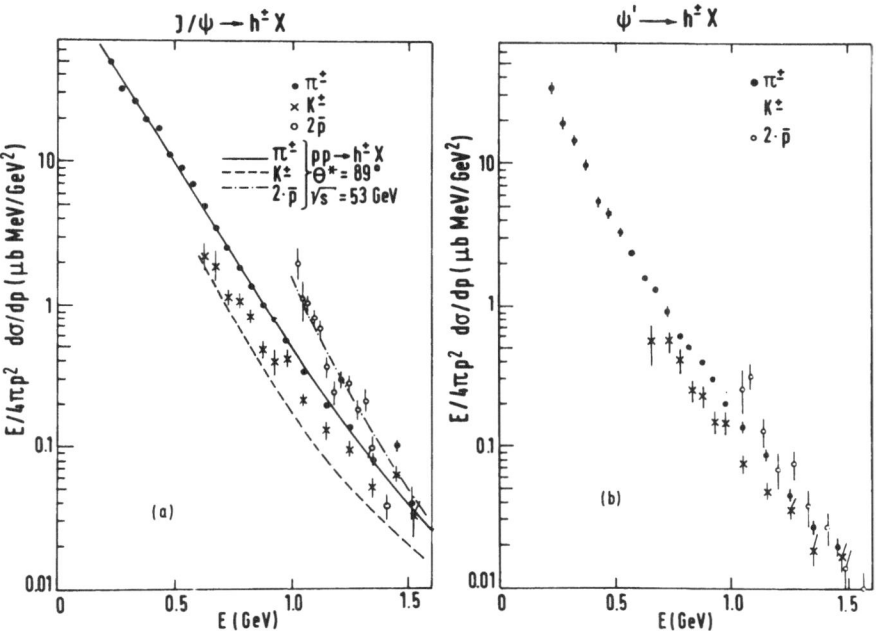

Fig. 7.17. Invariant cross sections, $(E/4\pi p^2)d\sigma/dp$, integrated over the resonance for $\pi^\pm$, $K^\pm$, and $\bar{p}$ for the J/ψ and ψ'. The $\bar{p}$ cross section has been multiplied by a factor of 2 /75/. The curves in (a) describe the inclusive particle yields from pp → $h^\pm$X /94/

Figure 7.17 presents the particle spectra as invariant cross sections $\left(E/4\pi p^2\right)d\sigma/dp$ versus particle energy. The π cross section for the J/ψ can be described by a single exponential. In the case of the ψ' there is a break near E = 0.4 GeV: the enhancement below this energy stems from the cascade decay $\psi' \rightarrow \pi^+\pi^- J/\psi$. The 2 $\cdot$ $\bar{p}$ cross sections lie above the $\pi^{\pm}$ cross section by the same factor for both resonances. Describing these data by a single exponential, $[(E/4\pi p^2)d\sigma/dp \sim \exp(-bE)]$ the following slope values (in GeV^{-1}) were found:

$$J/\psi : \quad b_\pi = 5.9 \pm 0.1 \qquad b_K = 5.2 \pm 0.3 \qquad b_{\bar{p}} = 7.2 \pm 0.6$$

$$b_\pi = 5.8 \pm 0.1 \qquad b_K = 4.7 \pm 0.3 \qquad b_{\bar{p}} = 6.3 \pm 0.6$$

$$(E_\pi > 0.5 \text{ GeV})$$

The behavior of the inclusive spectra observed for J/ψ and ψ' is very similar to that found in the nonresonant region (see Sect. 6.4). Even more surprising is the strong resemblance to the data on pp collisions obtained at ISR for the central region /94, 95/. The curves shown in Fig. 7.17 represent the invariant pp cross sections for $\sqrt{s}$ = 5.3 GeV and y $\approx$ 0 /94/ multiplied by a single common normalization factor. Instead of the particle energy E, the transverse energy $E_T = \sqrt{m^2 + P_T^2}$ was used. The pp data show a somewhat smaller relative kaon rate.

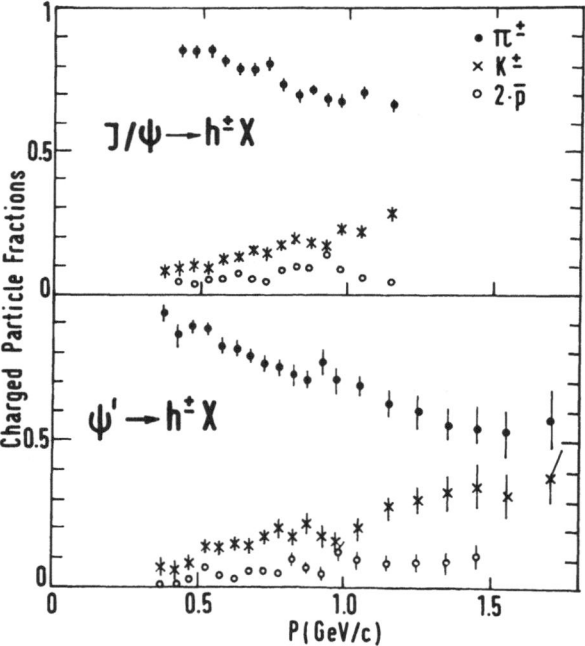

Fig. 7.18. Charged particle fractions as a function of momentum for the J/ψ and ψ'. The $\bar{p}$ fraction has been multiplied by a factor of 2 /75/

The particle ratios, e.g., R_π = (number of $\pi^\pm$)/(sum of $\pi^\pm$, $K^\pm$ and 2 · $\bar{p}$) are
plotted as function of momentum in Fig. 7.18. The K and $\bar{p}$ fractions increase with
momentum, and the π fraction decreases correspondingly. R_π and R_K seem to approach
a common value of ~0.4 at p ≳ 2 GeV/c. Integration over all momenta gives the following average particle ratios:

	$R_{\pi^\pm}$	$R_{K^\pm}$	$R_{2\bar{p}}$
J/ψ:	87.5 ± 1.3 %	8.9 ± 1.0 %	3.6 ± 0.9 %
ψ' :	90.8 ± 1.0 %	6.9 ± 0.9 %	2.3 ± 0.7 %

No significant difference between J/ψ and ψ' is observed; in particular, the K
fraction for the ψ' is not larger than for the J/ψ. However, 57 ± 8 % of the ψ'
decays proceed via the cascade transition, ψ' → J/ψX. This contribution should be
subtracted when comparing J/ψ and ψ'. The subtraction was made by DASP and led to
the following result:

ψ' (without cascade decay):
$R_{\pi^\pm}$ = 90.0 ± 3 % $R_{K^\pm}$ = 8.5 ± 2.7 % $R_{p,\bar{p}}$ = 1.5 ± 1.9 %

within errors, the particle ratios are the same as for the J/ψ.

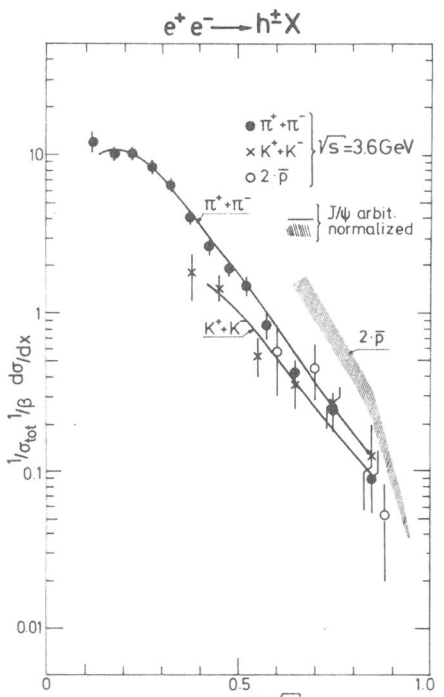

Fig. 7.19. The cross section $(1/\sigma_{tot})(1/\beta)\frac{d\sigma}{dx}$,
$x=2E/\sqrt{s}$, for the sum $\pi^+ + \pi^-$, $K^+ + K^-$ and
twice the $\bar{p}$ yield at 3.6 GeV and from J/ψ
decay. The J/ψ cross sections are shown by
the curves and the hatched band. They have
been normalized by a common factor. (Data
from DASP, /96/)

Annihilation into hadrons at high energies appears to proceed via a primary quark-antiquark pair. The hadronic decay of the J/ψ on the other hand is believed to have three gluons as an intermediate state. For this reason one might expect the J/ψ to yield a steeper falling x spectrum (i.e., more low momentum particles) as compared to nonresonant hadron production. In Figure 7.19 the quantity $(1/\sigma_{tot})$ $(1/\beta)$ $d\sigma/dx$ is shown for pions from J/ψ decay and at a c.m. energy of 3.6 GeV /96/. Within errors no difference is observed in the shape of the two spectra. Most likely the energy of ~3 GeV is too low for the two mechanisms to produce noticeably different spectra. This may be seen from the fact that at 3 GeV the sphericity distributions from phase space and jet formation are nearly the same.

8. Radiative Decays of J/ψ and ψ'

The experimental data strongly favor the hypothesis that J/ψ and ψ' are bound states of a fourth type of quark, the charm quark c, J/ψ (ψ') = $c\bar{c}$. The suppression of decays into ordinary hadrons is attributed to the OKUBO-ZWEIG-IIZUKA /97/ mechanism, which states that disconnected quark diagrams are suppressed compared to connected ones:

Since the physical $c\bar{q}$ states are heavier than $\frac{1}{2}$ $M_{J/\psi}$ the hadronic decays of the J/ψ can proceed only via disconnected diagrams. The comparison of the J/ψ decay width with those of ordinary mesons gives a OZI suppression factor on the order of 10^{-4} (see Sect. 7.1). Because of this strong suppression radiative decays which are of order α can be expected to play an important role.

8.1 The Decays $J/\psi \rightarrow \gamma\pi^0$, $\gamma\eta$, and $\gamma\eta'$

In the following we shall discuss the experimental data on the following decays:

$$J/\psi(\psi') \rightarrow \gamma\pi^0 \tag{8.1}$$
$$\gamma\eta \tag{8.2}$$
$$\gamma\eta' \tag{8.3}$$

Since these are two-body decays they lead to monochromatic γ rays which might show up in a measurement of the photon energy spectrum from J/ψ decay. Figure 8.1 shows the data from SLAC-LBL /98/ who observed photons which converted in the beam pipe, constructing the photon energy from the e^+,e^- momenta. No evidence is found for the channels (8.1 - 3) which should show up at E = 1.54, 1.44, and 1.23 GeV, respectively. The DASP collaboration sought these decays in final states involving only photons. In a first attempt events with two collinear or nearly collinear photons were analyzed /99/. Collinear photon pairs are produced in the process $e^+e^- \rightarrow \gamma\gamma$ which can be calculated from QED (3.7). Superimposed on this cross section,

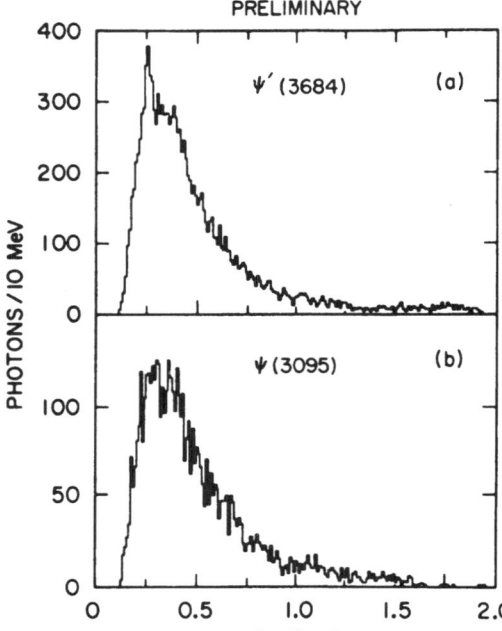

Fig. 8.1. Photon energy spectra from J/ψ and ψ' decay, respectively /98/

which decreases smoothly with s, there could be a peak near the mass of the J/ψ (or ψ') from a decay of the type J/ψ → γX⁰ where X⁰ in turn decays into two photons. (The decay J/ψ or ψ' → γγ is strictly forbidden for a spin 1 particle.) If the mass of X is small (e.g., X = π^0) or close to the mass of the J/ψ (or ψ') the events will look like nearly collinear photon pairs.

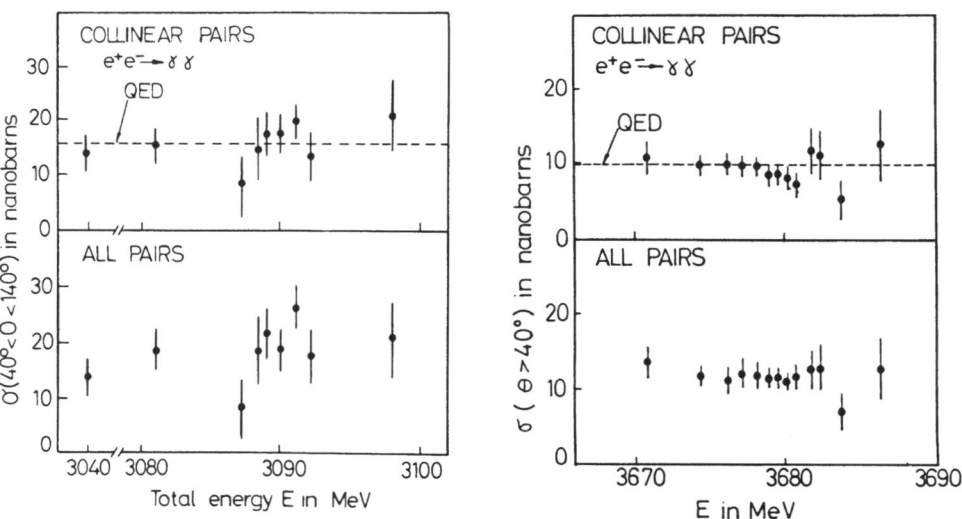

Fig. 8.2. Cross section for $e^+e^- \rightarrow \gamma\gamma$ in the vicinity of the J/ψ and ψ' /99/

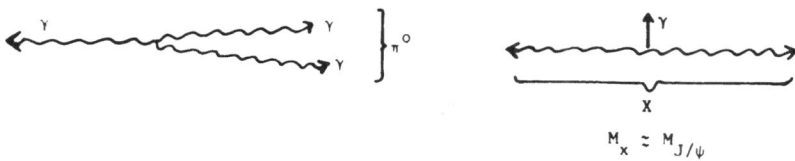

The cross sections measured at the J/ψ and ψ' for collinear (to within 6°) and nearly collinear photon pairs are shown in Figure 8.2. They are in good agreement

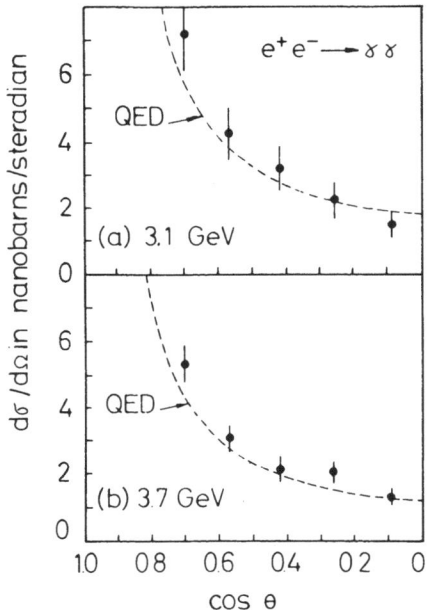

Fig. 8.3. Angular distributions of $e^+e^- \to \gamma\gamma$ for c.m. energies in the vicinity of the J/ψ and ψ' /99/

with the QED prediction. The photon angular distributions agree with QED as well (Figure 8.3). No enhancements are found at the resonances. Table 8.1 summarizes the measured upper limits.

Table 8.1. Upper limits (90 % C.L.) on branching ratios from two-photon events /99/

Decay mode	J/ψ	ψ'
$\Gamma_{\gamma\gamma}/\Gamma$	< 0.003	< 0.008
$\Gamma_{\pi^0\gamma}/\Gamma$	< 0.01	< 0.01
$\frac{\Gamma_{\chi^0\gamma}}{\Gamma} \frac{\Gamma_{\chi \to \gamma\gamma}}{\Gamma_{\chi \to all}}$	< 0.003 2.99 < M_χ < 3.09 GeV	< 0.008 3.58 < M_χ < 3.68 GeV

Three-Photon Events

The three-photon channel was investigated by the DASP collaboration /100/ and by the DESY-Heidelberg experiment /101/. We shall discuss the DASP analysis in some detail. Processes that can give rise to three-photon events are

a) $J/\psi \rightarrow \gamma X$,
        $\rightarrow \gamma\gamma$

where $X = \pi^0$ (branching ratio to $\gamma\gamma$ = 100 %), η (38 %), η' (1.9 %),... .

b) A direct $J/\psi \rightarrow \gamma\gamma\gamma$ decay. This decay does not lead to a narrow enhancement in the $\gamma\gamma$ mass distribution.

c) QED processes $e^+e^- \rightarrow 3\gamma$ of the type depicted by the following diagram:

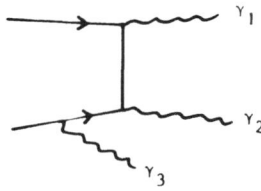

The energy spectrum of the radiated photon (γ_3) is bremsstrahlunglike, $P(K)dK \rightarrow K^{-1}$ and leads to a $\gamma_1\gamma_2$ mass spectrum that peaks at the kinematical limit.

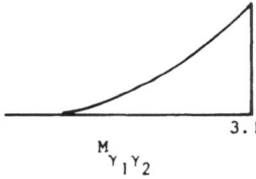

d) Multi π^0 production, e.g. $J/\psi \rightarrow \pi^0\pi^0\gamma$ where two of the final state photons are not detected in the apparatus.

The event selection was done in the following way. From a total of $\sim 10^6$ J/ψ decays $\sim$250 events were found with exactly three photons detected. For each photon the production angles were measured taking the position of the interaction point

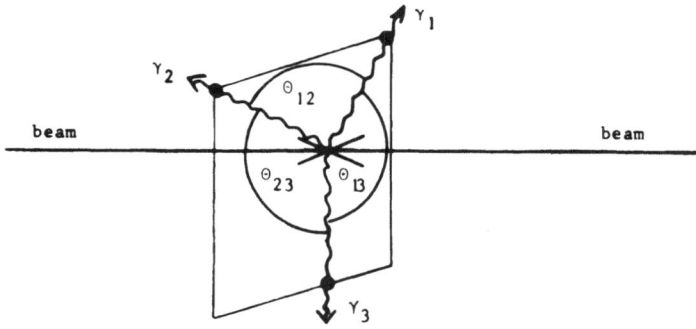

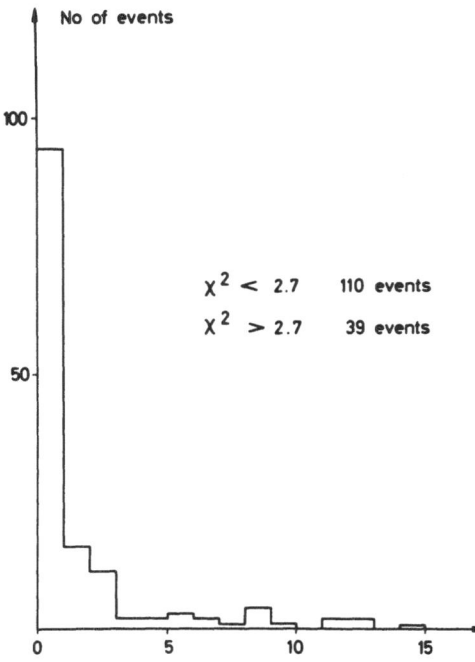

No of events

$\chi^2 < 2.7$ 110 events

$\chi^2 > 2.7$ 39 events

Fig. 8.4. Number of three-photon events at the J/ψ plotted as a function of χ^2 as computed from the hypothesis that the photons are coplanar /100/

from Bhabha events observed in the same runs. Genuine 3γ event have to lie in a plane. The test for coplanarity was made by a 1-constraint fit, which gave the χ^2 distribution shown in Figure 8.4. Candidate for the decay J/ψ → 3γ were required

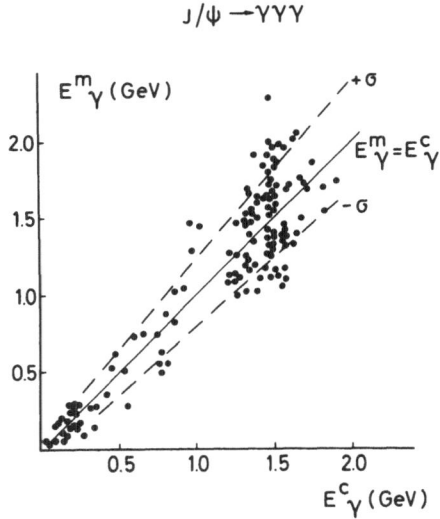

J/ψ ⟶ γγγ

E^m_γ (GeV)

$+\sigma$

$E^m_\gamma = E^c_\gamma$

$-\sigma$

E^c_γ (GeV)

Fig. 8.5. Distribution of measured versus computed photon energy for coplanar three-photon events at the J/ψ /100/

to have $\chi^2 < 2.7$. One hundred ninety events satisfied this cut; approximately 10 % genuine 3γ events are lost. Using energy and momentum conservation the energies of the three-photons were computed and compared to those measured by the detector using shower counter pulse height. The comparison shows good agreement (Figure 8.5). Since the measured γ energies are imprecise the computed values were used for further analysis. The angular resolution of $\pm 1^{\circ}$ leads to a two-photon mass resolution of about 20 MeV nearly independent of mass.

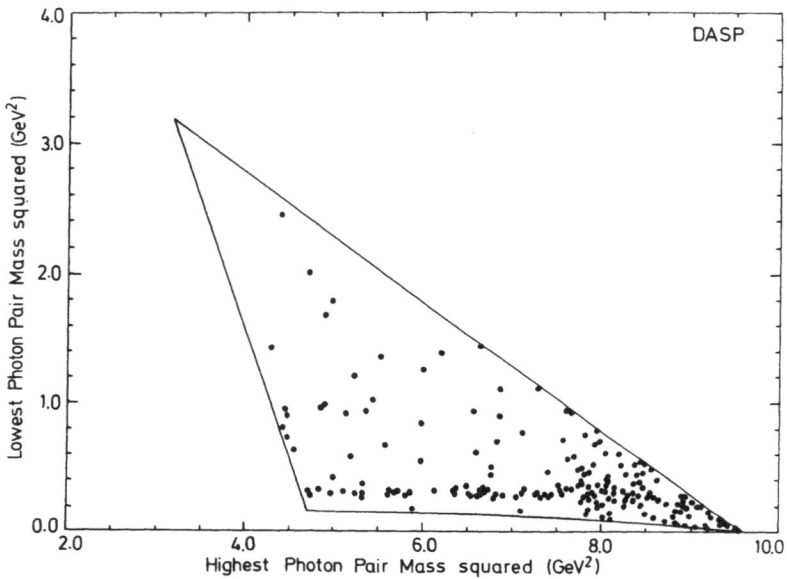

Fig. 8.6. Dalitz plot of coplanar three-photon events at the J/ψ /100/

Of the three possible $\gamma\gamma$ mass combinations only two are independent. Figure 8.6 shows a Dalitz plot of the low- and high-mass combinations. A clustering of events around the η mass is clearly visible. Figure 8.7 shows the projection onto the low-mass combination. The curve marked QED indicates the background expected from the QED processes. Apart from the η signal some η' production may be present.

In the same event sample a search was made for the J/$\psi \rightarrow \gamma\pi^{\circ}$ decay and a few candidates were found. A very clear signal for the J/ψ decay into $\gamma\eta'$ was observed by the DESY-Heidelberg collaboration analyzing the $\pi^+\pi^-\gamma\gamma$ final state /104/.

Figure 8.8 shows the $\pi^{\pm}\gamma\gamma$ mass distribution with a strong ρ signal and the $\pi^+\pi^-\gamma$ mass distribution with a peak at the position of the η'.

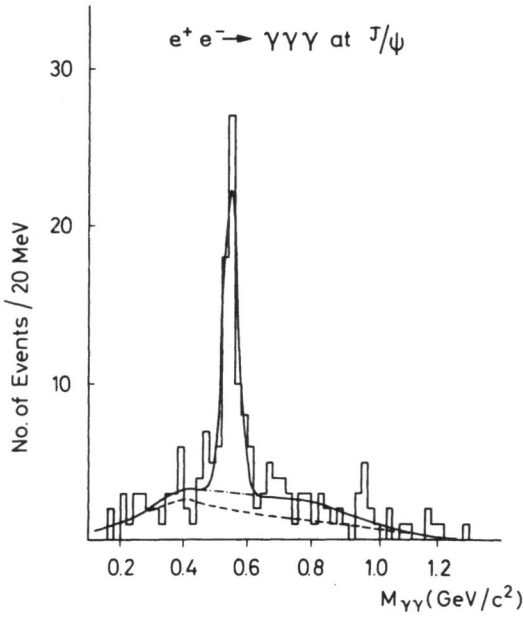

Fig. 8.7. Distribution of the lowest photon pair mass for three-photon events at the J/ψ /100/

The DASP collaboration has also analyzed three-photon events in the ψ' region. The Dalitz-plot distribution is shown in Figure 8.9. No evidence for ψ → γη or γη' decays was observed.

Table 8.2 summarized the branching ratios measured for the decays (8.1 - 3).

8.2 The Radiative Decays J/ψ → fγ, f'γ and $\pi^+\pi^-\gamma$, K^+K^-

PLUTO /102/ and DASP /103/ have observed the decay J/ψ → f(1270)γ. It was detected in the final state $\pi^+\pi^-\gamma$. Some difficulty was caused by copious $\pi^+\pi^-\pi^0$ production from πρ decay.

90

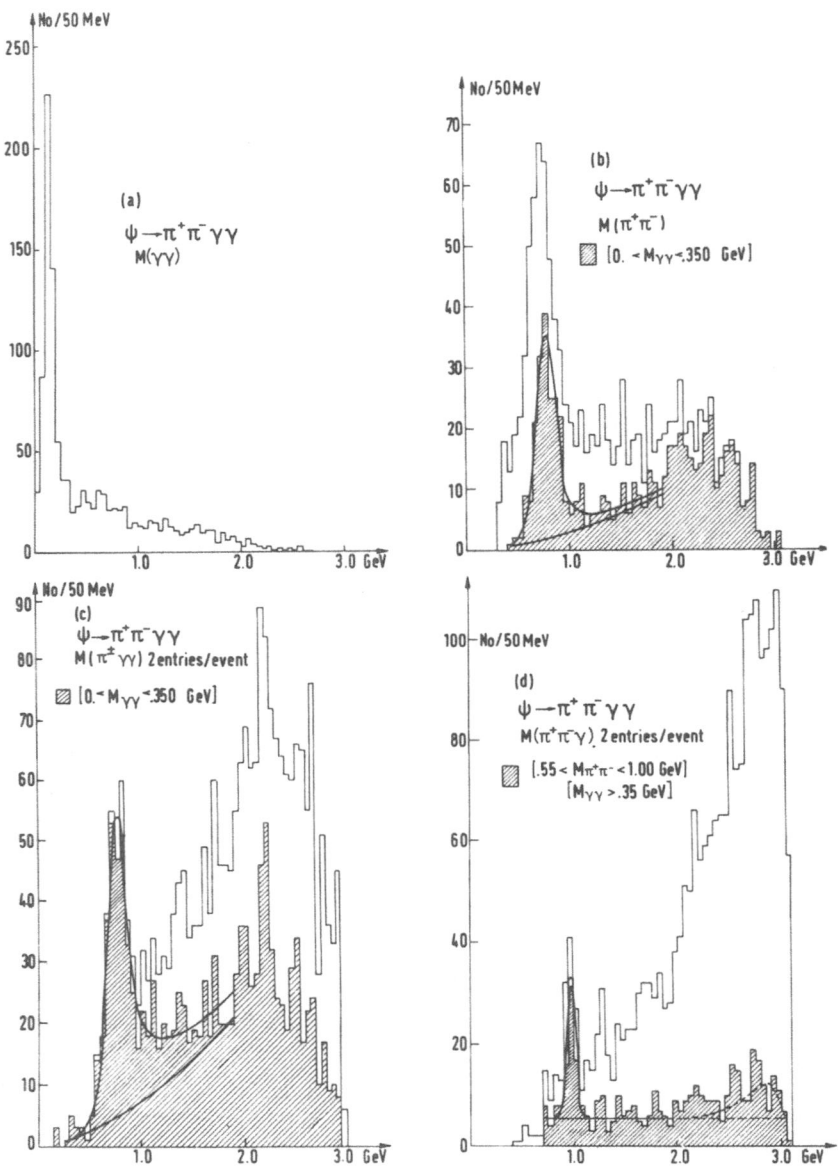

Fig. 8.8. Invariant mass distributions of two- and three-particle combinations from J/ψ decay /91/. (a) M(γγ). (b) M($\pi^+\pi^-$); Hatched histogram: events with $M_{\gamma\gamma}$ < 0.35 GeV. (c) M($\pi^{\pm}\gamma\gamma$); Hatched histogram as in (a). (d) M($\pi^+\pi^-\gamma$); Hatched histogram: events with M(γγ) $\geq$ 0.35 GeV, 0.55 < M($\pi^+\pi^-$) < 1 GeV

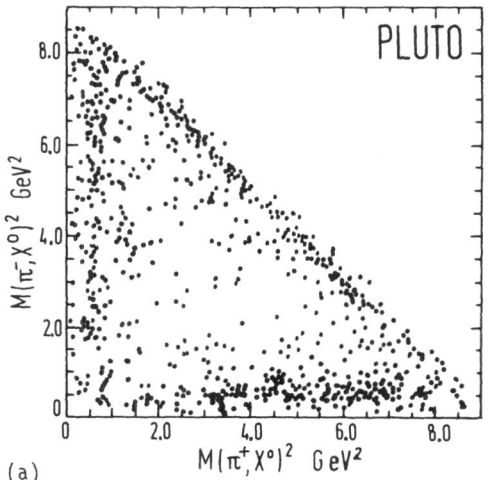

(a)

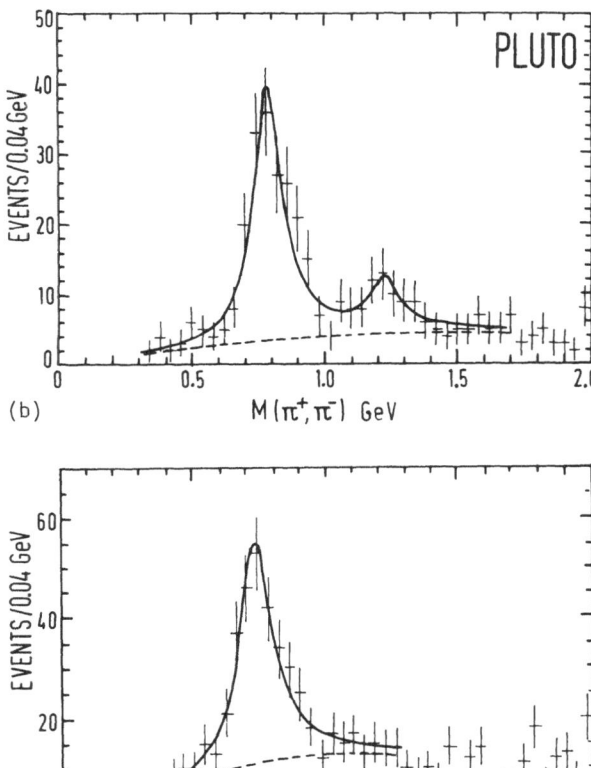

(b)

(c)

Fig. 8.9. (a) Dalitz plot of M^2 (π^-X^0) versus $M^2(\pi^+X^0)$ where X^0 stands for either a photon or a π^0-meson, (b) The $M(\pi^+\pi^-)$ distribution of the $\pi^+\pi^-\gamma$ sample excluding events lying in the ρ^+ and ρ^- mass bands, (c) Combined $M(\pi^+X^0)$ and $M(\pi^-X^0)$ distribution for $\pi^+\pi^-\gamma$ sample where X^0 represents either a photon or a π^0-meson. Events lying within the ρ^0, ρ^- and π^0, ρ^+ mass bands, respectively, were removed. The solid lines represent the best fit to the data, the dashed lines represent the estimated backgrounds /102/

<u>Table 8.2.</u> Radiative decays of J/ψ and ψ': Branching ratios Γ_f/Γ_{tot}

Decay Mode	J/ψ		ψ'	
$\gamma\pi^0$	$7.3 \pm 4.7 \cdot 10^{-5}$	/100/	$< 10^{-2}$	/99/
$\gamma\eta$	$0.80 \pm 0.18 \cdot 10^{-3}$	/100/	$< 0.4 \cdot 10^{-3}$	/100/
	$1.3 \pm 0.4 \cdot 10^{-3}$	/101/		
$\gamma\eta'$	$2.2 \pm 1.7 \cdot 10^{-3}$	/100/	$< 7 \cdot 10^{-3}$	/100/
	$1.8 \pm 0.8 \cdot 10^{-3}$	/101/		
$\gamma f(1270)$	$2.0 \pm 0.7 \cdot 10^{-3}$	/102/		
	$(0.9 \pm 0.3)-(1.5 \pm 0.3) \cdot 10^{-3}$ [a]	/103/		
$\gamma f'(1516)$	$\leq 0.34 \cdot 10^{-3}$	/103/		
$\gamma\pi^+\pi^-(M_{\pi\pi} < 1 \text{ GeV})$	$< 0.7 \pm 0.2 \cdot 10^{-3}$	/103/		
$\gamma\pi^+\pi^-(M_{\pi\pi} > 1.6 \text{ GeV})$	$0.17 \pm 0.11 \cdot 10^{-3}$	/103/		
$\gamma K^+ K^-(M_{KK} > 1.6 \text{ GeV})$	$\leq 0.25 \cdot 10^{-3}$	/103/		
$\rho^0\pi^0$	$4.5 \pm 0.3 \cdot 10^{-3}$	/82, 83, 102, 103/		
$\Phi\eta$	$1.0 \pm 0.6 \cdot 10^{-3}$	/83/		
	$0.8 \pm 0.9 \cdot 10^{-3}$	/64/		
$\Phi\eta'$	$< 1.3 \cdot 10^{-3}$	/83/		
Φf	$< 0.37 \cdot 10^{-3}$	/83/		
ωf	$2.8 \pm 1.1 \cdot 10^{-3}$	/83, 102/		
$\omega f'$	$< 0.16 \cdot 10^{-3}$	/83/		

[a] The exact value depends on the form assumed for the f production and decay angular distribution.

In the PLUTO experiment events with two charged tracks were fitted to the hypothesis $\pi^+\pi^-X^0$ where X^0 is a photon. Figure 8.9 shows the Dalitz plot and the mass distributions for events which gave an acceptable fit. Apart from prominent ρ peaks in the $\pi^\pm X^0$ and $\pi^+\pi^-$ mass spectra an accumulation of events around $M_{\pi^+\pi^-} \sim 1.25$ GeV is seen. Of the possible sources for this bump: $\varepsilon(1200)$, $f(1270)$, $\rho'(1250)$, only the $f(1270)$ was found to be consistent with the data. By C invariance the missing neutral then has to be a photon (and not a π^0).

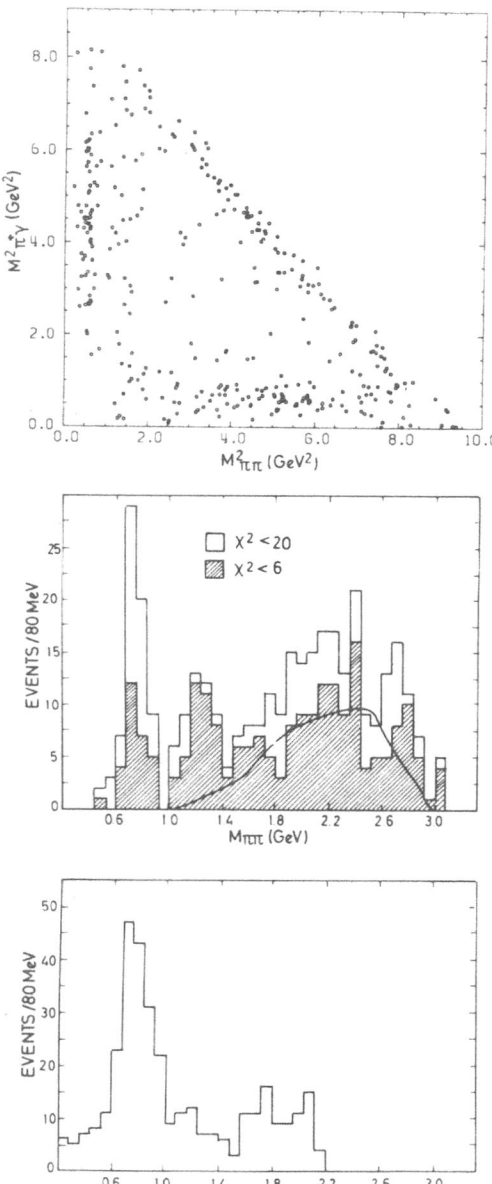

Invariant mass distribution for $\pi^+\pi^-\gamma$ events with $\chi^2 < 20$. Only the low-mass $\pi\gamma$ combination is shown /103/

$\pi^+\pi^-$ invariant mass distribution for $\pi^+\pi^-\gamma$ with $\chi^2 < 20$. The shaded area represents events with $\chi^2 < 6$. The line represents reflection from $\rho^{\pm}\pi^{\pm}$ events as calculated by Monte Carlo ($\chi^2 < 6$)

Fig. 8.10a. Dalitz plot for e^+e^-
$e^+e^- \to J/\psi \to \pi^+\pi^-\gamma$ events with $\chi^2 < 20$

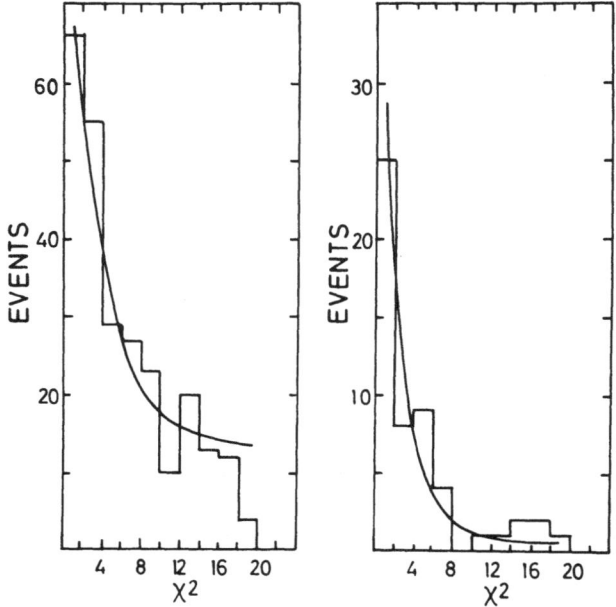

Fig. 8.10b. x^2 distributions for $\pi^+\pi^-\gamma$ events; left: in the ρ^0 band ($m^2_{\pi\pi} < 1.0$ GeV^2) or in the $\pi^\pm$ band ($m^2_{\pi\gamma} < 1.0$ GeV^2); right: in the f band (1.0 GeV^2 < $m^2_{\pi\pi}$ < 2.0 GeV^2). The lines represent the expected distributions in these bands as calculated by Monte Carlo /103/

   The DASP group selected events with a charged pion identified in the magnetic spectrometers plus an additional charged track and a single photon. In Figure 8.10 the Dalitz plot and the mass spectra are shown for events fitting the $\pi^+\pi^-\gamma$ hypothesis with x^2 < 20. The result is similar to that from PLUTO: a strong population of the ρ bands and a signal in the f region. The x^2 distribution in the f region is considerably narrower than for ρ events (see Figure 8.10b) which gives supporting evidence that the events in the f region are indeed $\pi^+\pi^-\gamma$ events. Table 8.2 lists the branching ratios determined in the two experiments. Due to the limited acceptance the DASP result depends on the assumed form of the f production and decay angular distribution.

   Similar to the measurement just described DASP /103/ searched for the decay $J/\psi \rightarrow f'(1516)\gamma$ with f' decaying into K^+K^-. No clear signal was found (Figure 8.11). An upper limit on the decay rate is given in Table 8.2

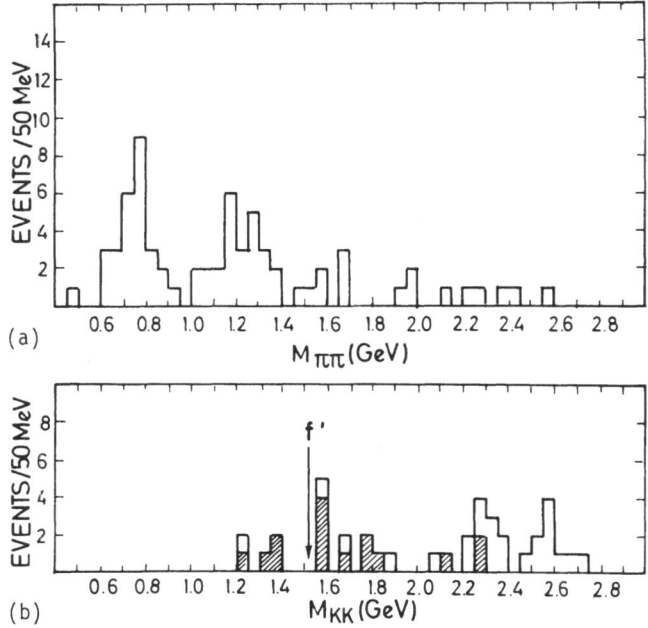

Fig. 8.11. (a) $\pi^+\pi^-$ invariant mass distribution for $\pi^+\pi^-\gamma$ events with $\chi^2 < 6$ and $m^2_{\pi\gamma} > 1.2$ GeV^2 for both $\pi\gamma$ combinations, (b) K^+K^- invariant mass distribution for $K^+K^-\gamma$ events with $\chi^2 < 6$ and $m^2_{K\gamma} > 1.2$ GeV^2 for both $K\gamma$ combinations /103/

8.3 Discussion[1]

Table 8.2 shows the radiative decays $J/\psi \to \gamma\eta$, $\gamma\eta'$ and γf to have roughly the same strength as the hadronic decays $J/\psi \to \phi\eta$ or ωf, and $\Gamma_{\gamma\pi^0}$ to be much smaller than $\Gamma_{\gamma\eta}$, $\Gamma_{\gamma\eta'}$ or $\Gamma_{\gamma f}$. Three different quark diagrams can contribute to the radiative decays /104/[2].

J/ψ c ————— g ∿∿∿ q   c ∿∿∿ g ————— q   c ————— c
          c̄ ————— ∿∿∿ q̄   c̄ ————— ∿∿∿ q̄   c̄ ————— c̄ M
              (a)   (b)   (c)

---

[1] We are grateful to T.F. Walsh for a detailed discussion on this subject.

[2] A possible mixing of the J/ψ with ordinary quarks will be neglected (see FRITZSCH and JACKSON /104/).

In the first two diagrams c and $\bar{c}$ couple to ordinary quarks $q,\bar{q}$ through gluon exchange. In the third diagram the meson $M(= SU_3$ singlet) is assumed to have a $c\bar{c}$ component[3].

For diagram a the $q\bar{q}$ system is in an SU_3 octet state since the $c\bar{c}$ system is a singlet and the photon couples only to the octet part of $q\bar{q}$.

In diagram b the $q\bar{q}$ system is in a singlet state.

The π^0 meson, being a pure SU_3 octet state, can only be produced via diagram a. Invoking VDM the $\gamma\pi^0$ branching ratio can be related to $\Gamma_{\rho^0\pi^0}$,

$$B(J/\psi \to \gamma\pi^0) \cong \frac{\alpha\pi}{\gamma_\rho^2} B(J/\psi \to \rho^0\pi^0) \approx 2 \cdot 10^{-5}, \tag{8.4}$$

which agrees with the measured value of $7.3 \pm 4.7 \cdot 10^{-5}$.

The η and η' mesons have singlet and octet components

$$\begin{aligned} \eta &= \eta_8 \quad \cos\theta + \eta_1 \sin\theta \\ \eta' &= -\eta_8 \sin\theta + \eta_1' \cos\theta, \end{aligned} \tag{8.5}$$

where

$$\begin{aligned} \eta_8 &= \sqrt{\frac{2}{3}} \ s\bar{s} - \sqrt{\frac{1}{6}} \ (u\bar{u} + d\bar{d}) \\ \eta_1' &= \sqrt{\frac{1}{3}} \ (u\bar{u} + d\bar{d} + s\bar{s}). \end{aligned} \tag{8.6}$$

By observing that the γ-q coupling is proportional to the quark charge, diagram a yields (up to phase space corrections)

$$\Gamma_{\gamma\pi^0} : \Gamma_{\gamma\eta} : \Gamma_{\gamma\eta'} = 3 : \cos^2\theta : \sin^2\theta \tag{8.7}$$

in disagreement with the experimental result: $\Gamma_{\gamma\pi^0} \ll \Gamma_{\gamma\eta}, \Gamma_{\gamma\eta'}$. We therefore can dispense with diagram a for the $\gamma\eta$ and $\gamma\eta'$ decays. Assuming SU_3 invariance diagram b predicts

$$\Gamma_{\gamma\pi^0} : \Gamma_{\gamma\eta} : \Gamma_{\gamma\eta'} = 0 : \sin^2\theta : \cos^2\theta \tag{8.8}$$

or (including corrections for phase space)

$$\Gamma_{\gamma\eta'} : \Gamma_{\gamma\eta} = \cos^2\theta = \begin{cases} 26 & |\theta| = 11° \quad \text{(quadratic mass formula)} \\ 5 & |\theta| = 24° \quad \text{(linear mass formula).} \end{cases}$$

---

[3] The mixing may be caused by $q\bar{q}$ annihilation into gluons. In this case diagrams b and c could be the same.

The predicted ratio is clearly at variance with the measured value of 2.6 ± 0.8. The absolute rates predicted by diagram b are difficult to estimate. One may endeavor to relate diagram b to the decay $J/\psi \to \gamma\pi^0$ via diagram a. Diagram b leads to one factor of α_s less where α_s is the strong interaction coupling constant. The result is

$$B(J/\psi \to \gamma\eta') \simeq \frac{\alpha_s^2(s = M_{\eta'}^2)}{\alpha_s^3(s = M_{J/\psi}^2)} B(J/\psi \to \gamma\pi^0). \tag{8.9}$$

The value of α_s near $s = M_{J/\psi}^2$ can be inferred from the J/ψ decays: $\alpha_s \cong 0.2$. No reliable estimate is available for α_s near the η' mass. Asymptotic freedom requires $\alpha_s(s_1) > \alpha_s(s_2)$ for $s_1 < s_2$; therefore

$$B(J/\psi \to \gamma\eta') > B(J/\psi \to \gamma\pi^0).$$

The γf decay rate is of the same order of magnitude as $\Gamma_{\gamma\eta}$ or $\Gamma_{\gamma\eta'}$. Assuming f to be a pure SU_3 singlet, diagram a cannot contribute. Comparing $\Gamma_{\gamma f}$ with the strong decay width $\Gamma_{\omega f}$ one finds

$$\Gamma(J/\psi \to \gamma f')/\Gamma(J/\psi \to \gamma f) = 1/2.$$

The experimental value is less than 1/3. Hence we observe for f and f' a behavior similar to that of η and η'.

FRITZSCH and JACKSON /104/ have investigated the possibility that the wave functions of η and η' possess a small $c\bar{c}$ component of size ε (diagram c). The decay width is of the following form:

$$\Gamma(J/\psi \to \gamma\eta) = \varepsilon_\eta^2 \frac{4\alpha}{3} \left(\frac{Q_c}{m_c}\right)^2 K^3 \Omega^2, \tag{8.10}$$

where $Q_c = 2/3$, m_c charge and mass of c quark

$\dfrac{eQ_c}{m_c}$ transition magnetic moment

K photon momentum in J/ψ rest system

Ω overlap integral.

Assuming the overlap integrals to be the same one finds for the relative sizes of the $c\bar{c}$ admixture

$$|\varepsilon_{\eta'}|^2 : |\varepsilon_\eta|^2 = \frac{\Gamma_{\gamma\eta'}}{K_{\eta'}^3} : \frac{\Gamma_{\gamma\eta}}{K_\eta^3} = (0.9 \pm 0.3) : (0.25 \pm 0.05) \text{ and} \tag{8.11}$$

$$|\varepsilon_{f'}|^2 \; : \; |\varepsilon_f|^2 = \frac{\Gamma_{\gamma f'}}{K_{f'}} \; : \; \frac{\Gamma_{\gamma f}}{K_f} = (< 0.29) \; : \; (1.0 \pm 0.3). \qquad (8.11)$$

In the approach of /104/ the $c\bar{c}$ admixture is generated by $q\bar{q}$ annihilation into two gluons. Furthermore, SU_3 symmetry breaking of the $gq\bar{q}$ coupling is allowed. For the ε's the following values are found: $\varepsilon_\eta = 1 \cdot 10^{-2}$, $\varepsilon_{\eta'} = 2.2 \cdot 10^{-2}$. As a consequence, the ratio $\Gamma_{\gamma\eta'}/\Gamma_{\gamma\eta}$ is [see (8.8)] around 3.9 which is close to the experimental result of 2.6 ± 0.8. The absolute magnitudes of the decay widths are also in satisfactory agreement with the data, viz

	experiment	theory
$\Gamma_{\gamma\eta}$	$= 56 \pm 12$ eV	60 eV
$\Gamma_{\gamma\eta'}$	$= 165 \pm 48$ eV	220 eV

Finally, we discuss the radiative decay widths for $\pi^+\pi^-$ and K^+K^- pairs outside of the f. The DASP experiment yielded $B(J/\psi \rightarrow \gamma\pi^+\pi^-, M_{\pi\pi} < 1$ GeV$) < 0.7 \pm 0.2 \cdot 10^{-3}$ under the extreme pessimistic assumption that all $\pi\pi$ pairs are genuine radiative decays. Correcting for isospin one find $B(J/\psi \rightarrow \gamma\varepsilon) < 1 \cdot 10^{-3}$. This is smaller than the value of $B(J/\psi \rightarrow \gamma\varepsilon) \approx 5 \cdot 10^{-3}$ predicted through a VDM analysis of the cascade decay $\psi' \rightarrow \pi^+\pi^- J/\psi$ /105/. Note also that the branching ratios for radiative decays into $\pi^+\pi^-$ or K^+K^- masses above 1.6 GeV are smaller than $B(J/\psi \rightarrow \gamma f)$ by almost a factor of ten.

8.4 J/ψ,ψ' Contributions to Photo- and Electroproduction

The following remarks are due to WALSH /106/. We first consider the form factors of light hadrons (i.e., without a $c\bar{c}$ component). The J/ψ contribution $F_{J/\psi}$ to the form factor relative to that of the low lying vector mesons V is

$$\frac{F_{J/\psi}}{F_V} \sim \underbrace{\frac{\gamma_V}{\gamma_{J/\psi}}}_{0(1)} \cdot \underbrace{\sqrt{\frac{\Gamma_{J/\psi hh}}{\Gamma_{Vhh}}}}_{\sim 10^{-2}} \underbrace{\frac{1 - s/m_V^2}{1 - s/m_{J/\psi}^2}}_{1 - 10} = 0(0.1) \qquad (8.12)$$

We see that $J/\psi, \psi' \ldots$ make a negligible contribution to the form factors of light hadron. Similarly, their contribution to photo- or electroproduction of light hadrons can in general be neglected. The J/ψ contribution, $\sigma_{J/\psi}$, to photoproduction of hadrons can be estimated from the cross-section data on $\gamma p \to J/\psi p$ using VDM /88/

$$\sigma_{J/\psi}(\gamma p \to \text{hadrons}) = \frac{\alpha}{\gamma^2_{J/\psi}} \, \sigma^{tot}(J/\psi p \to \text{hadrons})$$

$$\approx 0.6 \cdot 10^{-3} \cdot 1 \text{ mb} \approx 0.6 \text{ μb} \approx 0.6 \text{ % } \sigma^{tot}_{\gamma p} \qquad (8.13)$$

The total cross section for virtual photons ($Q^2 < 0$) on nucleons below the J/ψ threshold can be described by

$$\sigma^{old}_{\gamma p}(Q^2) \approx \sigma^{old}_{\gamma p}(0) \, \frac{m^2_\rho}{m^2_\rho + Q^2} \, . \qquad (8.14)$$

Assuming a similar Q^2 mass relation for the J/ψ we expect the J/ψ contribution to be of the order of

$$\frac{\sigma_{J/\psi}(Q^2)}{\sigma^{old}_{\gamma p}(Q^2)} \approx \sigma_{J/\psi}(0) \, \frac{m^2_{J/\psi}}{m^2_{J/\psi} + Q^2} \qquad (8.15)$$

or

$$\frac{\sigma_{J/\psi}(Q^2)}{\sigma^{old}_{p}(Q^2)} \approx 0.6 \cdot 10^{-2} \, \frac{1 + Q^2/m^2_\rho}{1 + Q^2/m^2_{J/\psi}} \, . \qquad (8.16)$$

The relative J/ψ contribution increases with Q^2 to a maximum value of ~10 % for $Q^2 \gg m^2_{J/\psi} = 9.6$ GeV^2. At smaller Q^2 the contribution will be much less, viz.

$$\frac{\sigma_{J/\psi}}{\sigma^{old}_{\gamma p}} \approx$$

$Q^2 = 0$	1	5	20 GeV^2
0.006	0.01	0.04	0.07

The J/ψ contribution is small and difficult to detect experimentally.

9. Search for Other Narrow Vector States

Before a classification of J/ψ and ψ' can be made one has to know whether other narrow ($\Gamma \approx$ few MeV) vector states exist, and, in particular, whether J/ψ is the ground state of the new family of vector particles.

   The mass region below the J/ψ has been scanned in different ways. One experiment /107/ measured lepton pair production

$$\gamma Be \rightarrow e^+e^-X,$$

using a bremsstrahlung beam of 7.2 GeV endpoint energy. The resulting lepton-mass spectrum does not show any narrow signal (Figure 9.1). The forward cross section

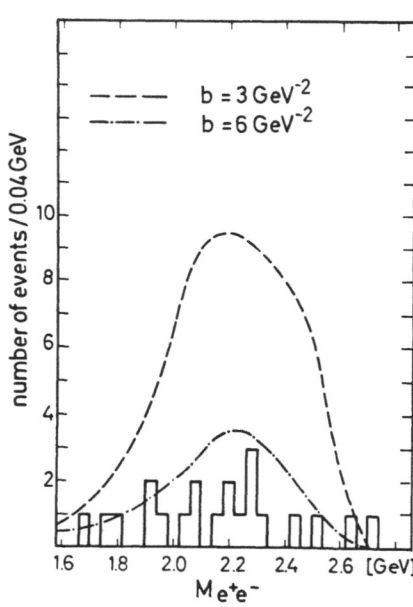

Fig. 9.1. e^+e^- mass spectrum measured in the reaction $\gamma Be \rightarrow e^+e^-X$ /107/

times branching ratio for a narrow (Γ < 40 MeV) state decaying into e^+e^- is found to be

$$\left.\frac{d\sigma}{dt}\right|_{t = 0} \cdot B(V \to e^+e^-) < 0.3 \text{ nb GeV}^{-2}$$

for a mass between 2.1 and 2.6 GeV. This may be compared with Φ production where $\left.\frac{d\sigma}{dt}\right|_{t = 0} \cdot B = 1.1$ nb GeV^{-2}.

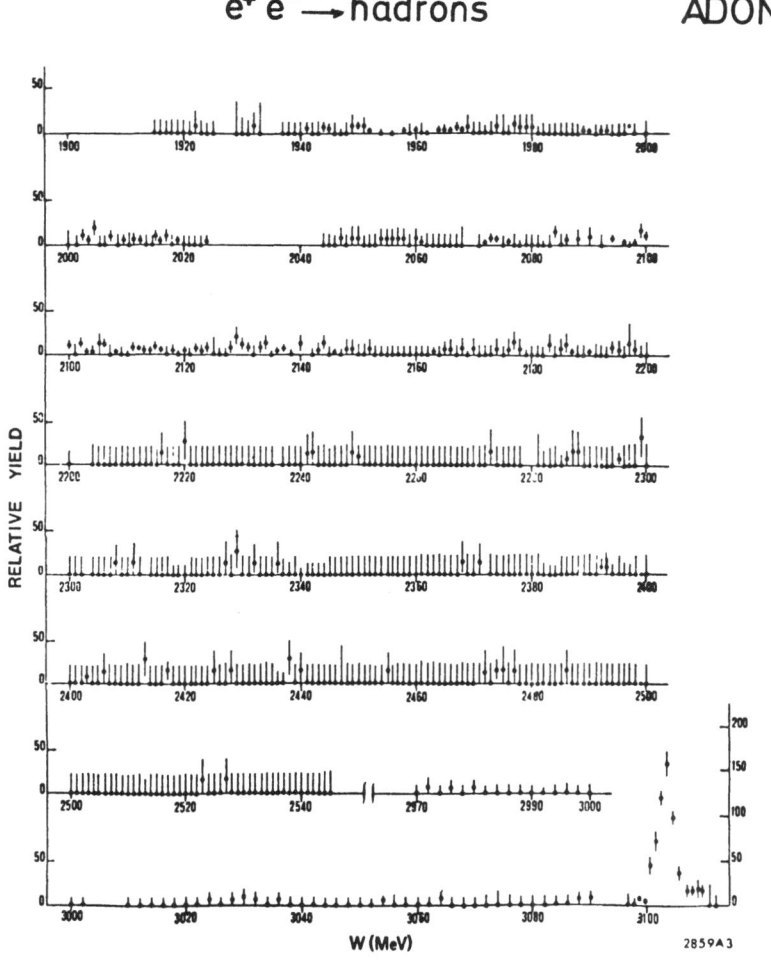

Fig. 9.2. Relative yield for the reaction $e^+e^- \to$ hadrons as a function of the total c.m. energy W /108/

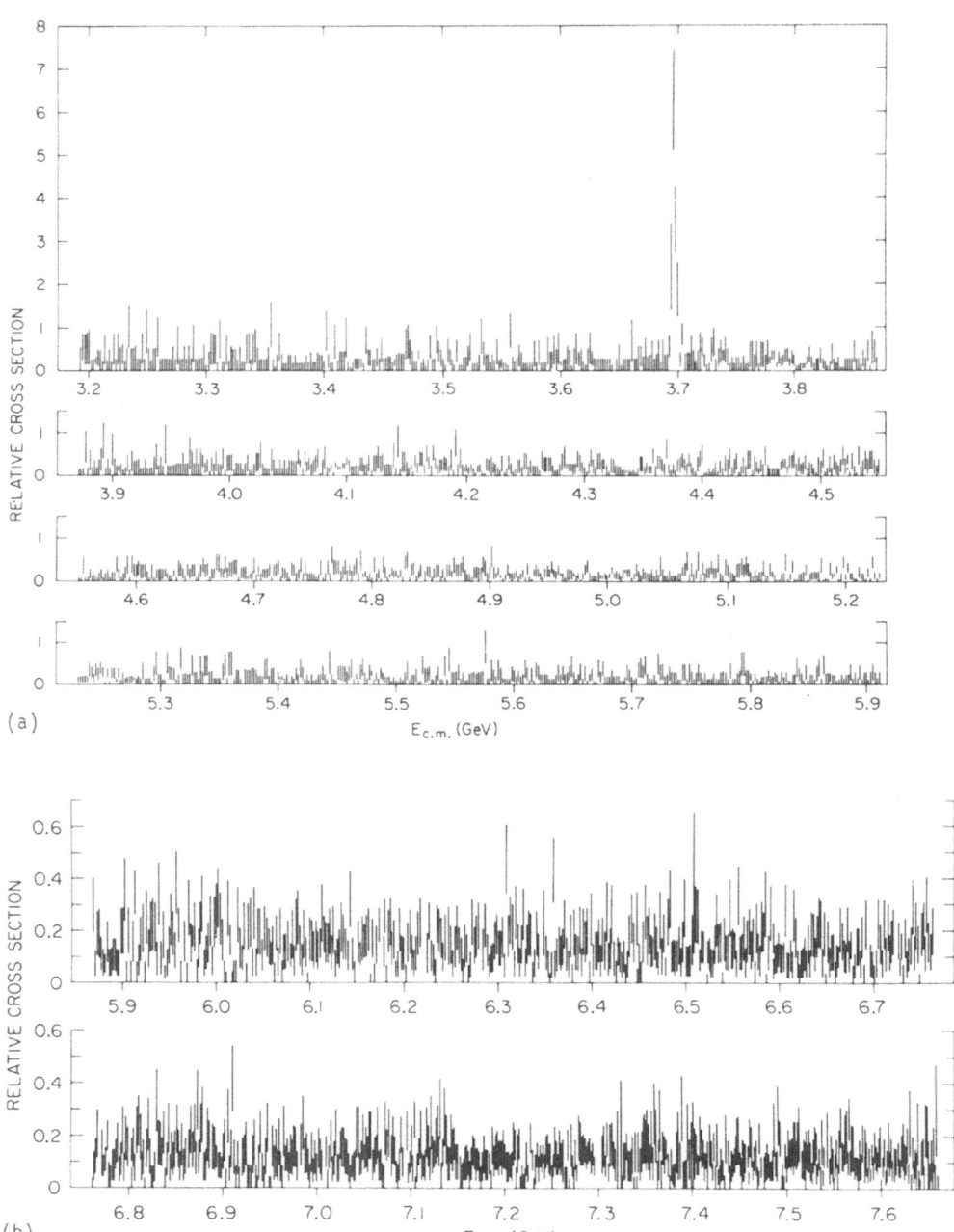

Fig. 9.3. (a) Relative yield for the reaction $e^+e^- \rightarrow$ hadrons as a function of the total c.m. energy E_{cm} /109/, (b) Relative yield for the reaction $e^+e^- \rightarrow$ hadrons as a function of the total c.m. energy E_{cm} /109/

At ADONE, three different experiments investigated the mass region between 1.91 and 3.09 GeV /56, 108/. Figure 9.2 shows one of these scans. No further narrow state was found. The upper limits on the cross section for $e^+e^- \to$ hadrons integrated over a possible narrow state of width Γ are given in Table 9.1.

Table 9.1. Search for narrow vector states in $e^+e^- \to$ hadrons

	Mass range [GeV]	Assumed width [MeV]	$\int \sigma_{hadron} d\sqrt{s}$ [nb MeV]	Percentage of $\int \sigma_{J/\psi} d\sqrt{s}$
ADONE /55, 108/	1.91 - 2.2	2	< 950	< 9 %
	2.2 - 2.54	2	< 660	< 6 %
	2.97 - 3.09	4	< 830	< 7 %
SPEAR /109/	3.2 - 5.9	2-4	< 1000	< 9 %
	4.9 - 7.6	2-4	< 450	< 4 %
SPEAR /111/	5.7 - 6.4			$\leq$ 2 %
	7.0 - 7.4			$\leq$ 1 %

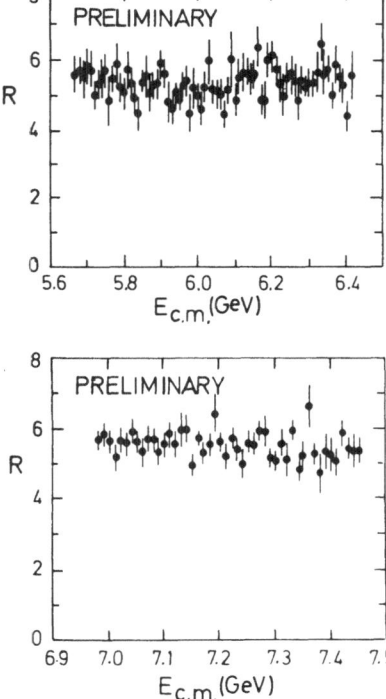

Fig. 9.4. Relative yield for the reaction $e^+e^- \to$ hadrons as a function of the c.m. energy E_{cm} /111/

At SPEAR a similar scan was made for energies between 3.2 and 7.6 GeV (see Figure 9.3) /109/. The data did not reveal any narrow signal at the level given in Table 9.1. Another high-mass scan with higher precision around 6 and 7.2 GeV was made by a Columbia-FNAL-Stonybrook collaboration studying 400 GeV pBe → e^+e^-X /110/. Their preliminary data indicated a narrow signal in the e^+e^- mass distribution near 6 GeV. No evidence was found for this state in e^+e^- annihilation (see Figure 9.4) /111/. The results of this scan are also quoted in Table 9.1.

Conclusions

The mesons J/ψ and ψ' are the only narrow vector states that exist between 2 and 7.6 GeV with a width $\Gamma \leq 4$ MeV and $\Gamma_{ee} \gtrsim 0.4$ keV. The J/ψ is the lowest lying vector state of the new family of particles.

10. The Quark Model Interpretation of J/ψ and ψ'

An intuitively simple and by now accepted explanation for J/ψ and ψ' was offered by the quark model. It assumed that a) besides the u, d, and s quarks a fourth quark Q exists which carries a new quantum number conserved by strong and electromagnetic interactions; b) the J/ψ and ψ' are $Q\bar{Q}$ bound states.

Based on these assumptions the model predicts the existence of further $Q\bar{Q}$ mesons and of mixed states of the type $(Q\bar{q})$ and $(\bar{Q}q)$. Since the widths of J/ψ and ψ' is small the masses of the $(Q\bar{q})$, $(\bar{Q}q)$ mesons were assumed to be $M_{Q\bar{q}} > M_{\psi'}/2$ which prevents decays of the type $\psi' \rightarrow (Q\bar{q})(\bar{Q}q)$.

In a next step the new quark Q was taken to be identical to the charm quark c which had been introduced earlier by GLASHOW et al. /112/ in order to understand certain weak interaction phenomena such as the smallness of the decay rate for $K_L^0 \rightarrow \mu^+\mu^-$ and the absence of strangeness changing neutral currents. The properties of the hypothesized charm quark c were

charge +2/3
strangeness 0
charm 1, where charm is a new quantum number
c decays only weakly into d or s quarks according to $c \rightarrow -d \sin\Theta_c + s \cos\Theta_c$, where Θ_c is the Cabibbo angle, $\Theta_c = 0.23$ (GIM mechanism).

10.1 $c\bar{c}$ Spectroscopy

The variety of $c\bar{c}$ states in terms of J^{PC} follows directly from the spin 1/2 nature of the quark and is equivalent to that of positronium. For instance, orbital angular momentum L = 0 leads to vector states (both quark spins parallel, also called orthocharmonium), and to pseudoscalars (spins antiparallel, paracharmonium). In order to compute level spacing, transition rates, etc., more assumptions about the $c\bar{c}$ system are necessary. In most calculations the c quark is assumed to be sufficiently heavy so that the $c\bar{c}$ system can be described by a nonrelativistic Schrödinger equation.

The forces acting between $c\bar{c}$ quarks in nonrelativistic calculations are approximated by a steeply rising attractive potential /113/. At short distances this force might be represented by gluon exchange. This will give rise to a term $-\frac{4}{3}\alpha_s/r$ in the potential where α_s is the strong interaction coupling constant. A linear term is added to the potential in order to ensure quark confinement:

$$V(r) = -\frac{4\alpha_s}{3\,r} + ar + V_0 \qquad (10.1)$$

A potential of this type /113-115/ will lead to the level scheme shown in Figure 10.1. The levels are labeled by J^{PC} with $P = (-1)^{L+1}$ and $C = (-1)^{L+S}$. For each value of L there are two bands of radial excitations with opposite charge conjugation depending on whether $S = 0$ or 1. The spectroscopic notation $n^{2S+1}L_J$ where $n - 1$ is the number of radial nodes is used to name the levels.

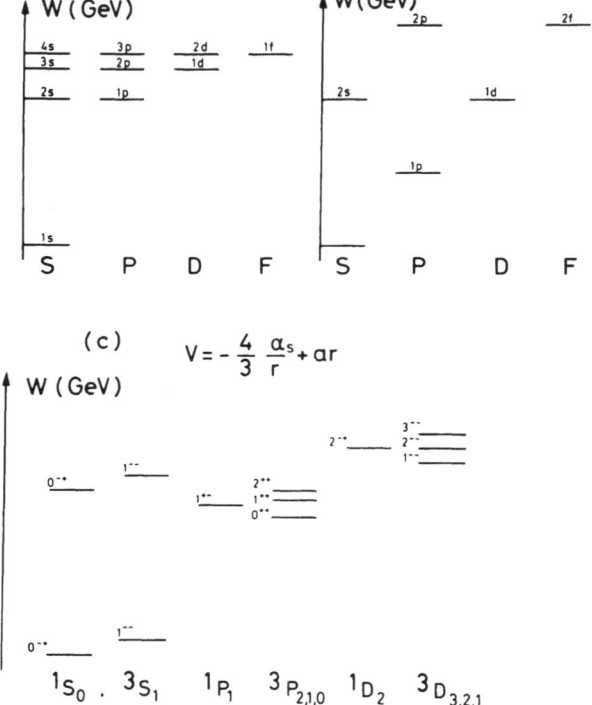

Fig. 10.1. Level scheme predicted for two fermions bound in a rising attractive potential, (a) pure Coulomb potential $V = \alpha/r$, (b) Harmonic oscillator potential $V = br^2$, (c) $V = -\frac{4}{3}\alpha_s\, m^2/r + a\, r$

The triplet S states 1^3S_1 and 2^3S_1 are identified with the J/ψ and ψ', respectively. The potential (10.1) will lead to a small spin-spin force and hence to a small mass splitting between triplet ($^3S_1$) and singlet ($^1S_0$) states. The P levels will split into one state ($^1P_2$) with odd and three states ($^3P_{2,1,0}$) with even charge conjugation. In a pure Coulombic potential the first set of P levels would be degenerate in mass with the 2^3S_1 level. The addition of a confining potential pushes the mass of the 1P levels below the mass of the 2^3S_1 level.

The first L = 2 level splits into one state ($^2D_1$) with even and three states ($^3D_{3,2,1}$) with odd charge conjugation. The 1^3D_1 state has the quantum number of the photon. By mixing with the nearby 2^3S_1 state its wave function acquires a finite value at the origin /15/. As a consequence the 1^3D_1 can show up, e.g., in the total e^+e^- cross section. The observation of this state was one of the major triumphs of charmonium spectroscopy which had predicted (EICHTEN et al. /114/) the ψ'' two years before its discovery /116, 117/.

The number and the quantum numbers of the predicted $c\bar{c}$ levels are mainly a consequence of the spin-1/2 nature of the c quark. The level spacing and the level widths on the other hand are strongly model dependent and cannot be predicted firmly.

10.2 The Vector States

The leptonic and hadronic widths of the J/ψ and the J/ψ - ψ' spacing can be used to fix the parameters of the potential (10.1). The $^3S_1$ state decays to lowest order into hadrons via a 3 gluon intermediate state. The hadronic width is given by

$$\Gamma(^3S_1 \to ggg \to \text{hadrons}) = \frac{160}{81} (\pi^2 - 9) \, \alpha_S^3 \, \frac{\left|^3S_1(0)\right|^2}{M^2} \,, \tag{10.2}$$

where $^3S_1(0)$ describes the wave function at the origin. The width for the decay into a pair of leptons is given by

$$\Gamma(^3S_1 \to e^+e^-) = 16\pi\alpha^2 e_Q^2 \, \frac{\left|^3S_1(0)\right|^2}{M^2} \,, \tag{10.3}$$

where M is the mass of the vector state and e_Q the charge of the quark. Applying these relations to the J/ψ yields $\alpha_S \approx 0.19$. From the J/ψ - ψ' spacing the slope of the confining term is found to be a = 0.25 GeV^2. These parameter values should be considered as an order of magnitude estimate only since higher gluonic corrections could be sizeable. The scale violations observed in neutrino interactions for instance prefer $\alpha_S \approx 0.4$. The R dependence of the potential computed for $\alpha_S = 0.2$,

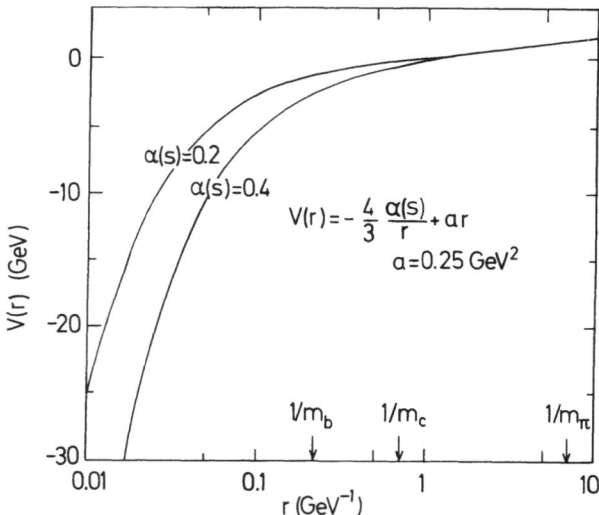

Fig. 10.2. The radius dependence of the charmonium potential

0.4, a = 0.25 GeV^2 is shown in Figure 10.2. It is evident that due to the large
De Broglie wavelength of the charm quark the charmonium spectrum as well as the decay
rates will be controlled primarily by the linear part of the potential.

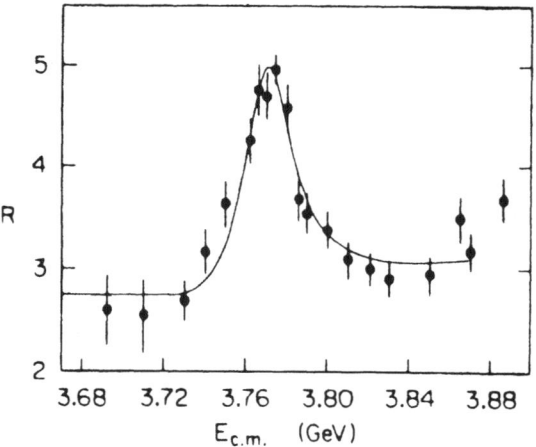

Fig. 10.3. R = $\sigma^{tot}/\sigma_{\mu\mu}$ near the ψ' after radiative corrections /116/

With the α_s and a parameters determined from J/ψ and ψ' the third vector state $\psi'' = 1^3D$, was predicted to be at 3.755 GeV /115/. Figures 10.3 and 10.4 show the excitation of the ψ'' in e^+e^- annihilation as measured by LBL-SLAC /116/ and the DELCO group /117/ at SPEAR.

The properties of the ψ'' are summarized in Table 10.1.

Table 10.1. Properties of the ψ''(3771)

Mass [MeV]	Γ [MeV]	Γ_{ee} [eV]	
3772 ± 6	28 ± 5	370 ± 90	LBL-SLAC /116/
3770 ± 6	24 ± 5	180 ± 60	DELCO /117/

branching ratios: $\psi'' \to D^0\overline{D^0}$ 49 ± 25 %

                      $\to D^+D^-$ 50 ± 38 %

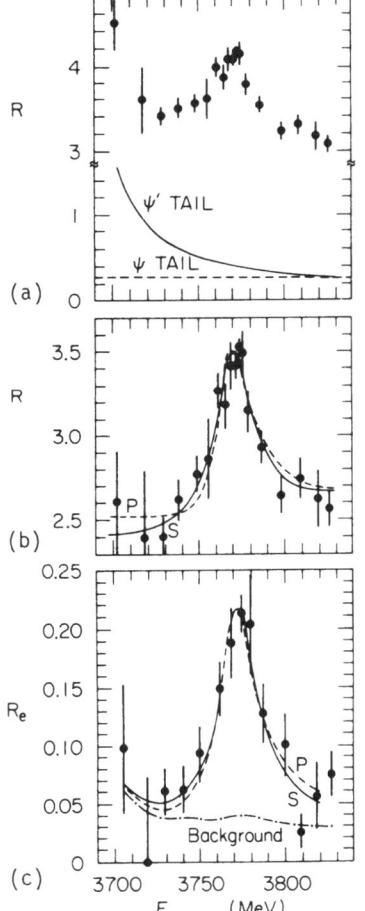

Fig. 10.4. R = $\sigma^{tot}/\sigma_{\mu\mu}$ near the ψ'' /117/. (a) measured R, (b) after radiative corrections, (c) $R_e = \sigma(e^+e^- \to eX)/\sigma_{\mu\mu}$

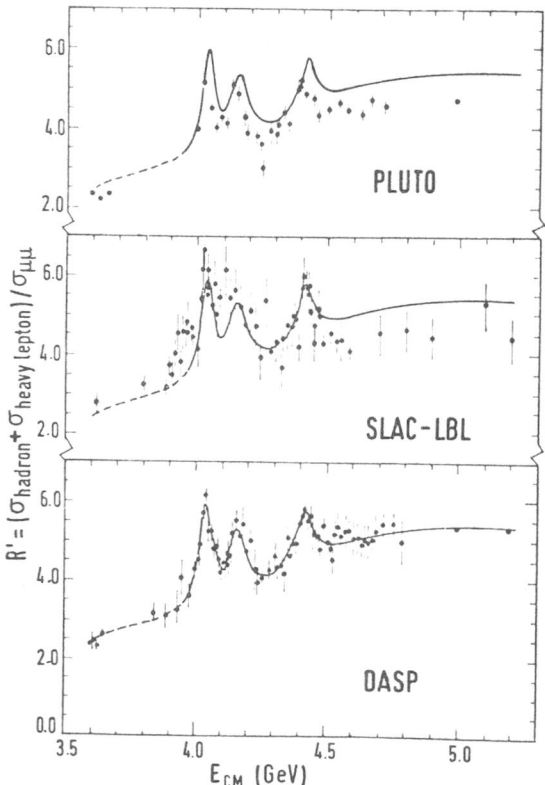

Fig. 10.5. Total hadron production: Plotted is the ratio $R'=(\sigma_{hadron}+\sigma_{heavy\ lepton})/$ $\sigma_{\mu\mu}$ from the PLUTO group /119/, SLAC-LBL /118/, and DASP /120/. The curve represents the resonance fit to the DASP data. (From /120/)

The ψ'' being ~ 45 MeV above the $D\bar{D}$ threshold is found to decay almost exclusively into a pair of D mesons and is therefore an ideal place to study the D properties (see Sect. 11.1).

Evidence for still heavier vector states has been found in studies of the total e^+e^- cross section above 4 GeV. Figure 10.5 shows the measurements of SLAC-LBL /118/, PLUTO /119/, and DASP /120/ in the region from 3.6 to 5.2 GeV. Preliminary data in this energy region are also available from DELCO /117/. The error bars shown in Figure 10.5 do not include systematic uncertainties which are primarily due to the limited solid angle.

The systematic errors are estimated to be ±15 % for DASP. The cross-section values shown in Figure 10.5 include the contribution from the heavy lepton τ. The three data sets agree with each other at the level of one unit in R which is of the order of the systematic errors. The cross-section data of PLUTO and DASP exhibit structures at 4.04, 4.16, and 4.4 GeV. The first and the third one were seen first by SLAC-LBL. Since many charmed particle channels open up in this energy region (D*D*, D**D*, FF, ...) an interpretation of these structures is not straightforward. If they are assumed to be resonances one find the mass and width values given in Table 10.2. They should be considered as preliminary since a precise determination will probably require a coupled channel analysis /123/. The result of the resonance fit to the DASP data is shown by the solid curve in Figure 10.5.

Table 10.2. Resonance parameters of vector states between 3 and 4.5 GeV

	Mass [MeV]	Γ [MeV]	Γ_{ee} [keV]	$B_{ee} = \dfrac{\Gamma_{ee}}{\Gamma}$	
J/ψ	3095±2	0.069±0.015	4.8±0.6	$6.9\pm0.9\cdot10^{-2}$	MIT /121/, SLAC-LBL /122/
ψ'	3686±2	0.228±0.056	2.1±0.3	$9.3\pm1.6\cdot10^{-3}$	SLAC-LBL /122/
ψ''(3771)		28±5	0.37±0.09	$(1.3\pm0.2)\cdot10^{-5}$	LBL-SLAC /116/
		24±5	0.18±0.06	$0.7\pm0.2\cdot10^{-5}$	DELCO /117/
4.04	4040±10	52±10	0.75±0.10	$1.4\pm0.4\cdot10^{-5}$	DASP /120/
4.16	4159±20	78±20	0.77±0.20	$0.9\pm0.3\cdot10^{-5}$	DASP /120/
4.41	4414±7	33±10	0.44±0.14	$1.3\pm0.3\cdot10^{-5}$	SLAC-LBL /118/
		66±15	0.40±0.10	$0.7\pm0.2\cdot10^{-5}$	DASP /120/

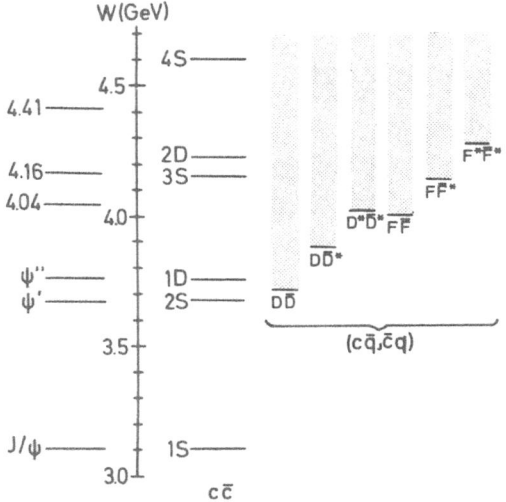

Fig. 10.6. The $c\bar{c}$ levels, the lowest lying charmed meson channels and the measured position of vector mesons /123/

In Figure 10.6 the observed vector mesons are compared to vector state levels predicted by a charmonium model. Between J/ψ and ~4.6 GeV the number of vector states known at present agrees with the number of states expected by theory. Whether the interpretation of the observed states as suggested by Figure 10.6 is the correct one, namely $\psi(4.04) = 3^3S_1$, $\psi(4.16) = 2^3D_1$, etc., is an open question.

10.3 P-Wave States

As mentioned above, three P-wave states $^3P_{2,1,0}$ are predicted between J/ψ and ψ' with even parity and even charge conjugation (Figure 10.1). If the ψ' - ^3P_J mass difference is smaller than $2 \cdot m_\pi$ these states can be reached from the ψ' by photon emission only,

$$\psi' \to \gamma \, ^3P_J.$$

The ^3P_J decay can either be purely hadronic via gluon annihilation (diagrams a,b) or by radiative decay into the J/ψ (diagram c).

The decay widths for (a,b) are proportional to the wave functions squared at small distances. These widths have been estimated by BARBIERI et al. /124/ as $\Gamma(2^{++})/\Gamma(0^{++}) = 4/15$ and $\Gamma(0^{++}) = 2.4$ MeV. The 1^{++} state is expected to be even narrower.

The decays $2^3S_1 \to \gamma \, ^3P_J$ can proceed to lowest order by an electric dipole (E1) transition. The rate for an E1 transition is given by

$$\Gamma(2^3S_1 \to \gamma 1^3P_J) = \left(\frac{16}{243}\right) \alpha(2J+1) \, k^3 \, |< 1P|r|2S >|^2. \tag{10.4}$$

The rate depends on the overlap between the radial wave functions for the S and P levels. For an E1 transition the angular distribution of the photon with respect to the beam axis is of the form

$$
\begin{aligned}
2^3S_1 \to \gamma 1^3P_0 &\qquad 1 + \cos^2\theta \\
\to \gamma 1^3P_1 &\qquad 1 - (1/3)\cos^2\theta \\
\to \gamma 1^3P_2 &\qquad 1 + (1/13)\cos^2\theta
\end{aligned} \tag{10.5}
$$

For the latter two transitions also higher multipole amplitudes could contribute. For this reason only the prediction for $^3S_1 \to \gamma \, ^3P_0$ is unique.

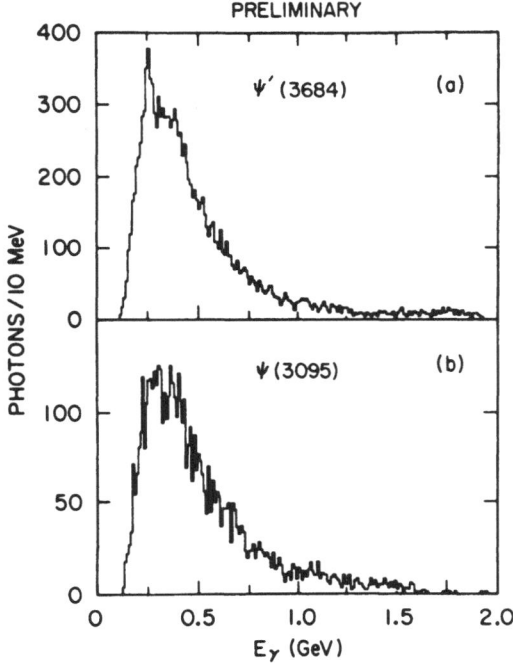

Fig. 10.7. The inclusive photon energy spectra /126/. (a) From J/ψ decay, (b) from ψ' decay

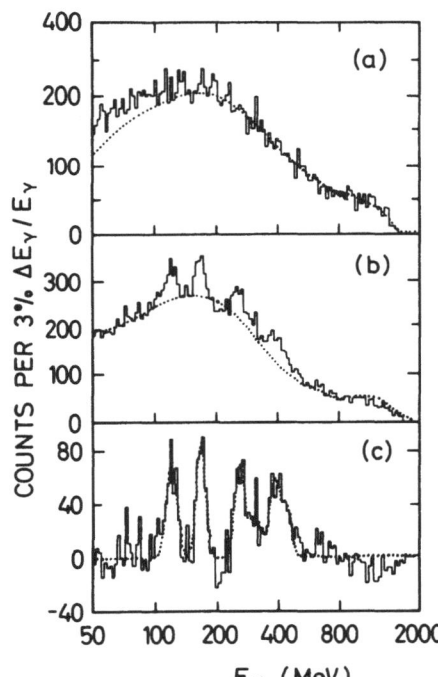

Fig. 10.8. The inclusive photon distribution measured by the Maryland, Pavia, Princeton, UC-San Diego, SLAC, and Stanford collaboration /127/ as a function of E_γ for: (a) the J/ψ, (b) ψ', (c) shows the difference between the data and the continuum in (b)

The ^3P_J states were detected in three different ways: as discrete lines in the photon spectrum, through the cascade decay $\psi' \to \gamma\,^3P_J$, $^3P_J \to \gamma J/\psi$ and as peaks in $\pi\pi$, 4π ... mass distributions. The study of the cascade decay led to the first discovery of an intermediate state ($P_c(3510)$) by DASP /125/.

a) Discrete Photon Lines

Monochromatic photons from ψ' decay were observed by SLAC-LBL /126/ and by the Maryland, Pavia, Princeton, UC-San Diego, SLAC, and Stanford collaboration /127/. The photon spectra obtained by the two experiments from J/ψ and ψ' decay are displayed in Figures 10.7 and 10.8. The ψ' spectrum shows several maxima superimposed on a smoothly varying background. No structures are observed in the J/ψ spectrum. The photon lines are centered at 121, 168, 256, and 383 MeV. The first three transitions correspond to intermediate states with masses of 3561, 3512, and 3481 MeV now called $\chi(3561)$, $P_c(3512)$, and $\chi(3418)$. The fourth line results from the decay $P_c(3512) \to \gamma J/\psi$. The branching ratios for the $\psi' \to \gamma\,^3P_J$ decays are given in Table 10.3. They account for more than one fourth of all ψ' decays. The theoretically predicted decay widths are consistently lower by a factor of 2 - 3.

Table 10.3. P_c/χ states

Decay	Branching ratio %	Γ [keV]	Reference	Theory [keV]
$\psi' \to \gamma\,\chi(3.41)$	7.2 ± 2.0	16 ± 5	/127/	44
	6.5 ± 2.2	15 ± 5	/126/	
$\psi' \to \gamma\,\chi(3.45)$	< 2.5	< 5.7	/127/	18
$\psi' \to \gamma\,P_c(3.51)$	7.1 ± 2.0	16 ± 5	/127/	38
$\psi' \to \gamma\,\chi(3.55)$	7.0 ± 2.3	16 ± 5	/127/	27

b) $\psi' \to \gamma\,^3P_J \to \gamma\gamma J/\psi$

Next we discuss the observation of the P states via the cascade decay. The J/ψ is identified by its $\mu^+\mu^-$ decay. Figure 10.9 shows the μ-pair mass spectrum from ψ' decay as measured by DASP /125/. Two narrow signals are observed; the higher one corresponds to the $\psi' \to \mu^+\mu^-$ decay and to μ-pair production by QED; the peak at 3.1 GeV results from $J/\psi \to \mu^+\mu^-$ decay. Events with a $\mu^+\mu^-$ mass in the J/ψ region and two photons are selected and fitted to the hypothesis $\psi' \to \gamma\gamma J/\psi$
$$\raisebox{0pt}{$\hookrightarrow \mu^+\mu^-$}.$$

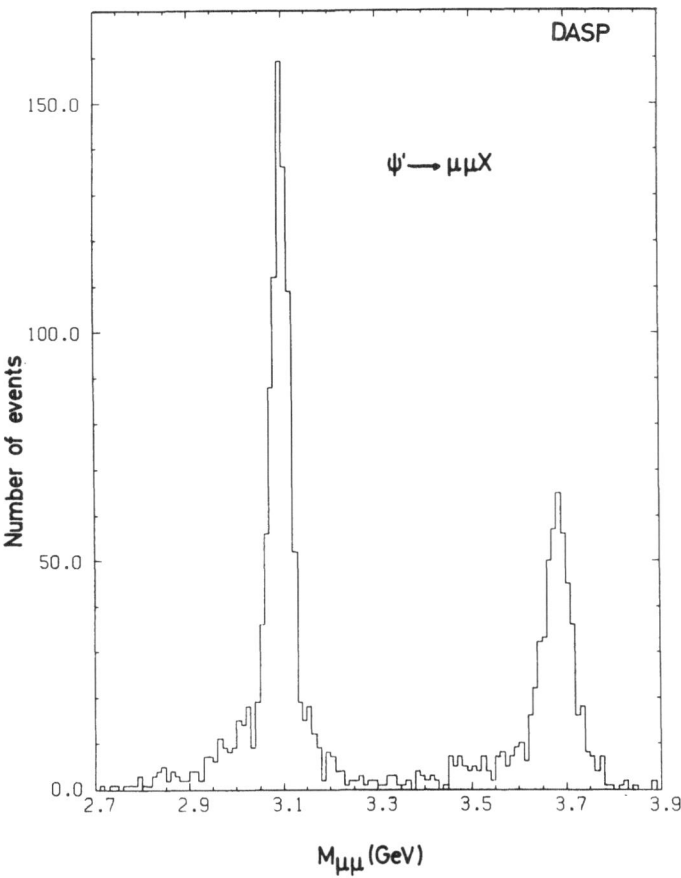

<u>Fig. 10.9.</u> Distribution of the $\mu^+\mu^-$ effective mass measured at the ψ' /125/

A large fraction of the events is due to the decay $\psi' \rightarrow \eta J/\psi$. They are removed by a cut in the $\gamma\gamma$ mass (e.g., by requiring $M_{\gamma\gamma}$ < 0.52 GeV in the DASP analysis).

Figure 10.10 shows the remaining events obtained in the experiments of DASP /125/, PLUTO /128/, and SLAC-LBL /129/. The low ($\gamma J/\psi$) mass solution is plotted versus the high ($\gamma J/\psi$) mass solution. A small percentage of these events is due to back-ground from $\psi' \rightarrow \pi^0\pi^0 J/\psi$. Two distinct clusters of events are observed around 3.51 and 3.56 GeV corresponding to the $P_c(3.51)$ and $\chi(3.56)$. Note that the mass spread is larger for the low-mass solution due to the Doppler broadening. Clear signals for the $P_c(3.51)$ and $\chi(3.56)$ have also been obtained by the DESY-Heidelberg group /130/ (see Figure 10.11). In addition to the two clusters at 3.51 and 3.56 GeV Figure 10.10 shows two groups of events centered at 3.45 and 3.41 GeV. The latter one can be identified with the $\chi(3418)$ seen in the photon spectrum and in the hadronic mass spectra.

116

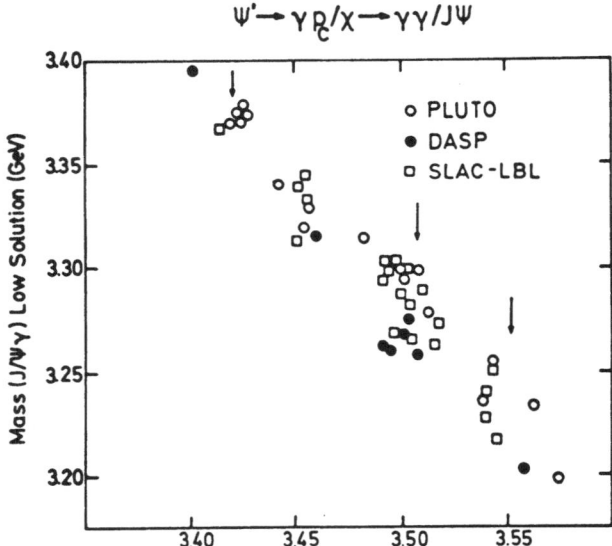

$\psi' \rightarrow \gamma \underset{c}{R}/X \rightarrow \gamma\gamma/J\psi$

Fig. 10.10. The decay $\psi' \rightarrow \gamma\gamma J/\psi$. The data are plotted as a function of the low mass solution versus the high mass solution. The plot shows the world data /125, 128, 129/ as of summer 1977

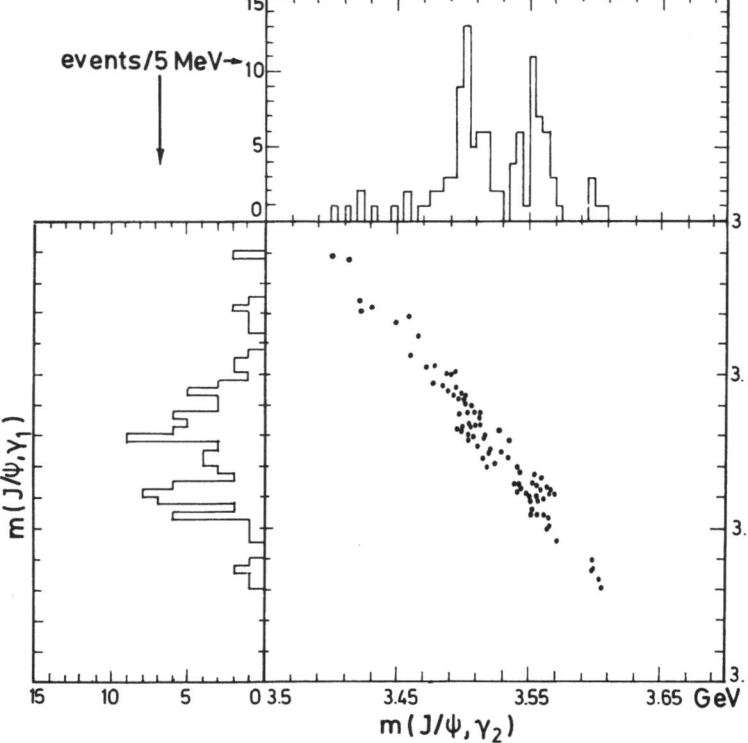

Fig. 10.11. The decay $\psi' \rightarrow \gamma\gamma J/\psi$: scatter plot of mass $M(\gamma_1 J/\psi)$ vs mass $M(\gamma_2 J/\psi)$ as measured by DESY-Heidelberg /130/

The DESY-Heidelberg group recently found evidence for a state near 3.60 GeV (see Figure 10.11) /130/. More data are needed to firmly establish the existence of this particle.

The 3.45 GeV group has not been observed in any other decay mode. We will come back to it later.

The products of the branching ratios are given in Table 10.4.

Table 10.4. Cascade decays $(\psi' \to \gamma P_c/\psi)$ $(P_c/\chi \to \gamma J/\psi)$

Intermediate State	BR$(\psi' \to \gamma P_c/\chi) \cdot$ BR$(P_c/\chi \to \gamma J/\psi)$ %	Reference
$\chi(3.41)$	3.3 ± 1.7	/127/
	0.3 ± 0.2	/125/
	0.14 ± 0.09	/130/
	0.2 ± 0.2	/129/
	(0.22 ± 0.08)[a]	
$\chi(3.45)$	< 0.61 90 % CL	/125/
	< 0.25 90 % CL	/130/
	0.8 ± 0.4	/129/
$P_c(3.51)$	5.0 ± 1.5	/127/
	2.1 ± 0.4	/125/
	2.5 ± 0.4	/130/
	2.4 ± 0.8	/129/
	(2.4 ± 0.3)[a]	
$\chi(3.56)$	2.2 ± 1.0	/127/
	1.6 ± 0.6	/125/
	1.0 ± 0.2	/130/
	1.0 ± 0.6	/129/
	(1.9 ± 0.2)[a]	
(3.59)	0.18 ± 0.06	/130/

[a] The combined branching ratio obtained by averaging the results of /125, 126, 130/

c) $\psi' \to \gamma^3 P_J \to \gamma$ hadrons

Hadronic decays of the intermediate states were observed by SLAC-LBL /126, 131, 132/ and recently by DASP /64, 76, 125/ analyzing the decays

118

$$\psi' \to \gamma P_c \; \chi \to \gamma \; \pi^+\pi^-, \; K^+K^-$$
$$\gamma \; 2\pi^+ \; 2\pi^-$$
$$\gamma \; \pi^+\pi^- K^+K^-$$
$$\gamma \; 3\pi^+ \; 3\pi^- \; .$$

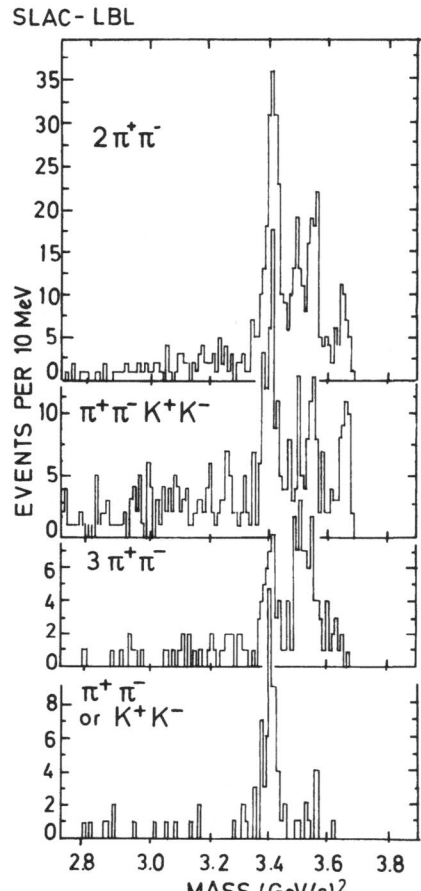

SLAC-LBL

Fig. 10.12. The decay $\psi' \to \gamma$ + charged hadrons. Mass distributions of the hadron system measured by SLAC-LBL /126, 131, 132/

Figure 10.12 shows the fitted mass distributions of the hadronic systems. The $2\pi^+ \; 2\pi^-$ and $\pi^+\pi^- K^+K^-$ spectra show three peaks with masses of 3.415 and 3.550 GeV. The peak at 3.68 GeV results from the direct decay of the ψ'. In the $3\pi^+ \; 3\pi^-$ distribution the two upper states are not resolved. There is a clear $\pi^+\pi^-$ (or K^+K^-) signal for the decay of the 3415 MeV state. Eight events are observed with a mass of 3.55 GeV. No signal is found at 3.50 GeV. The branching ratios are listed in Table 10.5. Note that none of the mass spectra give evidence for a 3.45 GeV state.

Table 10.5. Properties of the intermediate states

Mass [MeV]	Decay	$BR(\psi' \to \gamma P_c/\chi) \cdot BR(P_c/\psi \text{ decay})$ %	Reference	$BR(P_c/\psi \text{ decay})$[a] %
3413 ± 5	$\pi^+\pi^-$	0.07 ± 0.02	/131, 132/	1.0 ± 0.3
		0.06 ± 0.02	/64, 125/	
	K^+K^-	0.07 ± 0.02	/131, 132/	1.0 ± 0.3
$p\bar{p} < 0.0095$		0.055 ± 0.025	/64, 125/	
	$2\pi^+2\pi^-$	0.32 ± 0.06	/131, 132/	4.7 ± 0.9
	$\pi^+\pi^-K^+K^-$	0.27 ± 0.07	/131, 132/	3.9 ± 1.0
	$p\bar{p}\,\pi^+\pi^-$	0.04 ± 0.013	/131, 132/	0.6 ± 0.2
	$3\pi^+\pi^-$	0.14 ± 0.05	/131, 132/	2.0 ± 0.7
	$\gamma J/\psi$	0.22 ± 0.08	b	3.2 ± 1.2
	$\gamma\gamma$	< 0.014	/125/	< 0.2
	$\rho^0\pi^+\pi^-$	0.12 ± 0.04	/131, 132/	1.8 ± 0.6
	$K^{0*}K^+\pi^-$	0.17 ± 0.06	/131, 132/	2.5 ± 0.9
3.508 ± 4	$\pi^+\pi^- + K^+K^-$	< 0.015	/131, 132/	< 0.21
	$2\pi^+2\pi^-$	0.11 ± 0.04	/131, 132/	1.5 ± 0.6
	$\pi^+\pi^+K^+K^-$	0.06 ± 0.03	/131, 132/	0.85 ± 0.42
	$p\bar{p}\,\pi^+\pi^-$	0.01 ± 0.008	/131, 132/	0.14 ± 0.11
	$3\pi^+\pi^-$	0.17 ± 0.06	/131, 132/	2.4 ± 0.8
	$\gamma J/\psi$	2.4 ± 0.3	b	34 ± 4
	$\gamma\gamma$	< 0.013	/125/	< 0.18
	$\rho^0\pi^+\pi^-$	0.026 ± 0.022	/131, 132/	0.37 ± 0.31
	$K^{0*}K^+\pi^-$	0.31 ± 0.022	/131, 132/	0.44 ± 0.31
3.552 ± 6	$\pi^+\pi^- + K^+K^-$	0.02 ± 0.01	/131, 132/	0.29 ± 0.14
	$\pi^+\pi^-$	0.015 ± 0.008	/64, 125/	
	K^+K^-	0.012 ± 0.009	/64, 125/	
	$p\bar{p}$	< 0.0075	/64, 125/	
	$2\pi^+2\pi^-$	0.16 ± 0.04	/131, 132/	2.3 ± 0.6
	$\pi^+\pi^-K^+K^-$	0.14 ± 0.04	/131, 132/	2.0 ± 0.6
	$p\bar{p}\,\pi^+\pi^-$	0.02 ± 0.01	/131, 132/	0.29 ± 0.14
	$3\pi^+3\pi^-$	0.08 ± 0.05	/131, 132/	1.1 ± 0.7
	$\gamma J/\psi$	1.9 ± 0.2	b	27 ± 3
	$\gamma\gamma$	< 0.004	/125/	< 0.06
	$\rho^0\pi^+\pi^-$	0.05 ± 0.03	/131, 132/	0.71 ± 0.43
	$K^{0*}K^+\pi^-$	0.052 ± 0.031	/131, 132/	0.74 ± 0.44

[a] The values listed in Table 10.3 for $\psi' \to P_c/\chi$ were used to extract the branching ratios.

[b] The average values from Table 10.4 were used.

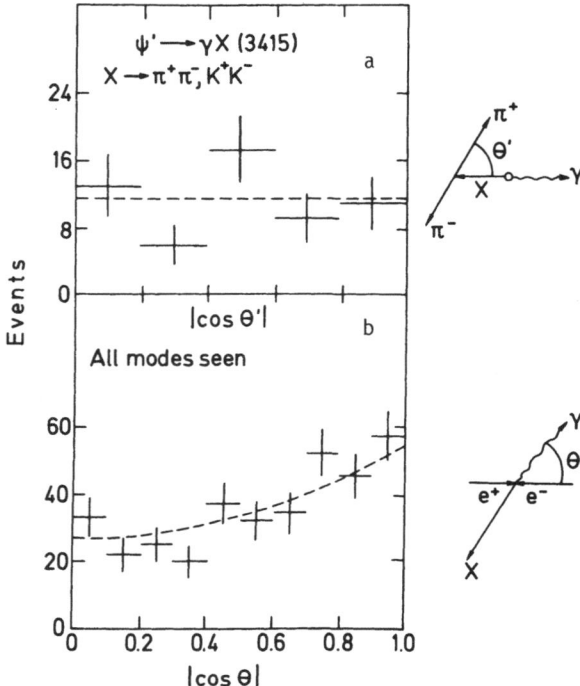

Fig. 10.13. The decay $\psi' \rightarrow \gamma X(3.41)$ measured by SLAC-LBL /126, 131, 132/. (a) Decay $X(3.41) \rightarrow \pi^+\pi^-$: distribution of the π^+ polar angle measured in the X system with respect to the X distribution of flight. (b) Photon production angular distribution measured with respect to the incoming e^+ beam

d) Quantum Numbers of the Intermediate States

The distributions in Θ, the angle of the photon with respect to the beam axis, has been obtained by summing the various decay channels. The result is shown in Figure 10.13 for the $X(3.41)$. The values of a obtained from a fit to the data of the form $1 + a \cos^2\Theta$ are as follows:

State:	$X(3.41)$	$P_c(3.51)$	$X(3.55)$
a	1.4 ± 0.4	0.1 ± 0.4	0.3 ± 0.4

Remember that for a spin 0 state a = 1. The $X(3.41)$ is therefore consistent with J = 0, whereas a J = 0 assignment for the $P_c(3.51)$ and the $X(3.55)$ is excluded on the 2σ level.

In the level scheme depicted in Figure 10.1 there are four levels with even charge conjugation between the ψ' and the J/ψ, one pseudoscalar $^1S_0$ with $J^{PC} = 0^{-+}$ and three ^3P states with $J^{PC} = 2^{++}$, 1^{++} and 0^{++}. Are the levels found consistent with these quantum numbers?

All levels have even charge conjugation since they are populated via $\psi' \to \gamma P_c/\chi$.

$\chi(3.413)$ It follows from C = + and the observed decay into $\pi^+\pi^-$ and/or K^+K^- that the state is an isoscalar and has positive parity. The angular distribution is consistent with J = 0. We can therefore safely assign this state to $J^{PC} = 0^{++}$.

$\chi(3.454)$ Needs to be confirmed.

$P_c(3.508)$ The absence of $\pi^+\pi^-$ and K^+K^- decays are consistent with the state belonging to an unnatural spin-parity sequence 0^-, 1^+, ... The angular distribution of the photon suggests J = 0 and isospin must be even since the resonance decays into an even number of pions.

$\chi(3.552)$ It follows from the observed decay into $\pi^+\pi^-$ and/or K^+K^- that the level is an isoscalar and belongs to a natural spin-parity sequence. The angular distribution of the photon excludes J = 0 with 2σ.

(3.59) Needs to be confirmed.

This is all that can be deduced from the data alone. However, comparing the available information with the levels expected in the charmonium model (Figure 10.1) leads to a unique assignment using the following reasoning. The 3.413 GeV level must be the 0^{++} state. This leaves only one natural spin-parity level - the $^3P_2(2^{++})$ level - which then must be associated with the 3.552 GeV level. The 3.508 GeV level has $J \neq 0$ and must therefore be the $^3P_1(1^{++})$ level. This assignment is also consistent with the fact that $P_c(3.51) \to \pi^+\pi^-$, K^+K^- has not been observed.

The decays $\psi' \to \gamma \, ^3P_{2,1,0}$ are to lowest order electric dipole transitions. The measured widths are listed in Table 10.3 and compared to model calculations assuming E1. Note that the predicted rates are considerably larger than the measured values. If the matrix elements are independent of J then the relative rates for E1 transitions are given by $\Gamma \sim k^3 (2J + 1)$. In this case we expect

$$\Gamma(2^3S_1 \to \gamma \, ^3P_0) \; : \; \Gamma(2^3S_1 \to \gamma \, ^3P_1) \; : \; \Gamma(2^3S_1 \to \gamma \, ^3P_2) =$$

$$1 \qquad : \qquad 0.9 \qquad : \qquad 0.6$$

compared to the observed ratios 1 : 1.04 : 1.02. Reversing the order of the levels, i.e., $^3P_0 > \, ^3P_1 > \, ^3P_2$ leads to a gross disagreement between the predicted and the observed rates.

In QCD the 0^{++} and the 2^{++} states can decay to lowest order via two-gluon emission into hadrons whereas the 1^{++} state must emit three gluons. The total hadronic width for 0^{++} and 2^{++} is therefore proportional to α_s^2 compared to α_s^3 for the 1^{++} state. Since α_s is less than one we naively expect the 1^{++} state to have a smaller hadronic width than the other two. This assertion is supported by detailed calculations /124, 133/ which find

$$\Gamma(0^{++}) : \Gamma(2^{++}) : \Gamma(1^{++}) = 15 : 4 : 1,$$

where the first ratio is the most reliable one. Experimentally we would therefore expect the 0^{++} level to have a smaller branching ratio for the radiative decay into the J/ψ than the other two and this is indeed supported by experiments.

The ratios of the hadronic widths can be estimated from the radiative transition with the following assumptions:

1) $\Gamma_{tot} = \Gamma(^3P \rightarrow \text{hadrons}) + \Gamma(^3P \rightarrow \gamma \; ^3S_1) = \Gamma_h + \Gamma_\gamma$

2) $\Gamma(^3P \rightarrow \gamma \; ^3S_1) \sim k^3$, i.e., independent of the matrix elements.

With these assumptions we can write

$$\Gamma_h = (\Gamma_{tot} - \Gamma_\gamma) = (\frac{1}{B_\gamma} - 1) \; \Gamma_\gamma \quad \text{where } B_\gamma = \Gamma_\gamma/\Gamma_{tot}. \tag{10.6}$$

With the radiative branching ratios listed in Table 10.5 we find

$$\Gamma_h(3.41) : \Gamma_h(3.55) : \Gamma_h(3.51) = (14.2\pm6.0) : (3.8\pm1.3) : (0.5\pm0.16)$$

consistent with the predictions.

The $^1P_1$ State

Besides the three ^3P_J states the potential (10.1) predicts a singlet P state with $J^{PC} = 1^{+-}$ (Figure 10.1). Because of odd C parity this level cannot be reached from the ψ' by a γ transition. The decay $\psi' \rightarrow \pi^0 \; ^1P_1$ is suppressed by isospin conservation. The strong decay $\psi' \rightarrow \pi\pi \; ^1P_1$ is probably forbidden by kinematic. No evidence so far was found for the existence of the $^1P_1$.

10.4 Pseudoscalar States

The potential (10.1) predicts the pseudoscalar states $(n_c, n'_{c3}...) \; 1 \; ^1S_0, \; 2 \; ^1S_0 \; ...$ to lie below the corresponding vector states (n^3S_1). The pseudoscalar states can decay into ordinary hadrons by two-gluon exchange, i.e.,

$$\Gamma(n_c \rightarrow \text{hadrons}) \sim \alpha_s^2.$$

Hence one expects the $^1S_0$ states to have a larger width than the vector states for which $\Gamma \sim \alpha_s^3$. More precisely

$$\Gamma(^1S_0 \to gg \to \text{hadrons}) = \frac{32\pi}{3} \alpha_s^2 \frac{\left|^1S_0(0)\right|^2}{M^2} .$$ (10.7)

The $^1S_0$ states having $C = +$ can also decay into two photons; the corresponding width is given by

$$\Gamma(^1S_0 \to \gamma\gamma) = \frac{256\pi}{27} \alpha_s^2 \frac{\left|^1S_0(0)\right|^2}{M^2} .$$ (10.8)

A comparison of (10.7) and (10.8) leads to

$$\frac{\Gamma(^1S_0 \to \gamma\gamma)}{\Gamma(^1S_0 \to \text{hadrons})} = \frac{8}{9} \frac{\alpha^2}{\alpha_s^2} \cong 1.3 \cdot 10^{-3}$$ (10.9)

for $\alpha_s = 0.2$. Under the assumption that $^3S_1$ and $^1S_0$ have the same radial wave functions (10.2) and (10.7) predict for the $^1S_0$ width

$$\Gamma(^1S_0 \to \text{hadrons}) = \frac{27}{5} \frac{\pi}{(\pi^2 - 9)\alpha_s} \Gamma(^3S_1 \to \text{hadrons}) \approx 5 \text{ MeV}.$$

If the triplet-singlet level splitting is small, the $^1S_0$ can be reached from the corresponding $^3S_1$ state by a radiative transition only, $^3S_1 \to \gamma \,^1S_0$. This is a pure spin-flip or M1 transition. The amplitude is proportional to the magnetic moment of the c quark. If this is assumed to be of the Dirac type,

$$\mu_C = \frac{(2/3)e}{2 \, m_C}$$ (10.10)

m_C = mass of c quark

the transition rate is given by

$$\Gamma(n_i \,^3S_1 \to n_f \,^1S_0) = \frac{16}{27} \alpha \frac{k^3}{m_C^2} |\langle n_f | n_i \rangle|^2.$$ (10.11)

k is the photon energy.

The overlap integral $|\langle n_f | n_i \rangle|^2$ is expected to be unity for $n_f = n_i$ and small for $n_f \neq n_i$. Table 10.6 shows the predicted width for $J/\psi \to \gamma\eta_C$ for various values of the η_C mass.

Table 10.6. Predicted decay width Γ and branching ratio B for the decay $J/\psi \to \gamma\eta_c$ as a function of the η_c mass

M_{η_c} [GeV]	Photon energy [MeV]	Γ [keV]	B [%]
3.04	50	0.2	0.3
2.99	100	1.6	2
2.89	200	14	20
2.79	300	45	70

A search of the photon spectrum from J/ψ decay failed to produce evidence for the $J/\psi \to \gamma\eta_c$ transition /127/ (see Figures 10.7, 10.8).

Evidence for a heavy and narrow resonance below the J/ψ was reported by the DASP group /134/ in the decay $J/\psi \to \gamma X \to \gamma\gamma\gamma$.

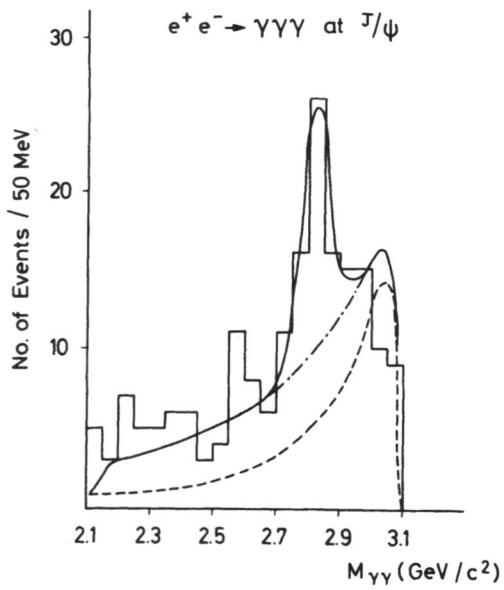

Preliminary DASP

$e^+ e^- \to \gamma\gamma\gamma$ at J/ψ

No. of Events / 50 MeV

$M_{\gamma\gamma}$ (GeV/c^2)

Highest Photon Pair Mass

---- QED
—·—·— QED + Reflection from η and η'
——— QED + Reflection from η and η' + X(2.82)

Fig. 10.14. $J/\psi \to \gamma\gamma\gamma$ measured by DASP /134/. Distribution of the highest photon pair mass

Events with three photons were selected using the DASP inner detector. For each photon the production angles were measured taking the position of the interaction point from Bhabha events observed concurrently. A 1C fit was made to the hypothesis $J/\psi \rightarrow \gamma\gamma\gamma$ and events with a $\chi^2 < 2.7$ were retained. The photon energies for the accepted events were computed. Of the three possible $\gamma\gamma$ mass combinations only two are independent. The events are plotted in Figure 10.14 as a function of the highest $\gamma\gamma$ mass in the event. The low-mass plot shows a clear η signal and an indication for η' production (see Figure 8.7). The position of the $\eta(547.1 \pm 4.2$ MeV) agrees with its known mass value; the observed η width ($\sigma = 24 \pm 4$ MeV) is consistent with the expected resolution of 20 MeV.

The high-mass data are shown in Figure 10.14 together with the contribution from QED and $\eta + \eta'$. The QED curve was computed from the matrix elements evaluated by BERENDS and GASTMANN /135/. The QED curve is an absolute prediction. The calculation was checked by measuring $e^+e^- \rightarrow \gamma\gamma\gamma$ at a c.m. energy of 3.6 GeV, i.e., away from resonances.

The high mass data show a peak near 2.82 GeV, called the $X(2.8)$[1]. In the mass interval 2.82 ± 0.04 GeV, 41 events are observed, compared to 19 events expected from QED and $\eta + \eta'$. This corresponds to a 5 standard deviation effect. The fitted mass is 2.82 ± 0.014 GeV; the value of the width is 0.04 ± 0.014 GeV. Considering the experimental resolution the measured width is consistent with that of a zero width resonance. The product of the branching ratios was determined to

$$B(J/\psi \rightarrow \gamma X) \cdot B(X \rightarrow \gamma\gamma) = 1.4 \pm 0.4 \cdot 10^{-4}.$$

The DESY-Heidelberg group searched also for the decay $J/\psi \rightarrow \gamma X \rightarrow \gamma\gamma\gamma$ /130, 136/. No peak was observed and a 90 % confidence upper limit was found at $B(J/\psi \rightarrow \gamma X) \cdot B(X \rightarrow \gamma\gamma) < 3.2 \cdot 10^{-4}$. This limit is about a factor of two higher than the value measured by DASP.

Further evidence for the $X(2.8)$ was recently reported by a group working at SERPUKOV /137/. The experiment using a 40 GeV π^- beam incident on a hydrogen target searched for events of the type

$\pi^- p \rightarrow \gamma\gamma$ n.

The $\gamma\gamma$ mass distribution (Figure 10.15) shows an enhancement at 2.88 ± 0.06 GeV with a systematic uncertainty of 0.03 GeV. The width of 0.17 GeV is consistent with the experimental resolution. No other enhancement is seen in the mass spectrum and the authors identify the peak with the $X(2.82)$. The photons are emitted isotropically in the X rest system as expected for a spin zero particle. Additional confirmation of the existence of the $X(2.82)$ by other experiments appears to be highly desirable.

---

[1] Note added in print: The Caltech-Harvard-Princeton-Stanford-SLAC collaboration using the Crystal Ball detector at SPEAR with larger statistics than in the DASP experiment failed to observe the $X(2.82)$. There is some evidence from this experiment for a state at 2.976 GeV observed in the decay $\psi' \rightarrow \gamma$ + anything.

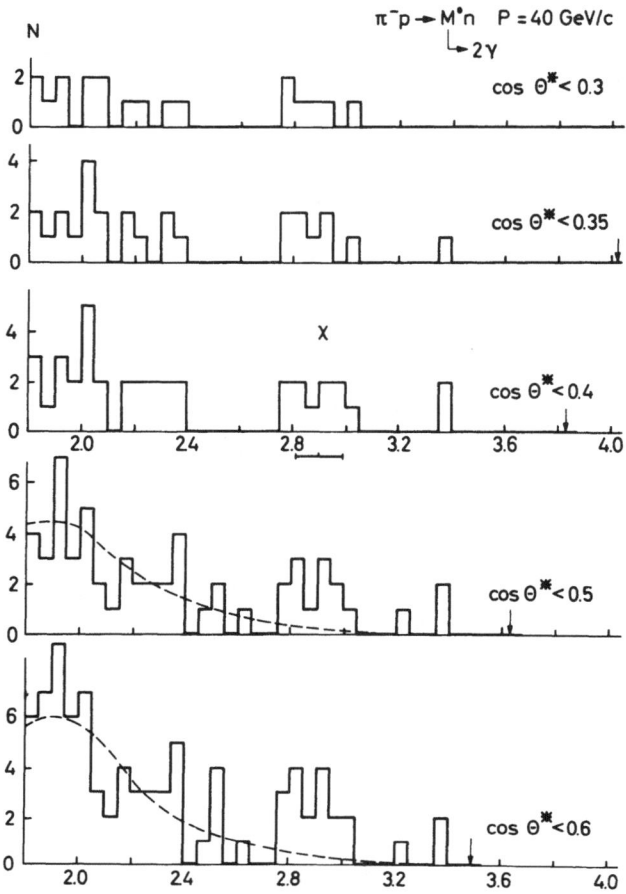

Fig. 10.15. The reaction $\pi^- p \to \gamma\gamma n$ at 40 GeV/c as measured at SERPUKHOV /137/. Plotted are the number of events versus the photon pair mass with the cosinus of the c.m. decay angle Θ^* as a parameter. Events are detected with an efficiency of at last 0.8 for photon pair masses below the value shown by the arrow

We now discuss searches for the decay $\psi' \to \gamma X$. None of these searches so far yielded any clear evidence for the X. However, remember that the $1\ ^1S_0$ and $2\ ^3S_1$ wave functions are expected to be nearly orthogonal resulting in a strong suppression of the decay $\psi' \to \gamma X(2.82)$.

The Maryland, Pavia, Princeton, UC-San Diego, SLAC, and Stanford collaboration /127/ from the measured photon spectrum of ψ' decay (Figure 10.8) deduced an upper limit of $B(\psi' \to \gamma X(2.82)) < 1\ \%$ in agreement with earlier results from SLAC-LBL /129/.

The DASP group /134/ studied 3γ events from ψ' decay. A total of 190 events was observed for an integrated luminosity of 1610 nb^{-1}. The high $\gamma\gamma$ mass solution is

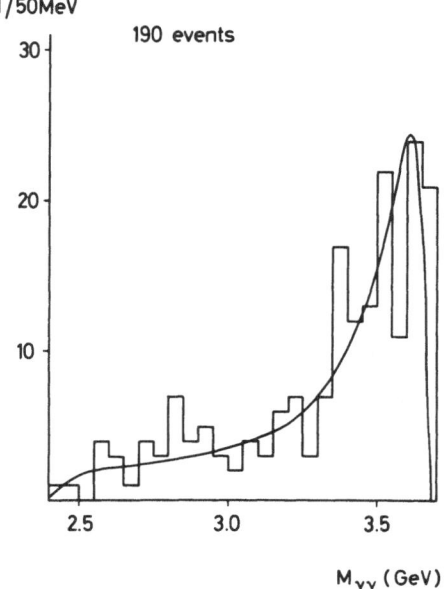

$e^+e^- \rightarrow \gamma\gamma\gamma$ at ψ'

Highest Photon pair mass Distribution

Fig. 10.16. $\psi' \rightarrow \gamma\gamma\gamma$ measured by DASP /134/. Distribution of the highest photon pair mass

plotted in Figure 10.16. No statistically significant peaks were observed and the absolute number of events are well explained by the QED process as shown by the solid line. The 90 % confidence upper limit is

$$B(\psi' \rightarrow \gamma X) \cdot B(X \rightarrow \gamma\gamma) < 1.4 \cdot 10^{-4}.$$

The properties of the X(2.82) are summarized in Table 10.7.

Table 10.7. The X(2.82)

		Ref.
Mass	2.82 ± 0.014 GeV	/134/
$B(J/\psi \rightarrow \gamma X) < 1.7$ %		/127/
$B(J/\psi \rightarrow \gamma X) \cdot B(X \rightarrow \gamma\gamma) = 1.4 \pm 0.4 \cdot 10^{-4}$		/134/
$B(\psi' \rightarrow \gamma X) \cdot B(X \rightarrow \gamma\gamma) < 1.4 \cdot 10^{-4}$		/134/

We now have all available data on hand to compare the X(2.82) with the predicted $1 \, ^1S_0$ state.

1) $\underline{B(J/\psi \to \gamma \, ^1S_0)}$

For a $^1S_0$ of mass 2.82 GeV theory predicts a branching ratio $B(J/\psi \to \gamma \, ^1S_0) \approx 50 \%$ compared to the measured upper limit of 1.7 %. Basic input into the theoretical calculation was the assumption that the singlet and triplet radial wave functions are the same and that the magnetic moment of the c quark is of the Dirac size. Several authors have speculated that the latter may not be true /138, 139/. A precise measurement of the $D^{*+} \to D^+\gamma$ and $D^{*0} \to D^0\gamma$ decay rates will shed light on this question /138/.

2) From the upper limit $B(J/\psi \to \gamma X) < 1.7 \%$ and from $B(J/\psi'' \to \gamma X)$, $B(X \to \gamma\gamma) = 1.4 \pm 0.4 \cdot 10^{-4}$ one finds $B(X \to \gamma\gamma) > 8 \cdot 10^{-3}$ which is to be compared with the predicted value of $1.3 \cdot 10^{-3}$, (10.8). The theoretical prediction is based on the value of $\alpha_s = 0.2$ deduced from $J/\psi \to ee$ decay.

3) Theory predicts the overlap integral between $2 \, ^3S_1$ and $1 \, ^1S_0$ to be small. This can be checked by a comparison of the J/ψ and ψ' decay rates into X(2.82). Removing the phase space factor one finds agreement with the theoretical expectation

$$\frac{\frac{1}{k'^3} \Gamma_{\psi' \to \gamma X(2.82)}}{\frac{1}{k^3} \Gamma_{J/\psi \to \gamma X(2.82)}} = \frac{|<X|\psi'>|^2}{|<X|J/\psi>|^2} < 0.11 \qquad (10.12)$$

k, k' are the corresponding γ energies.

4) According to BRADLEY and GAULT /140/ the observed features of the X(2.82) are consistent with those for the η_c provided relativistic spin-dependent corrections and higher order gluon corrections are taken into account. They predict the η_c' to have a mass of ~3.6 GeV.

5) Several authors /141/ speculated that the X(2.82) is not the lowest $c\bar{c}$ pseudoscalar state but rather a four-quark state $c\bar{c}q\bar{q}$. It was further assumed that this state is below the real $1 \, ^1S_0$ level and hence narrow. In this case the decays $\psi' \to \gamma X$ and $J/\psi \to \gamma X$ should both be electric dipole transitions and the widths, assuming equal matrix elements, will be in the ratio $\Gamma(\psi' \to \gamma X)/\Gamma(J/\psi \to \gamma X) = (k'/k)^3 \approx 25$. This is an order of magnitude larger than the experimental upper limit of 2.3.

Before closing the subject of three-photon events we want to mention an interesting exercise that can be made with events coming from the direct decay $J/\psi \to \gamma\gamma\gamma$. In the quark model the decay rate is proportional to the sixth power of the quark charge. To search for the direct decay $J/\psi \to \gamma\gamma\gamma$ the DASP group /134/ considered only events with a minimum photon pair mass larger than 1 GeV. This cut excludes the radiative decays via η, η' and X(2.82). A total of 5 events were observed compared to 5.7 events predicted by QED ($e^+e^- \to \gamma\gamma\gamma$). This leads to a 90 % confidence upper limit of $\Gamma(J/\psi \to \gamma\gamma\gamma) < 5.1$ eV. This limit is about a factor of two above the rate predicted using the charmonium model of a bound $c\bar{c}$ pair, with charge $e_c = 2/3$ for the c quark and therefore consistent with $e_c = 2/3$. The upper limit is well below the value of ~30 eV predicted for a quark with unit charge.

10.5 The 3.45 GeV Level

As discussed above, the 3.45 GeV state is not yet firmly established. It was observed only in the $\gamma J/\psi$ decay mode. CHANOWITZ and GILMAN /142/ first suggested identifying it with the $2\ ^1S_0$ state. However, this causes serious problems. From the branching ratio $B[\psi' \to \gamma\chi(3.45)] \cdot B(\chi \to \gamma J/\psi) = 0.8 \pm 0.4$ % determined by SLAC-LBL and the upper limit $B[\psi' \to \gamma\chi(3.45)]$ quoted by /127/ one finds $B[\chi(3.45) \to \gamma J/\psi] > 25$ %, in contradiction to the theoretical predictions that the $2\ ^1S_0 - 1\ ^3S_1$ overlap integral should be small [see (10.12)].

   Recently it was pointed out by HARARI /142/ that the large singlet-triplet splitting operative if the X(2.82) is identified with the $^1S_0$ may push the $^1D_2$ ($J^{PC} = 2^{-+}$) level below the ψ' mass /142/. Since this state can decay into $\gamma J/\psi$ the $\chi(3.45)$ could be the $^1D_2$ state.

Summary

The observed number of levels and their ordering is that expected for a pair of fermions bound in a steeply rising potential. The fine structure splitting of the P-levels, i.e., $^3P_2 > ^3P_1 > ^3P_0$, is that expected for a vector force. A pure scalar potential would reverse the order and give a level assignment in contradiction to the experiments. The level splittings are larger than predicted but the radiative rates are smaller by a factor of two. Indirect evidence shows that the ratio of the hadronic widths of the P states are in agreement with predictions based on QCD.

11. Charmed Mesons

The existence of a fourth quark besides the familiar u, d, and s quarks implies
that the SU(3) nonet of 8 + 1 mesons will be replaced /143/ by a hexadecimet of
15 + 1 states as shown in Figure 11.1. Besides ordinary SU(3) resonances with C = 0
the hexadecimet contains six states with open charm, C = ±1 and one $c\bar{c}$ state with
hidden charm. The quark content and the names given to the charmed pseudoscalar

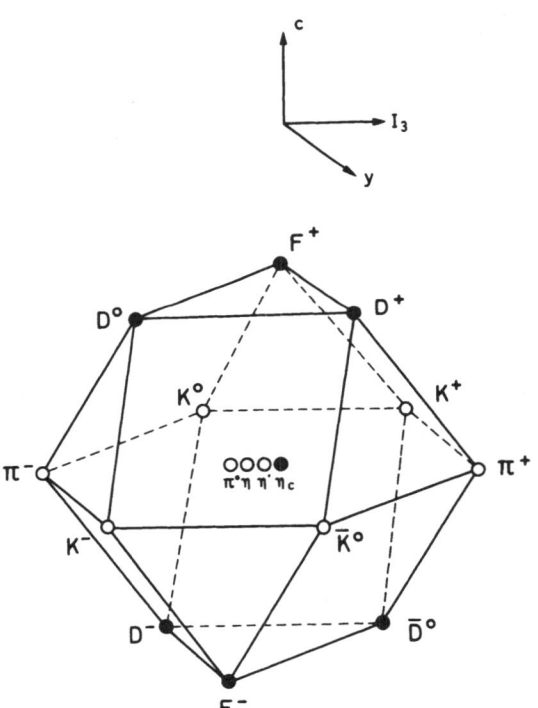

Fig. 11.1. The hexadecimet of the pseudoscalar mesons. Charm is plotted along the
z axis, Y and I_3 along, respectively, the y axis and the x axis. The π^0, η, and η'
mesons are denoted by the open circles at the origin, η_c by the black circle

131

mesons are

$$C = +1: D^+ = c\bar{d} \qquad D^0 = c\bar{u} \quad F^+ = c\bar{s}$$
$$= 0: \eta_c = 1\ ^1S_0 = c\bar{c}$$
$$= -1: D^- = \bar{c}d \qquad \bar{D}^0 = \bar{c}u \quad F^- = \bar{c}s.$$

Higher mass charmed mesons will cascade by strong and/or electromagnetic decays into these states, which then decay weakly with the following decay modes:

$$D(F) \to \ell\bar{\nu}_e$$
$$\to \ell\bar{\nu}_e + \text{hadrons}$$
$$\to \text{hadrons}.$$

The leptonic decay mode is suppressed by kinematics ($J_3^D = 0 \neq J_3^\ell + J_d^{\bar{\nu}} = 1$). Semi-leptonic decays are predicted with a branching ratio on the order of 20 % and most decays ($\approx$ 80 %) will therefore yield only hadrons in the final state /143, 144/.

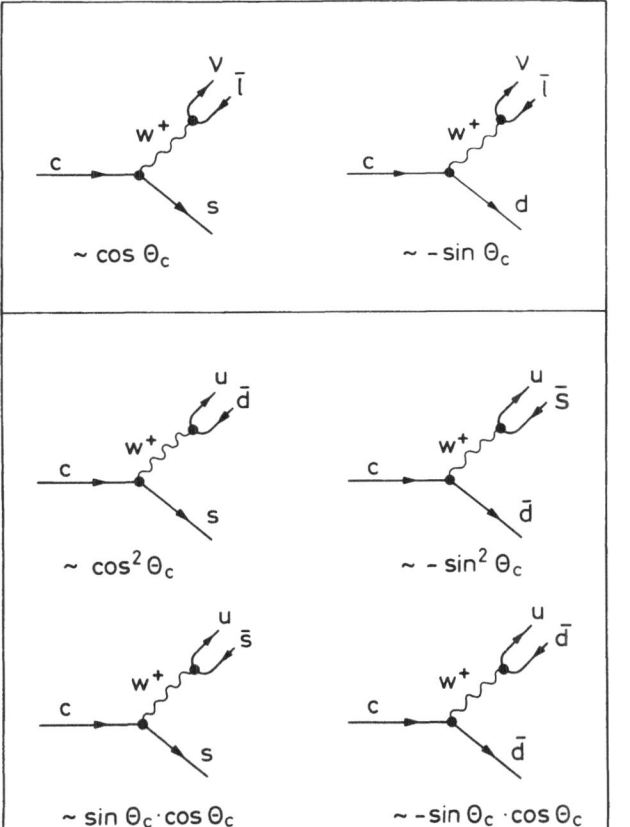

(a)

(b)

Fig. 11.2. (a) Schematic diagrams for the semileptonic decay of the charmed quark, (b) Schematic diagrams for the decay of the charmed quark into noncharmed quarks

132

Any new flavor will produce mixed lepton-hadron final states and show up as narrow resonances above threshold. However, the flavor can be identified by the properties of the final states. The GIM mechanism /112/ (Figure 11.2) leads for charmed mesons to the following Cabibbo favored decay modes:

$$D^0 \rightarrow (e^+\nu_e) \ (K^-...)$$
$$\rightarrow (\bar{K}n\pi)^0$$
$$D^+ \rightarrow (\ell^+\nu_e) \ (\bar{K^0}...)$$
$$\rightarrow (\bar{K}n\pi)^+$$
$$F^+ \rightarrow (\ell^+\nu_e) \ (\eta, \eta', K\bar{K},...)$$
$$\rightarrow (\eta n\pi)^+, (\eta'n\pi)^+, (K\bar{K}n\pi)^+.$$

According to the GIM mechanism associated production of D mesons will show up as an increase in the yield of kaons, a strong correlation between leptons and kaons in multiprong final states, and apparent exotic decays like $D^+ \rightarrow K^-\pi^+\pi^+$. The production of F mesons will produce an increase in the yield of particle with a large $s\bar{s}$ component like η, η', or ϕ.

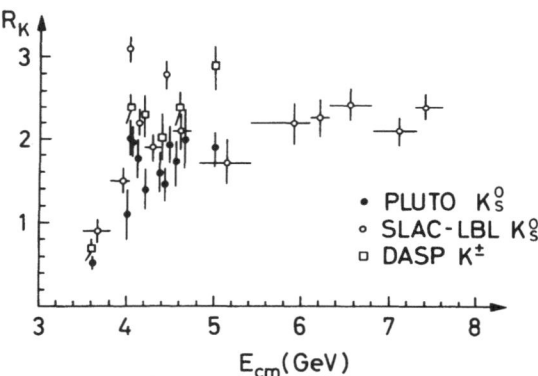

Fig. 11.3. Inclusive K_S^0 and $K^\pm$ production as a function of the c.m. energy as measured by PLUTO /145/, DASP /146/, and SLAC-LBL /147/. Plotted is

$$R_{K^0} = \frac{2\sigma(e^+e^- \rightarrow K^0X)}{\sigma(e^+e^- \rightarrow \mu^+\mu^-)} \quad \text{and} \quad R_{K^\pm} = \frac{\sigma(e^+e^- \rightarrow K^\pm X)}{\sigma(e^+e^- \rightarrow \mu^+\mu^-)}$$

as a function of the c.m. energy. [$\sigma(e^+e^- \rightarrow K^\pm X)$ is the sum of inclusive K^+ and K^- production]

A strong increase in the yield of neutral and charged kaons correlated with the step in the total cross section at 4.0 GeV was indeed observed, first by PLUTO /145/ and DASP /146/ and then confirmed by SLAC-LBL /147/. This is shown in Figure 11.3 where the ratios $R_{K^{\pm}} = \sigma(e^+e^- \to K^{\pm}X)/\sigma(e^+e^- \to \mu^+\mu^-)$ and $R_{K^0} = 2\sigma(e^+e^- \to K^0X)/\sigma(e^+e^- \to \mu^+\mu^-)$ are plotted as a function of the c.m. energy. The data are in rough agreement. The SLAC-LBL K^0 cross sections are in general larger than those of PLUTO; the PLUTO values have not been corrected for a loss (≈ 20 %) due to the cutoff in K^0 momentum ($P_K > 0.2$ GeV/c). The peak at 4.415 GeV is only reflected in the SLAC-LBL data; no clear evidence is seen for it in the DESY data. This might be due to a lack of statistics at exactly this energy.

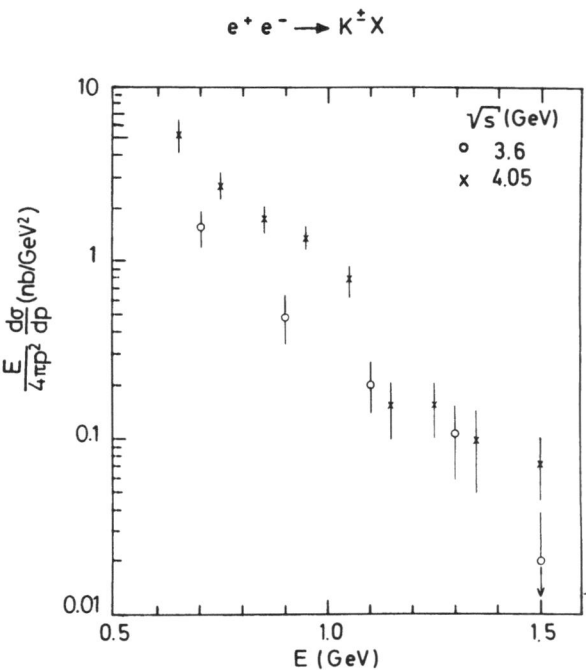

Fig. 11.4. The invariant cross sections $E/(4\pi p^2)$ dσ/dp for the sum of K^+ and K^- production at c.m. energies of 3.6 GeV and 4.05 GeV. (From DASP /146/)

A simple kinematical observation can be used to show that the increase in K production is indeed due to charm production. At threshold the charmed mesons are produced at rest, hence kaons resulting from their decays must have energies less than one-half of the beam energy. This is demonstrated by Figure 11.4, where the invariant cross section $(E/4\pi p^2)$ dσ/dp, measured by DASP /146/, is plotted as a

134

function of the kaon energy for c.m. energies of 3.6 GeV (below charm threshold) and 4.05 GeV (above charm threshold). The step in the kaon yield is caused by particles with an energy less than one half of the beam energy.

If the decay of each charmed hadron would yield one kaon then $\frac{1}{2} (\Delta R_{K^0,\overline{K}^0} + \Delta R_{K^\pm}) = \Delta R_C$ where $\Delta R_K \cdot \sigma_{\mu\mu}$ is the observed increase in K yield and $\Delta R_C \cdot \sigma_{\mu\mu}$ the contribution of charm production to σ^{tot}. From Figure 11.3 one finds $\frac{1}{2} (\Delta R_{K^0,\overline{K}^0} + \Delta R_{K^\pm}) \approx 1.7$ measured as the difference in R_K between 3.6 and 5.0 GeV. The DASP group reported for ΔR_C a value of 2.1 ± 0.3. Thus nearly every charmed hadron decays into final states containing a kaon as predicted by the GIM mechanism.

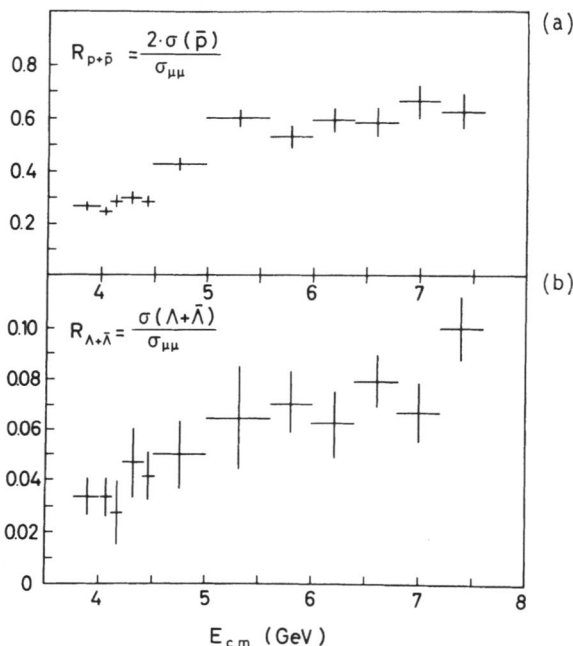

Fig. 11.5. (a) $R(p + \bar{p}) = \frac{2\sigma(\bar{p})}{\sigma_{\mu\mu}}$ versus c.m. energy, (b) $R(\Lambda + \bar{\Lambda}) = \frac{\sigma(\Lambda + \bar{\Lambda})}{\sigma_{\mu\mu}}$ versus c.m. energy. (Data from SLAC-LBL)

The SLAC-LBL group has measured the yield of antiprotons and Λ, $\bar{\Lambda}$ as a function of the c.m. energy between 3.7 and 7.6 GeV (Figure 11.5). The $\bar{p}$ yield is seen to rise above 4.5 GeV which is the threshold for charmed production.

11.1 The D Mesons

First evidence for charmed mesons was found by SLAC-LBL in a study of hadron events produced between 3.9 and 4.6 GeV and at 4.028 GeV /148/. In the $K^{\pm}\pi^{\mp}$, $K^{\pm}\pi^{\mp}\pi^{\mp}$, and $K^{\pm}\pi^{\pm}\pi^{-}\pi^{-}$ mass distributions narrow peaks were observed at a mass of 1.87 GeV. The small width ($\Gamma < 40$ MeV) of this state and the fact that the $K^{\pm}\pi^{\mp}\pi^{\mp}$ decay mode could not come from an ordinary $q\bar{q}'$ system strongly suggested the identification with the D^{0}, D^{+} mesons. Meanwhile the best place to study the D mesons was found to be the $\psi''(3771)$ /116, 117/ which decays nearly exclusively into $D\bar{D}$ /116/. The ψ'' is only 40 MeV above the $D\bar{D}$ threshold and this has the added advantage that the D's are almost at rest.

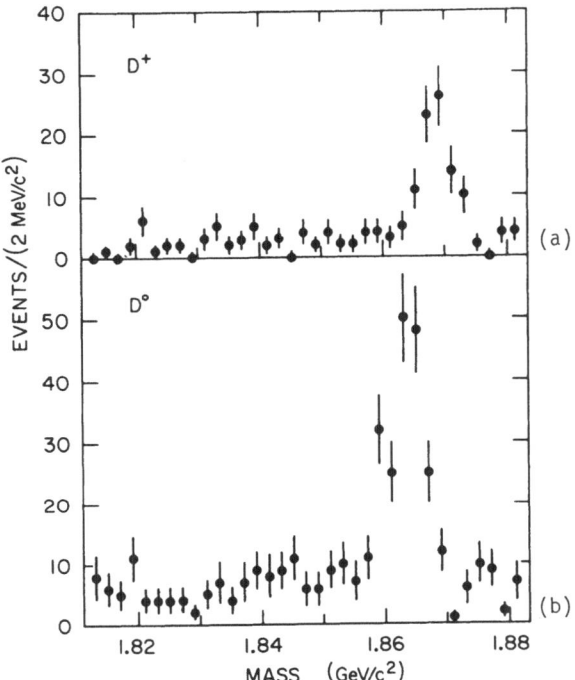

Fig. 11.6. ψ'' decay as measured by LBL-SLAC /116/: Invariant mass spectra for the sum of all observed (a) $D^{\pm}$ and (b) D^{0} decay modes

Figure 11.6 shows the sums of charged and neutral $K\pi$, $K\pi\pi$, and $K\pi\pi\pi$ mass spectra measured at the ψ''. Clear signals for the $D^{\pm}$ and D^{0} are observed. The favorable kinematics allow a precise determination of the $D^{\pm}$ and D^{0} masses. The result is listed in Table 11.1. The charged D is found to be heavier than the neutral D, $M_{D^{\pm}} - M_{D^{0}} = 5.1 \pm 0.8$ MeV

136

Table 11.1. D and D* mass values /116/

	M [MeV]
D^0	1863.3 ± 0.9
D^+	1868.4 ± 0.9
D^{*0}	2006.0 ± 1.5
D^{*+}	2008.6 ± 1.0
$M_{D^+} - M_{D^0}$	5.1 ± 0.8
$M_{D^*} - M_{D^{*0}}$	2.6 ± 1.8

D - Branching Ratios: Absolute branching ratios for D^0 and D^+ decays were determined by LBL-SLAC at the $\psi''(3771)$. Since the ψ'' is only ~40 MeV above $D\bar{D}$ threshold the only OZI allowed strong decay is into $D\bar{D}$. Noticing that the ψ'' has a strong decay width ($\Gamma \approx 25$ MeV), it is safe to assume that the ψ'' decays almost 100 percent into $D\bar{D}$. This assumption was checked by comparing the rates of events with one and two D's identified, respectively, in specific decay channels. The following numbers of events were found:

$\psi'' \rightarrow D^0 \quad X \qquad$ 194 events (one D identified)
$\qquad \llcorner_{\rightarrow K\pi}$

$\rightarrow D^0 \quad D^0 \qquad$ 8 events (two D's identified)
$\qquad \llcorner_{\rightarrow K\pi} \llcorner_{\rightarrow K\pi}$

$\psi'' \rightarrow D^\pm \quad X \qquad$ 82 events (one D identified)
$\qquad \llcorner_{\rightarrow K\pi}$

$\rightarrow D^+ \quad D^- \qquad$ 2 events (both D's identified)
$\qquad \llcorner_{\rightarrow K\pi} \llcorner_{\rightarrow K\pi}$

Since the one D event rate is proportional $B(\psi'' \rightarrow D\bar{D}) \cdot B(D \rightarrow K\pi)$ while the two D rate is $\sim B(\psi'' \rightarrow D\bar{D}) B(D \rightarrow K\pi)^2$ both ratios can be computed separately from the data. The result is

$B(\psi'' \rightarrow D^0 D^0) = 49 \pm 25$ %
$B(\psi'' \rightarrow D^+ D^-) = 50 \pm 38$ % and
$B(\psi'' \rightarrow D\bar{D}) = 99 \pm 48$ %,

which is consistent with $B(\psi'' \rightarrow D\bar{D}) = 1$.

Table 11.2 lists the measured D^0 and D^+ decay rates.

Table 11.2. D and D* branching ratios /116/

	$B(D \to f)$ [%]	Q [MeV]
$D^0 \to K^-\pi^+$	2.2 ± 0.6	
$K^0\pi^+\pi^-$	3.5 ± 1.1	
$K^-\pi^+\pi^+\pi^-$	2.7 ± 0.9	
$D^+ \to K^0\pi^+$	1.5 ± 0.6	
$K^-\pi^+\pi^+$	3.5 ± 0.9	
$D^{*0} \to D^0\pi^0$	55 ± 15	7.7 ± 1.7
$D^0\gamma$	45 ± 15	142.7 ± 1.7
$D^{*+} \to D^0\pi^+$	65 ± 9	5.7 ± 0.5
$D^+\pi^0$	29 ± 8	5.2 ± 0.9
$D^+\gamma$	$2 - 10$	140.2 ± 0.9

$\Gamma(D^{*0})/\Gamma(D^{*+}) = 0.95 \pm 0.4$

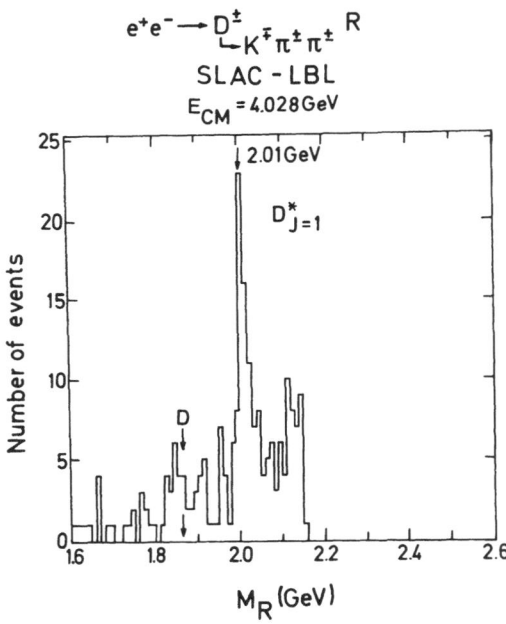

Fig. 11.7. Reaction $e^+e^- \to D^{\pm}R$ measured by SLAC-LBL /148, 149/. Mass spectrum
$\quad\quad\quad\quad\quad\quad\quad\quad\quad\quad\quad \lfloor_{\to k^+\pi^{\mp}\pi^{\pm}}$
of the recoiling system R

D* Meson: The data taken by SLAC-LBL at 4.028 GeV provided also evidence for the first excited D states, the D*^{0} and D*$^+$ /148, 149/. Figure 11.7 shows the mass spectrum of the system recoiling from D$^\pm$. A narrow signal is found at a mass of 2.008 GeV. Note that no clear evidence is seen for a recoiling D which indicates that the D$\bar{\text{D}}$ production cross section at this energy is considerably smaller than for DD* although the Q value for DD* is only ~150 MeV.

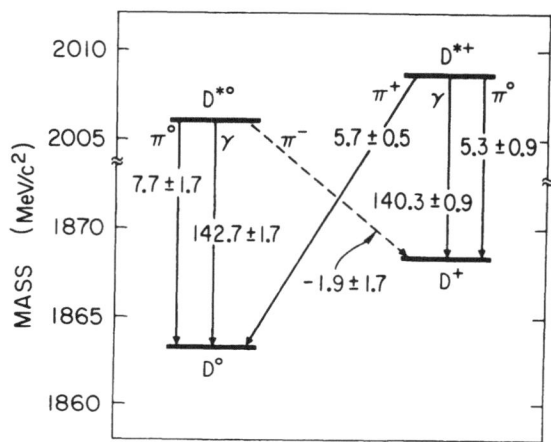

Fig. 11.8. Q values for the D* → D transitions as measured by LBL-SLAC /116/

The D* mass values are given in Table 11.1. The Q values for the D* → Dπ decays are only a few MeV or are negative (D*^{0} → D$^+\pi^-$). This is the reason for the small D* width. For the same reason radiative D* decays, D* → Dγ, play an important role. Figure 11.8 illustrates the possible D* → D transitions. The observed branching ratios are given in Table 11.2. From an analysis of the angular distribution for D mesons produced near threshold and of the D* decay channels, spin and parity of D and D* were found to be $J_D^P = 0^-$, $J_{D*}^P = 1^-$ where the D was chosen to have negative parity /149/.

A detailed study of D production at 4.028 GeV besides DD* revealed a sizeable fraction of D*$\bar{\text{D}}$* production although the Q value is only 16 MeV for the D^0*$\bar{\text{D}}^0$* and 12 MeV for the D$^\pm$*D$^\mp$* channels /150/. Correcting for phase space effects by a factor p^3 the relative cross-section ratios for the D production channels were measured to be D$^0\bar{\text{D}}^0$: D$^0\bar{\text{D}}^0$* + $\bar{\text{D}}^0$D*^{0} : D*$^0\bar{\text{D}}^0$* = 0.2 ± 0.1 : 4.0 ± 0.8 : 128 ± 40.

Spin counting predicts the ratios 1 : 4 : 7. Hence the 4.04 structure is coupled mainly to the D*$\bar{\text{D}}$* channel. This has led to the speculation that the structures seen in σ^{tot} above 4.0 GeV and in particular the 4.04 structure are c$\bar{\text{q}}$ $\bar{\text{c}}$q molecule states /151/. As a consequence a large rate (several percent) for J/ψ + X production at 4.04 GeV was expected. The PLUTO group /152/ has searched for this process and measured an upper limit of ~10^{-3} σ^{tot} for J/ψX production.

11.2   The F Mesons

The charm model predicts mesons carrying both charm and strangeness. The ground
state is the $F^{\pm}$. The GIM favored decay of the F is into an $s\bar{s}$ system leading to
final states containing $K\bar{K}$, Φ, η, or η'.

$$F^+ \frac{\quad \bar{s} \qquad\qquad \bar{s} \quad}{\quad c \qquad\qquad s \quad} = K\bar{K}, K\bar{K}\pi, \phi, \eta, \eta'.$$

$$= \pi, \ldots$$

Sizeable η production is therefore an indication of F production. DASP /153/ searched
for F production by studying events of the type

$$e^+e^- \rightarrow \eta + \geq 2 \text{ charged tracks} + X.$$

The η was identified by its decay into two photons. A search of this type is hampered
by the $\gamma\gamma$ mass resolution which in the DASP experiment was 80 MeV, and by a large
$\gamma_i\gamma_j$ combinatorial background: near 4 GeV on the average two to three π^0's are produced leading to 4 - 6 photons or 6 - 15 two-photon mass combinations. The event
selection was done as follows. The events accepted were required to have at least
two photons and two charged tracks coming from the interaction region. The photons
were detected in the inner detector and their angles and energies were measured.
Photons with energies between 0.14 and 1.0 GeV were considered candidates in the
determination of the two-photon mass spectrum $M(\gamma_i\gamma_j)$. The vector sum of the
momentum of the two photons (γ_i,γ_j) was required to be between 0.3 and 1.4 GeV. The
photon detection efficiency was 50 % at 0.05 GeV, rising to 80 % at 0.1 GeV at
0.1 GeV and 95 % above 0.3 GeV.

   The data were grouped into five c.m. energy intervals, 3.99 to 4.10 = "4.03"
GeV (1178 nb^{-1}), 4.10 to 4.23 = "4.17" GeV (509 nb^{-1}), 4.23 to 4.36 = "4.36" GeV
(603 nb^{-1}), 4.36 to 4.49 = "4.42" GeV (2240 nb^{-1}), 4.5 to 4.99 = "4.60" GeV (727 nb^{-1})
and 5.0 GeV (1270 nb^{-1}). The numbers in quotes are the average energies weighted
by the luminosity for a given interval, and those in parentheses are integrated
luminosities for each data set.

   Figure 11.9 shows the two-photon effective mass distribution for the various
center-of-mass energies; while a clear π^0 peak is observed at all energies, η signals are clearly seen only at 4.17 and 4.42 GeV, and possibly at 4.60 GeV. The
lack of an η signal at the 4.03 GeV resonance, which is dominated by D$\bar{D}$* and D*$\bar{D}$*
production, indicates that the branching ratio for the decay D $\rightarrow \eta$ + anything is
small.

   In order to determine the η cross section acceptance and efficiency corrections
have to be applied which in this case depend sensitively on the details of the
production mechanism. The authors of /153/ considered two different models for η

140

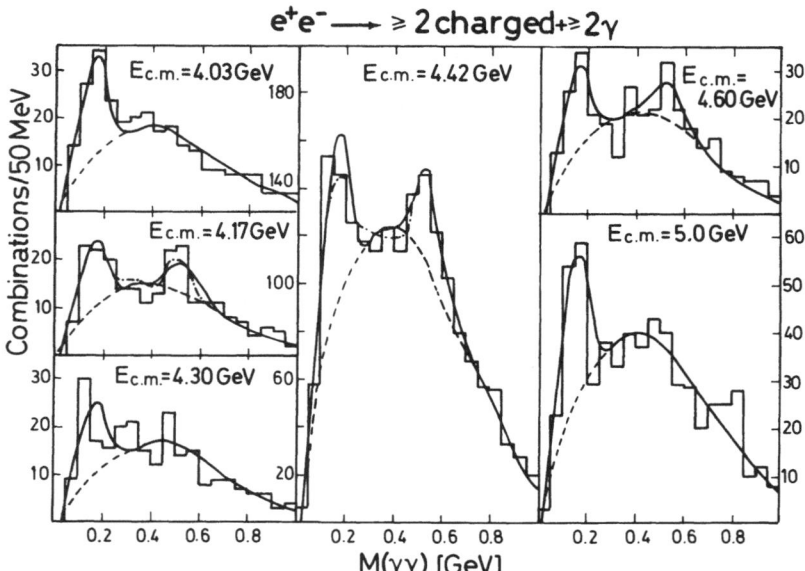

Fig. 11.9. M($\gamma\gamma$) mass distribution at the various c.m. energies. The solid lines are the results of a fit to a sum of background, π^0 and η contributions. The dashed lines correspond to the amount of background required by the fit under the η and π^0 peaks (method 1). The dashed-dotted lines at 4.17 and 4.42 GeV are the results of a fit corresponding to the sum of F production and the background, described by the M($\gamma\gamma$) mass distribution at $E_{c.m.}$ = 4.03 GeV (method 2). /153/

production. Model 1 is the statistical isospin model for F decay by QUIGG and ROSNER /154/. In the calculation a 16 % semileptonic decay ratio, 38 % hadronic decays with an η and 9 % with an η' in the various final states were assumed. Model 2 is a phase space model which assumes that on the average 4.2 charged particles and 2.8 π^0's are produced together with an η. At 4.17 GeV model 1 gives an η acceptance of 4.2 % whereas model 2 yields 3.5 %. For the π^0 acceptance the discrepancy between the two models is larger: 0.33 % according to model 1 and 0.72 % according to model 2.

The reliability of the analysis was checked by computing from the observed number of π^0's the total π^0 cross section. At all energies the π^0 cross sections calculated with model 1 were consistent within 50 % with [$\sigma(\pi^0) = \sigma(\pi^+) + \sigma(\pi^-)$]/2 as measured in the same experiment /155/.

Figure 11.10 shows the total inclusive η cross section $\sigma(\eta)$, as a function of the c.m. energy, calculated using model 1.

While no η signal is seen at 4.03 GeV, a significant production is seen between 4.10 and 4.70 GeV. Note that η production below 4.10 GeV is less than 0.5 nb.

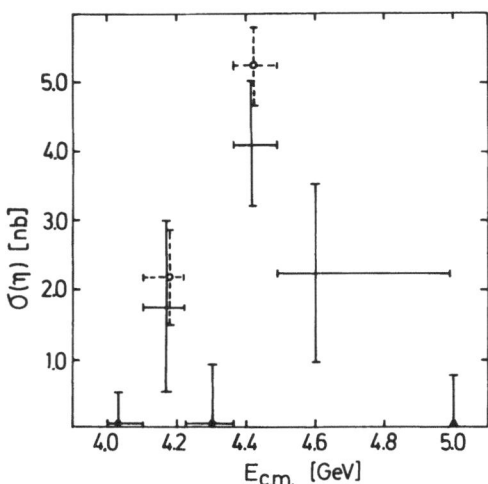

Fig. 11.10. η inclusive cross section as a function of the c.m. energy, calculated according to method 1 (solid lines) and to method 2 (dashed lines). /153/

This is surprisingly small: it is smaller than pion production (i.e., π^+ or π^-) by a factor of 50 - 80 and smaller than kaon production (i.e., K^+ or K^-) by a factor of ten.

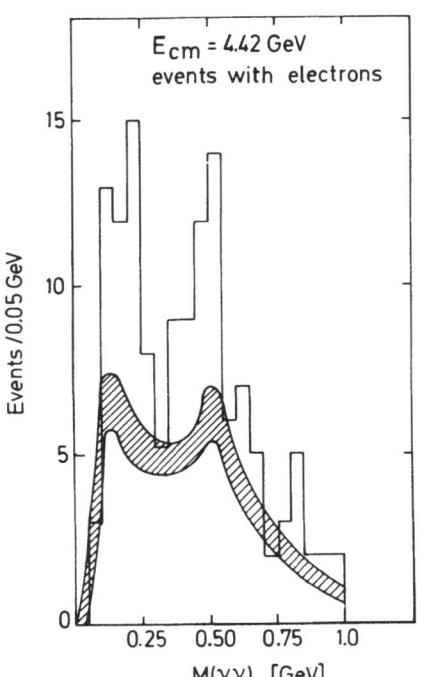

Fig. 11.11. $M(\gamma\gamma)$ mass distribution at $E_{c.m.} = 4.42$ GeV, for events having an electron in the DASP inner detector. The shaded band corresponds to a $\pm 1\sigma$ uncertainty in the expected background from hadrons simulating an electron and photons converting in the beam pipe /153/

The following upper limit was deduced in /156/: B.R. (D → η + anything) < 2 %. Using the D cross section of /156/ at 4.42 GeV, it would imply that σ(η; from D decay) ≤ 0.4 nb, which is more than a factor 10 smaller than σ(η) at 4.42 GeV, implying that η's at 4.42 GeV are produced by a different source. Using σ(η) at 4.42 GeV, and assuming that all of the 4.42 GeV resonance structure is due to F production, the following lower limit can be obtained: B.R. (F → ηX) ≥ 34 %.

Figure 11.11 shows the M(γγ) mass distribution at 4.42 GeV for events having an electron in the DASP inner detector. The contamination due to hadrons simulating an electromagnetic shower was estimated by looking at the process e^+e^- → J/ψ → ρπ → $\pi^+\pi^-\pi^0$, and it was found to be 1.2 ± 0.5 % per charged track. The contamination due to photons converting in the beam pipe and Dalitz π^0 decay was estimated to be less than 2.5 % of the events. Folding those numbers with the events that contributed to the M(γγ) mass distribution in Figure 11.9 gives rise to the background distribution shown in Figure 11.11 as a ±1σ band. Clear η and π^0 peaks can be seen above this background, indicating that both are the decay products of weakly decaying states. While the π^0's have their source in τ, D, and F production, the η signal cannot be due to the first two sources due to the lack of η signal at 4.03 GeV. Therefore the weakly decaying F meson becomes the most natural source of η production in this energy region.

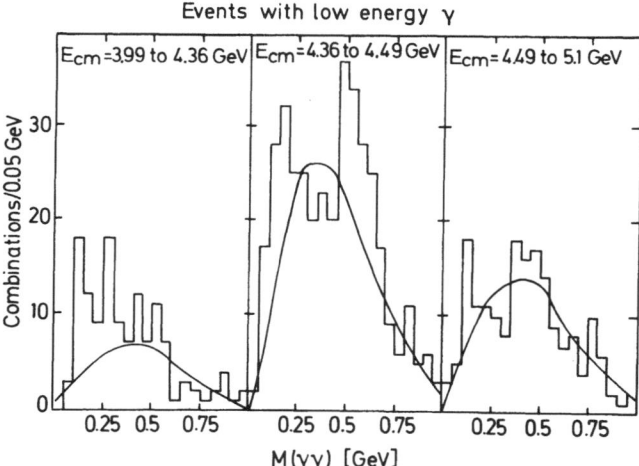

Fig. 11.12. M(γγ) mass distribution for events having a low-energy photon (E_γ < 140 MeV), below, at, and above the 4.42 GeV $E_{c.m.}$ region. The solid lines are estimates of uncorrelated photon background, normalized for M(γγ) > 0.7 GeV /153/

In an earlier publication the same group found that η's are produced at 4.4 GeV
in conjunction with a low-energy photon (E_γ < 140 MeV), indicating that the dominant
F production mechanism in this region occurs via the FF* and/or F*F* channels.
Figure 11.12 shows the M(γγ) distribution for events having a photon of less than
140 MeV momentum at c.m. energies below, in, and above the 4.42 GeV region. While
the first and last distributions do not show any strong η signal, the 4.42 GeV data
show a clear η peak. Moreover, this signal lies on a smaller background than the
one in Figure 11.9 for the same c.m. energy. This correlation of the η signal with

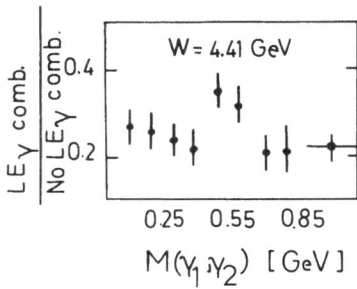

Fig. 11.13. Ratio of the number of combinations
for events having a low-energy photon to the number of combinations not having a low-energy photon
as a function of M(γγ), at $E_{c.m.}$ = 4.42 GeV /153/

events having a low-energy photon is shown more clearly in Figure 11.13, where the
ratio of events having a low-energy photon to those that do not have is shown as
a function of M(γγ). A 3.6 σ peak at the η mass is seen, indicating that η's in
the 4.42 GeV region are produced in conjunction with a low-energy photon. This con-
firms the previous observation that FF* or F*F* seem to be the dominant production
mechanism.

Since the largest amount of η signal is observed at the 4.42 GeV region, the
decay $F^\pm \to \eta\pi^\pm \to \gamma\gamma\pi^\pm$ was studied to determine the F meson mass. Candidates for the
two-body decay F → ηπ were sought using the DASP outer spectrometers, which allow
particle identification, and momentum measurement [$\Delta(p)/p$ = 0.02 p (GeV/c)] of
charged particles. To enter the fitting procedure a charged pion with momentum above
0.6 GeV/c was required, and at least two photons with energies above 0.1 GeV forming
an M(γγ) in the η region. One of the photons forming M(γγ) had to be in the inner
detector, while the second photon could be either in the inner detector or in the
shower counter of the spectrometer arms. Finally there had to be at least one or
more photons with an energy below 0.2 GeV (γ_{low}). A total of 43 events satisfied
these selection criteria at the 4.42 GeV region, and 79 events at all other energies.
These events were fitted to the reactions

$$e^+e^- \to (FF^* \text{ or } F^*F) \to F\ F\ \gamma_{low}$$
$$\phantom{e^+e^- \to (FF^* \text{ or } F^*F) \to F} \downarrow \eta\ \pi$$
$$\phantom{e^+e^- \to (FF^* \text{ or } F^*F) \to F \downarrow \eta} \downarrow \gamma\ \gamma$$

(11.1)

144

$$e^+e^- \to F^*F^*$$
$$\underset{\gamma_{low}}{\downarrow} \quad F$$
$$\qquad\quad \underset{\eta\ \pi}{\downarrow}$$
$$\qquad\quad \vert\!\to \gamma\ \gamma$$

(11.2)

These are 2C fits because of the mass constraint on $M(\gamma\gamma)$ and the requirement that $\pi\eta$ ($\pi\eta\ \gamma_{low}$) and the missing vector must have the same mass m_F (m_{F^*}) in the case of reaction (11.1), (11.2).

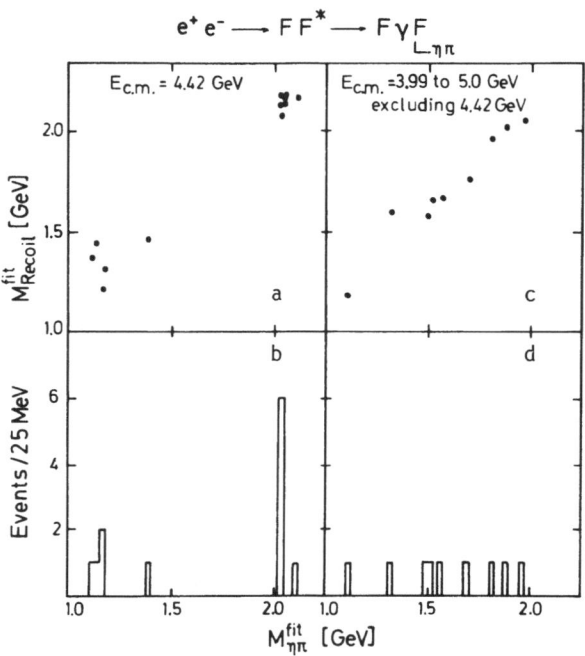

Fig. 11.14. Fitted $\eta\pi$ mass vs fitted recoil mass, assuming $e^+e^- \to FF^*$, where $F^* \to F\gamma$ and $F \to \eta\pi$ at (a) $E_{c.m.}$ = 4.42 GeV and (c) at all other energies excluding 4.42 GeV. Histograms (b) and (d) are the projections of (a) and (c), respectively, along the $M(\eta\pi)$ axis /153/

There were 15 events in the 4.42 GeV region and 11 events at all other energies that gave a fit to reaction (3.1) with $\chi^2 < 8$. By making the additional requirement $|M_{fit}^{(\eta\pi)} - M_{meas}^{(\eta\pi)}| \leq 250$ MeV in order to cut on badly measured events, those numbers reduced furthermore to 12 and 10 events, respectively. Figures 11.14a and c show the fitted $\eta\pi$ mass vs the fitted recoil mass for the two regions. At 4.42 GeV a clustering is seen at M($\eta\pi$) = 2.04 GeV and M$_{recoil}$ = 2.15 GeV (Figure 11.14a) while no such clustering is seen in Figure 11.14c for all other energies. All events in Figure 11.14c lie on a band given by the kinematics of the fit. Figures 11.14b and 11.14d show the projections of Figures 11.14a and 11.14c along the M($\gamma\gamma$) axis. A peak containing six events is seen at M($\eta\pi$) = 2.04 ± 0.01 GeV for the 4.42 GeV data, while no such a peak is seen for all other energies; this implies that the background under the peak at 4.42 GeV is less than 0.2 event. The events at 4.42 GeV also give an acceptable fit to reaction (3.2) with a lower value for the F mass [M($\eta\pi$) = 2.00 GeV]. The spread in the M($\eta\pi$) distribution is slightly larger than in the case of hypothesis (11.1), as expected due to the ambiguity in the determination of the F* → Fγ_{low} relation. Allowing for possible systematic uncertainties, the best estimate is m$_F$ = 2.03 ± 0.06 GeV.

The mass difference between F* and F was directly determined from the energy of the γ_{low} for the six events in the $\eta\pi$ mass peak. The result is m$_{F*}$ - m$_F$ = 0.11 ± .046 GeV. The cross section for those six events is found to be 0.41 ± 0.18 nb, giving

$$\frac{BR(F \rightarrow \eta\pi)}{BR(F \rightarrow \eta + anything)} = 0.09 \pm 0.05 \quad (\pm 30 \text{ \% syst.})$$

consistent with the assumption made in the Monte Carlo model (0.062) to compute the η acceptance.

Since a strong η signal is seen at the 4.17 GeV region, the authors of /153/ made a search for events that would fit the process

$$e^+e^- \rightarrow F^{\pm}F \qquad (11.3)$$
$$\vert_{\rightarrow \eta \pi^{\pm}}$$
$$\vert_{\rightarrow \gamma \gamma} .$$

The selection criteria were the same as those imposed for reactions (11.1) and (11.2) except that no requirement was made on the presence of a γ_{low}. After imposing a χ^2 cut at 8, five events survived; only one event was above M = 1.95 GeV. This event had a mass of 2.03 ± 0.02 GeV, which is consistent with the F mass value found at the 4.42 GeV region.

Another group /157/ reported a preliminary observation of a peak in the $K^+K^-\pi^{\pm}$ mass distribution at E$_{c.m.}$ = 4.16 GeV. The mass value for this peak is 2039 ± 1.0 MeV, consistent with the mass value for the $\eta\pi$ decay mode given by the DASP group.

A discussion of the F and F* properties expected by theory can be found, e.g., in /154, 158/.

11.3 Semileptonic Decays of Charmed Particles

11.3.1 Origin of Electron Inclusive Events

The pair production of new particles with large leptonic or semileptonic decay modes will lead to mixed final states containing leptons and hadrons. The observation of such final states at a level above the background expected from higher order electromagnetic interactions and semileptonic pion or kaon decays is direct evidence for the production of new particles. A new lepton or a hadron with a new flavor is but two examples of such particles. As will be discussed later there is now convincing evidence that besides charmed hadrons also a new lepton is pair produced at c.m. energies above 4 GeV.

<u>Table 11.3.</u> Properties of heavy sequential leptons and new hadrons

	L	H
Production	$e^+e^- \to L^+L^-$ (point cross section $e^+e^- \to L^+L^-$ + hadrons (negligibly small, less than α^2 of elastic production near threshold	$e^+e^- \to H\bar{H}$ (damped by form factors) $e^+e^- \to H\bar{H}$ + hadrons or $H^*\bar{H}^*$ (dominant at higher energies, cross section will have structure)
Decay modes	$L \to \ell\,\bar{\nu}_\ell\,\nu_L$ $\to \nu_L \cdot$ hadrons	$H \to \ell\,\bar{\nu}_\ell$ (suppressed if the lowest flavor state has spin 0) $\to \ell\,\bar{\nu}_\ell \cdot$ hadrons $\to$ hadrons
Final states: $e\mu$ + neutrinos	important, clear signature [e(μ) from three body decay]	negligible [e(μ) from a multibody decay]
$\ell\ell$ + neutrinos + hadrons	negligible, order α^2 at threshold	large [e(μ) from a multibody decay]
e(μ) + neutrino + hadrons	large, lepton spectrum computable and hard, hadrons have low multiplicity	large, lepton spectrum soft, soft, hadrons have high multiplicity

The anticipated features, of a new lepton and a new hadron are summarized in Table. 11.3. The low multiplicity expected for the decay of a pair of heavy leptons into a final state with an electron (muon) plus hadrons arises as follows. In the decay $L \to \nu_L$ + hadrons, the hadrons come from a low-mass current. If the multiplicity is related to that from a virtual photon of the same mass, it will be small. Specific calculations support this conjecture. The leptonic decay of its partner L contributes in general only one charged track. Inelastic production, i.e., $e^+e^- \to L\bar{L}$ + hadrons, is negligible at these energies. The high multiplicity expected in the decay of a pair of heavy hadrons will presumably lead to a multiplicity comparable to that observed in the decay of an ordinary hadron of the same mass -, i.e., on the average of 2 to 3 charged particles plus a few photons. From the semileptonic decay of its partner we might expect one or two charged particles and a few photons. Thus a cut on the total multiplicity can be used to separate the two classes with $e^+e^- \to L\bar{L} \to e(\mu) + X$ predominantly populating the two-prong class.

The momentum spectrum of the observed electron (muon) can also be used to classify inclusive lepton events. All particles in the leptonic decay $\tau \to \ell\nu\bar{\nu}$ are point-like and large values of Q^2, i.e., large electron energies, are therefore not suppressed by form factors. Also, at least two of the particles have negligible mass. In a semileptonic decay of a charmed particle the form factor will disfavor large values of Q^2. The mass of the hadron system will also lead to smaller electron momenta.

We will first discuss the data on semileptonic decay of charmed particles. Information on the τ particle is presented in the next section.

11.3.2 Results on the Semileptonic Decays of Charmed Hadrons

Semileptonic decays of charmed hadrons have been widely considered in the literature /143, 144, 159/. It follows directly from Figure 11.2 that the amplitude for the favored mode is proportional to $\cos\Theta_c$ with $\Delta S = \Delta Q = +1$ and $\Delta I = 0$. The suppressed mode has an amplitude proportional to $-\sin\Theta_c$ with $\Delta C = \Delta Q$, $\Delta S = 0$ and $\Delta I = 1/2$. A D meson should predominantly decay into final states like $e\bar{\nu}_e K$ or $e\bar{\nu}_e K^*(892)$. Disfavored modes like $D \to e\bar{\nu}_e \pi$ should be suppressed by $tg^2\Theta_c \cong 0.05$. The decay into $e \bar{\nu} K^*(1420)$ is suppressed by phase space and soft pion theorems predict that the decay $D \to e\bar{\nu}K(n\pi)$ where the $K(n\pi)$ system does not form a resonance should be small.

The semileptonic branching ratio of charmed hadrons can be determined from a measurement of mixed electron hadron final states

$$e^+e^- \to C\bar{C} \to (e \bar{\nu} \ldots) (X).$$

The average branching ratio BR(C $\to$ eX) is obtained from

$$B(C \to eX) = \frac{\text{number of eX-events}}{2\sigma(c\bar{c})}.$$

The total cross section for charmed hadron production is defined as $\sigma(e^+e^- \to c\bar{c}) = \sigma(\text{tot}) - R(3.6)\sigma(\mu\bar{\mu})$ where $R(3.6)\sigma(\mu\bar{\mu})$ is the cross section for noncharmed hadron production determined below charm threshold at 3.6 GeV.

In principle mixed muon hadron events can also be used. However, the lepton spectrum must be measured down to low momenta with a rejection factor of 10^3 against hadrons. Such a rejection factor is difficult to obtain for low-energy muons.

DASP has measured /160/ the single-electron spectrum for momenta above 200 MeV/c and c.m. energies between 3.6 GeV and 5.2 GeV for a total integrated luminosity of 6300 nb^{-1}. Electrons were defined as particles which gave a signal in the proper Cerenkov counter and had $\beta = 1$ (p < 0.35 GeV/c) or gave a large pulse height (p > 0.35 GeV/c) in the shower counter. A pion had a probability of 4×10^{-4} to pass these criteria. Electron pairs from Dalitz decay or pair conversion were rejected by pulse height cuts on the scintillation counters mounted before the magnet.

To reduce the background of electromagnetic origin the event was required to contain at least one nonshowering track. This could be a track in a spectrometer arm identified as a hadron or a muon. A charged track traversing the inner detector could also be called nonshowering if it fired less than 1.5 tubes per layer, averaged over all layers which had at least one tube set. Events containing only two charged particles were particularly sensitive to electromagnetic background. Here it was required that less than 1.25 tubes per plane be activated. Tests of these criteria using well-defined pions showed that 95 % of the pions but fewer than 5 % of the electrons satisfied the tight criteria.

The background to this sample from beam-gas events, Compton scattering on the material mounted in the front of the Čerenkov counter, or from two-photon processes was estimated and found to be small. The total background was (11.5 ± 3.5) % for the two-prong sample and (15 ± 5) % for the multiprong sample. These values are in agreement with measurements done at the ψ' resonance and at 3.6 GeV.

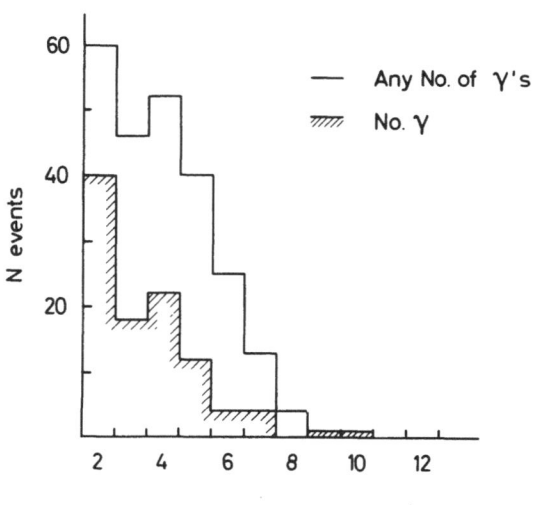

Observed Prong Distribution in $e^+ e^- \to e^{\pm} + X$

— Any No. of γ's

▨ No. γ

N events

N prong with $e^{\pm}$

Fig. 11.15. The charged track distribution observed by DASP /160/ for inclusive electron events. The electron is included in the prong number. The shaded distribution is for events without photons

After all cuts 60 two prongs and 182 multiprong events with an electron momentum above 0.2 GeV/c remained in this sample. The measured charged multiplicity distributions (including the electron) for all inclusive electron events and for those with only charged tracks are plotted in Figure 11.15. The distribution peaks for n_{ch} = 2 but it is rather wide with events up to a charged multiplicity of eleven. As was shown by DASP both charm and τ-pair production are needed to explain the observed multiplicity distribution. The observed events are grouped into two-prong (predominantly τ events) and multiprong events (predominantly charm).

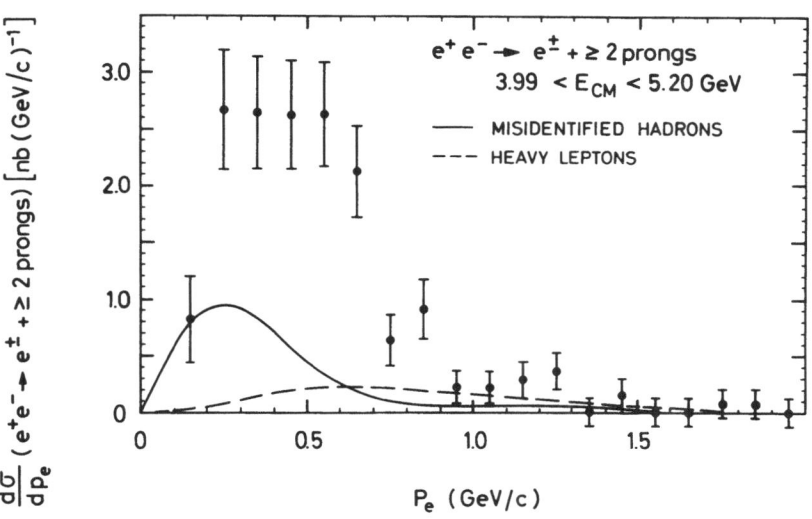

Fig. 11.16. The inclusive electron momentum spectrum measured by DASP /160/ between 3.99 and 5.2 GeV for multiprong events

The lepton spectrum associated with the multiprong sample is shown in Figure 11.16. The estimated background due to hadron misidentification or heavy lepton production is also plotted. The background was scaled from measurements below threshold. The heavy lepton contribution was estimated assuming a τ branching ratio of 30 % to decay into final states with three or more charged particles. DASP finds that less than 12 % of the events with $n_{ch} \geq 3$ can be explained as heavy lepton production. The simple cut on hadron multiplicity therefore yields a rather clean sample of charm decays.

The electron spectrum contains information on the semileptonic and the leptonic decay modes of the lowest mass charmed hadrons. Figure 11.16 demonstrates that semileptonic decays are much more important than leptonic decays because the latter, being two-body decays, would produce a peak in the electron spectrum around 1 GeV/c. This is in gross disagremment with the data which peak around an electron momentum of 0.5 GeV with only few events above 0.7 GeV/c. To study the observed momentum spectrum in more detail we consider the spectrum obtained for c.m. energies between 3.99 and 4.08 GeV. The charm cross section in this energy region is dominated by $D\bar{D}^*$ and $D^*\bar{D}^*$ production and is below the threshold for F production. The spectrum, corrected for the background and the heavy lepton contribution, is shown in Figure 11.17.

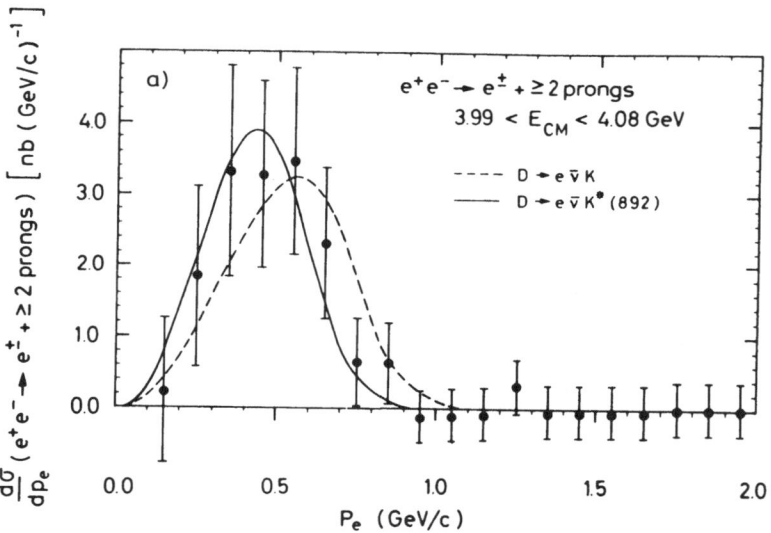

Fig. 11.17. The electron momentum spectrum for $D \to e\nu X$ as measured by DASP /160/. The momentum distributions expected for $D \to e\bar{\nu}_e K$ and $D \to e\bar{\nu}_e K^*(891)$ are shown by the dashed and the solid curves

The spectrum in Figure 11.17 was fitted to three possible channels: $D \to e\bar{\nu}_e \pi$, $D \to e\bar{\nu}_e K$, and $D \to e\bar{\nu}_e K^*(892)$. A V-A current was assumed and the form of the spectra was taken from a paper by ALI and YANG /159/. Note that the theoretical spectra are model dependent. These fits gave a χ^2 value for 10 degrees of freedom of: 29.6 for $D \to e\bar{\nu}_e \pi$, 6.3 for $D \to e\bar{\nu}_e K$, and 2.8 for $D \to e\bar{\nu}_e K^*(892)$. The decay $D \to e\bar{\nu}_e \pi$ can therefore be excluded as the sole semileptonic decay mode of the D. The data can be fitted with either $D \to e\bar{\nu}_e K$ or $D \to e\bar{\nu}_e K^*(892)$. A good fit to the spectrum can also be obtained assuming the charm changing weak current to be right handed in the decay $D \to e\bar{\nu}_e K^*(892)$.

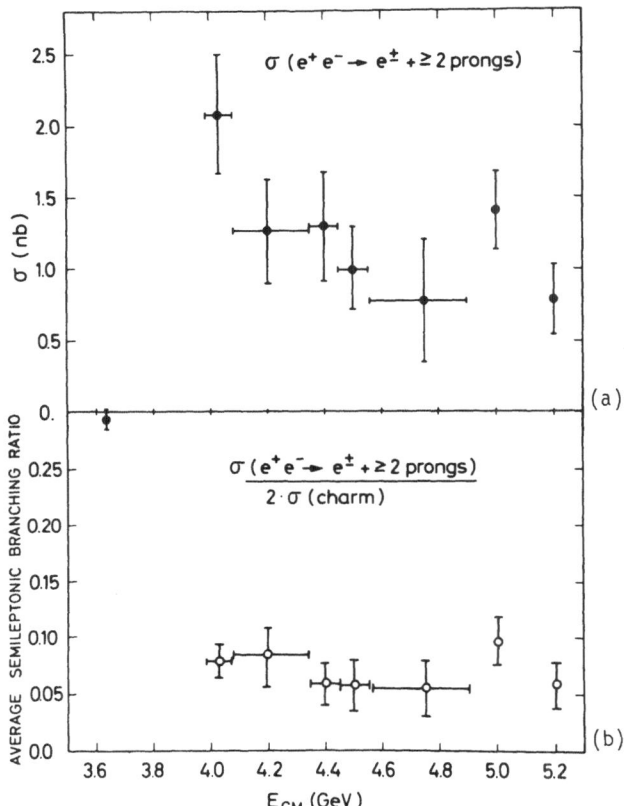

Fig. 11.18. The cross section measured by DASP /160/ for the inclusive production of electrons plus nonshowering track plus additional charged tracks as a function of c.m. energy, (b) the average semileptonic branching ratio for charmed hadrons as a function of energy. The error bars are statistical only

The absolute cross section for inclusive electron production $e^+e^- \rightarrow e^{\pm} + X$, where X contains at least two charged tracks and any number of photons, is plotted in Figure 11.18a as a function of energy. The data have been corrected for radiative effects. The background from hadron misidentification and the contribution from heavy leptonic production have been subtracted.

The inclusive cross section due to charmed particle production can be written as

$$\sigma(e^+e^- \rightarrow e^{\pm}X) = \Sigma_{i,j} \, \sigma(e^+e^- \rightarrow C_i\bar{C}_j) \cdot \{B(C_i \rightarrow e\bar{\nu}_eX) + B(C_j \rightarrow e\bar{\nu}_eX)\}.$$

Here $\sigma(e^+e^- \rightarrow C_i\bar{C}_j)$ denotes the effective cross section for producing the lightest charmed hadron stable against strong and electromagnetic decays. These particles might either be produced directly or result from the cascade decay of excited charmed hadrons. The cross section $\sigma(e^+e^- \to C_i\bar{C}_j)$ was obtained by subtracting the cross sections for "old" hadron production from the total hadronic cross section.

Near threshold, where only neutral and charged D production can contribute DASP finds

$$B(D \to e + X) = 0.08 \pm 0.02.$$

This should be compared to the value

$$B(C \to e + X) = 0.072 \pm 0.02$$

obtained by averaging over all energies between 3.9 and 5.2 GeV (see Figure 3.18b). These values were extracted using the DASP /120/ data for the total cross section and the error quoted is mainly systematic.

The semileptonic branching ratio can also be determined from the fraction of inclusive electron events containing a second electron. Using this method the DASP group finds $B(C \to e^-X) = 0.16 \pm 0.06$. Note that this value is independent of the charm cross section.

The DASP results are supported by the results from two recent experiments at SPEAR.

The DELCO group at SPEAR has measured /161/ the cross section for $e^+e^- \to e^\pm + \geq 2$ hadrons $+ \geq 0$ photons. Electrons are identified in 65 % of 4π using a Cerenkov counter sensitive only to electrons (pions have the threshold at 3.7 GeV/c). The Cerenkov counters are backed by an array of lead scintillator (2 r.l. divided into 3 layers) shower counters which cover 60 % of 4π. The on-line trigger is rather loose requiring only a charged track plus a signal from two shower counters. Electron inclusive multi-prong events were selected from this sample demanding that at least two shower counters should fire. The candidate track for an electron was required to give signals in the Cerenkov and the appropriate shower counter. To reduce the background from Dalitz pairs, photon conversions, and δ rays it was required that no other track should be within an angle of 18 mrad/p(GeV/c) (p is the momentum of the softer hadron) with respect to the electron candidate. Hadronic events at the J/ψ resonance passed these selection criteria with an efficiency of $(3.5 \pm 0.3) \cdot 10^{-3}$.

$$R_e = \frac{\sigma(e^+e^- \to e^\pm + \geq 2 \text{ hadrons} + \geq 0 \text{ photons})}{\sigma(e^+e^- \to \mu^+\mu^-)}$$

is plotted in Figure 11.19 as a function of c.m. energy. The dotted curve represents the estimated background. A clear peak is seen at 3.77 GeV produced by the decays of the $\psi''(3772)$ into pairs of charged and neutral D mesons. The ratio reflects the charm cross section with a rise at 4 GeV, a dip around 4.25 GeV, and presumably some structure around 4.4 GeV.

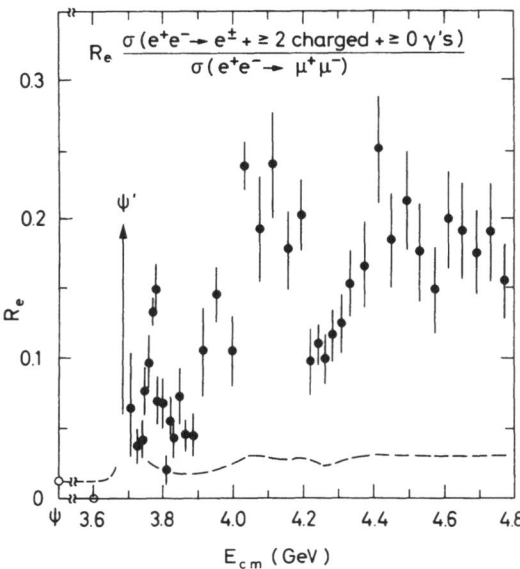

<u>Fig. 11.19.</u> R_e, the cross section for $e^+e^- \rightarrow e^{\pm} + \geq 2$ prongs $+ \geq 0$ photons normalized to the point cross section, plotted versus the c.m. energy. The dotted line indicates the background. The data are from DELCO /161/

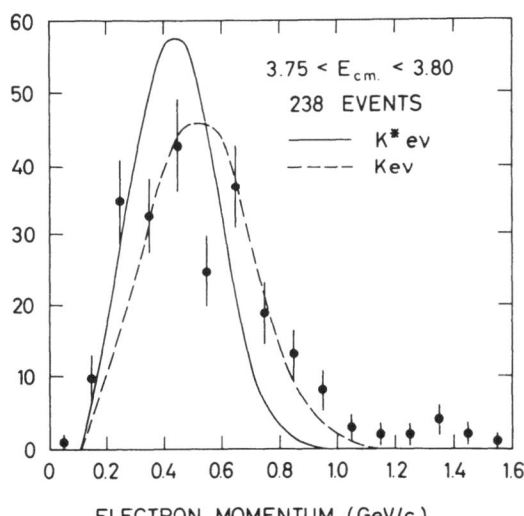

<u>Fig. 11.20.</u> Electron momentum spectrum measured by DELCO /161/ for the multiprong events at the ψ''. The dashed and the solid curves show the expected shape for the decay modes $D \rightarrow e\bar{\nu}_e K$ and $D \rightarrow e\bar{\nu}_e K^*(892)$ (evaluated for V-A couplings)

Normalizing to the cross section for $e^+e^- \to \psi''(3772)$ yields $B(D \to eX) = (0.11 \pm 0.03)$.

The electron momentum spectrum at the $\psi''(3772)$ is plotted in Figure 11.20. The dashed and the solid curves show the expected shapes for the decay modes $D \to e\bar{\nu}_e K$ and $D \to e\bar{\nu}_e K^*(892)$.

The DELCO group has collected data at the $\psi''(3772)$ with reduced magnetic field in order to be sensitive to electrons with very low momenta resulting from $D \to e\bar{\nu}_e K^*$ (1420). They find no evidence for this decay mode and quote

$$\frac{D \to e\bar{\nu}_e K^*(1420)}{D \to e\bar{\nu}_e + anything} < 0.1.$$

The LBL-SLAC group selects /162/ events of the type

$$e^+e^- \to e^{\pm} + \geq 2 \text{ charged tracks} + \geq 0 \text{ photons}$$

in the energy region between 3.7 and 7.4 GeV. Electrons are identified in the lead glass wall by demanding that the total energy deposited should be consistent with the momentum measured in the magnet. Furthermore the energy division between the active converters $(3.5 \cdot X_0)$ and the back blocks $(10.5 \cdot X_0)$ should be consistent with that of a showering particle and the time of flight should agree within 1 ns with that expected for an electron.

The probability that these criteria are satisfied by a hadron was measured at the J/ψ resonance. They find this probability to be 1.5 % for a particle of momentum 300 MeV/c decreasing to 0.4 % for a particle of 1.2 GeV/c. The probability averaged over the hadron spectrum is 1.1 % for momenta above 300 MeV/c.

As discussed above the resonance $\psi''(3772)$ is a clear and well-defined source of D mesons. The LBL-SLAC group found a total of 62 candidates for electron inclusive multiprong events at the $\psi''(3772)$. The background, not including τ pair production, was estimated to contribute 25 ± 5 events. Assuming that the excess signal results from semileptonic D decays they find

$$B(D \to eX) = (0.072 \pm 0.028).$$

The momentum spectrum is plotted in Figure 11.21 and compared to the spectra predicted for $D \to e\bar{\nu}_e$, $e\bar{\nu}_e K$ and $e\bar{\nu}_e K^*(892)$. The spectrum is not corrected for the contribution from τ pair production. The data are consistent with $D \to e\bar{\nu}_e K$ (confidence level 33 %) or $D \to e\bar{\nu}_e K^*$ (C.L. 13 %) but less consistent with $D \to e\bar{\nu}_e \pi$ ((C.L. 13 %). The data are inconsistent with $D \to e\pi$ which would produce a flat electron spectrum from 810 to 1080 MeV/c.

Data on inclusive electron multiprong events have also been collected at higher energies where not only D but also F and charmed baryon production are expected to be important. Preliminary results are listed in Tables 11.4 and 11.5.

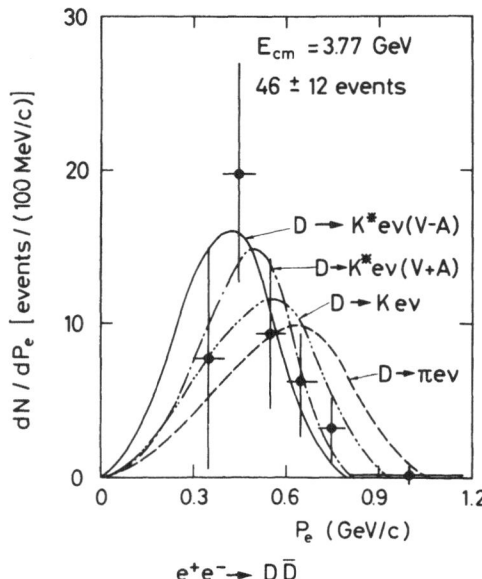

$e^+e^- \rightarrow D\bar{D}$

<u>Fig. 11.21.</u> Electron momentum spectrum measured by LBL-SLAC /162/ for electron in-clusive multiprong events. Predictions for $D \rightarrow e\bar{\nu}_e\pi, e\bar{\nu}_e K$ and $e\bar{\nu}_e K^*(892)$ are shown

<u>Table 11.4.</u> Preliminary results on inclusive electron multihadron events /161, 162/

E_{CM} [GeV]	R_e	R_{ch}	$B(C \rightarrow eX)$
4.1 - 4.2	0.26 ± 0.08	2.1 ± 0.5	0.077 ± 0.030
4.4 - 5.7	0.23 ± 0.06	1.9 ± 0.5	0.074 ± 0.028
6.4 - 7.4	0.26 ± 0.06	1.9 ± 0.4	0.087 ± 0.032

The data on the semileptonic charm decays are summarized in Table 11.5.

<u>Table 11.5.</u> Semileptonic branching ratio of D mesons

Experiment	$B(D \rightarrow eX)$
DASP	0.08 ± 0.02
DELCO	0.11 ± 0.03
LBL-SLAC	0.072 ± 0.028

The quoted values are valid for a mixture of charged and neutral D mesons. In principle they can be measured separately at the ψ''(3772) resonance by selecting events with one charged or one neutral D meson.

From dimuon production by neutrinos a preliminary value of B(C → μX) ≈ 0.11 was obtained /164/.

The semileptonic branching ratio is larger than the value of 0.04 predicted /143/ from the weak decays of strange particles. This indicates that the mechanism responsible for enhancing the nonleptonic channels in strange particle decays is less effective /144/ for charmed particle decays. In fact if none of the available channels is selectively enhanced one expects a semileptonic branching ratio of 0.20. This number is obtained by simple counting: the W decay can proceed in five different ways, W → eν, μν, and qq̄' times three because of three different colors. Assuming the same coupling strength, each channel has the probability 1 : 5 = 0.20.

One event was found by DASP with three electrons plus hadrons. This number is consistent with the expected background leading to an upper limit of

$$\sigma(e^+e^- \to 3e + X) < 0.1 \text{ nb},$$

with 90 % confidence. Events of that type could arise from a charm changing neutral current, which allows a charmed hadron to decay into two electrons plus hadrons /165/. A neutral lepton paired with the electron in a right-handed doublet would also yield events with three electrons and hadrons /166/.

DASP /160/ has determined the number of charged kaons emitted in electron multihadron events. This provides an independent consistency check on the nature of the weak current responsible for charm decay. If it is the GIM current then almost every electron event will have a KK̄ pair. The measurement was done with events that had an identified charged hadron (π, K or p̄) in the magnetic spectrometer, an electron in the inner detector, and possible other charged particles or photons. No ep̄X events were seen. From the observed K to π ratio and the measured charged multiplicity it was found that each multiprong event contained on the average 0.90 ± 0.18 charged kaons per event in agreement with the GIM prediction.

12. The Heavy Lepton τ

In 1975 PERL et al. reported evidence for events of the type

$$e^+e^- \rightarrow e^{\pm}\mu^{\mp} + \text{nothing,} \tag{12.1}$$

where "nothing" meant that no other particles were registered in the detector /167/. The analysis was hampered by background from purely hadronic processes: hadrons had an 18 % (20 %) probability to fake an electron (a muon). Of the 24 eμ events found 6-8 events were estimated to come from hadronic background. There was also the question whether or not other strongly or electromagnetically interacting particles had been produced together with e and μ. Since the detector covered only two-thirds of 4π additionally produced particles had a fair chance to escape detection.

Besides the experimental uncertainties there was the question of interpretation. Electron-muon events can arise, e.g., from the pair production of charmed particles or of a new lepton. The two mechanisms can be distinguished by their different production and decay patterns. Measurements on

1) $e^+e^- \rightarrow e\mu$ + nothing
2) $e^+e^- \rightarrow e(\mu)$ + minimum ionizing track
    + any number of photons

provided convincing evidence that above 4 GeV besides charmed particles a new type of weakly decaying particle, τ, is being produced /168/. The final proof that this particle indeed exists was given by the DASP collaboration which first observed τ pair production below charm threshold at the ψ' /169/. We first discuss the properties expected for a charged heavy lepton /170/.

12.1 Expected Properties of a Heavy Lepton

Assuming the lepton τ to be pointlike the e^+e^- production cross section is given by

$$\sigma_{\tau\tau} = \frac{4\pi}{3} \frac{\alpha^2}{s} \beta_\tau \left(1 + \frac{1 - \beta_\tau^2}{2}\right), \tag{12.2}$$

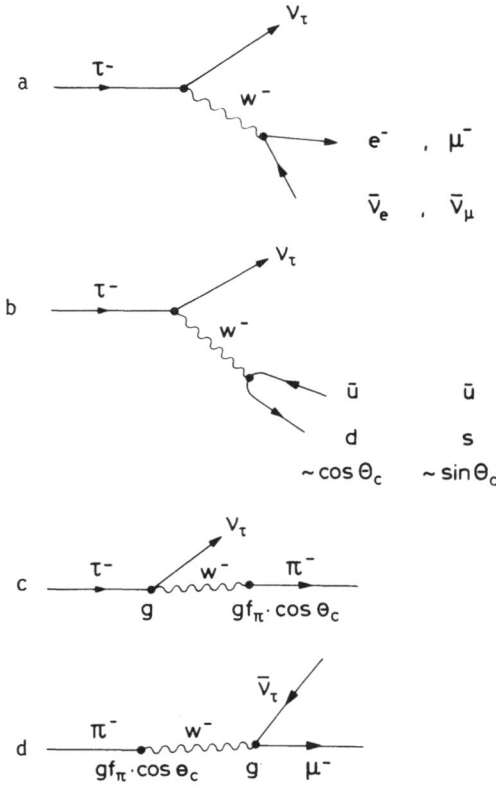

Fig. 12.1. (a,b) The graphs for leptonic and semihadronic decays of a heavy sequential lepton, (c,d) the relationship between $\pi^- \to \bar{\nu}_\mu \mu$ and $\tau^- \to \nu_\tau \pi^-$

where $\beta_\tau = p/E$ is the τ velocity. If the lepton decays weakly and the decay is mediated by the standard V-A weak current the partial decay widths can be calculated or estimated (see Figure 12.1).

The leptonic decay modes can be computed unambiguously

$$\Gamma_e \equiv \Gamma(\tau \to \nu_\tau\, e\bar{\nu}_e) = \frac{G_F^2 \cdot m_\tau^5}{192\pi^3}$$

$$\Gamma_\mu \equiv \Gamma(\tau \to \nu_\tau\, \mu\bar{\nu}_\mu) = \Gamma_e \cdot F(y)$$

$$G_F = \frac{1.02}{m_p^2} \times 10^{-5}\, (\text{GeV}^{-2}) \text{ and}$$

$$F(y) = (1-8y + 8y^3 - y^4 - 12y^2 \cdot \ln y). \tag{12.3}$$

$F(y = m_\mu^2/m_\tau^2)$ is a small phase space correction.

A lepton decays semihadronically as shown in Figure 12.1b. Conventional theory predicts that the τ decays into final states of low multiplicity and a small ratio of kaons to pions. This is indeed reproduced in actual calculations /170, 171/ and an estimate of various branching ratios is listed in Table 12.1.

Table 12.1. Branching ratios for a sequential lepton of mass 1.8 GeV /171/

Decay mode	Branching ratio	$\Gamma(\tau \to \nu_\tau X)/\Gamma(\tau \to \nu_\tau e\, \nu_e)$
$\nu_\tau e\, \bar\nu_e$	0.18	1
$\nu_\tau \mu \nu_\mu$	0.18	0.97
$\nu_\tau \pi$	0.10	0.60
$\nu_\tau \rho$	0.22	1.24
$\nu_\tau\, A_1$	~0.1	0.41
$\nu_\tau\, 4\pi$		0.44 ± 0.10
$\nu_\tau\, 5\pi$	0.2	~0.44
$\nu_\tau\, 6\pi$		~0.11
$\nu_\tau\, 7\pi$		~0.11
$\nu_\tau\, K$	<0.01	0.03
$\nu_\tau\, K^*(892)$	0.01	0.05
$\nu_\tau\, Q$	<0.01	0.02
$\nu_\tau (K \cdot n\pi), n > 3$	0.01	0.07

Some of the decay widths are rather well known:

$\tau \to \nu_\tau \pi$: The pion mode is directly related to the $\pi \to \mu\nu$ decay as shown in Figures 12.1c,d. Note that the mass squared of the weak current is the same in both cases ($Q^2 = m_\pi^2$). As a consequence the value of the form factor is the same for both decays. The partial width is given by

$$\frac{\Gamma(\tau^- \to \nu_\tau \pi^-)}{\Gamma_e} = \frac{12\pi^2 \, f_\pi^2 \, \cos^2\theta_c}{m_\tau^2} = \frac{2.1 \, m_p^2}{m_\tau^2} \qquad (12.4)$$

with $f_\pi = 0.137 \, m_p$ (m_p = proton mass).

$\tau \to \nu_\tau K$: This decay is directly related to $K \to \bar\nu_\mu \mu$.

$\tau \to \nu_\tau \rho$: This decay is related via CVC to $e^+e^- \to \rho$. The width for $\tau \to \nu_\tau \rho$ can be evaluated with an accuracy of 20 % using the measured values of the ρ-γ coupling constant.

$\tau \to \nu_\tau \cdot (n\pi)$: These decay modes are related via CVC to $e^+e^- \to (n\pi)$. (n even).

Pair production of new leptons will lead to mixed electron muon events via:

$$e^+e^- \to \tau^-\bar{\tau} \to (\nu_\tau e \bar{\nu}_e)(\nu_\tau \mu \bar{\nu}_\mu).$$

The number of eμ events can be written as

$$N_{e\mu} = 2\sigma_{\tau\bar{\tau}} \cdot L \cdot A_e \cdot A_\mu \cdot B_e \cdot B_\mu.$$

The $\tau\bar{\tau}$ cross section, (12.2), varies rapidly near threshold and a measurement in this energy region can be used to determine the τ mass. Well above threshold the cross section is not very sensitive to the precise value of the τ mass and is given by the point cross section. L is the luminosity, A_e and A_μ the acceptances for electrons and muons. The acceptances depend on the shape of the decay lepton spectrum, i.e., on the form of the weak coupling. The difference in acceptance introduced by a V-A or a V+A coupling is small as long as the leptons are measured down to low momenta.

A measurement of the eμ yield at high energies can therefore be used to determine $B_e \cdot B_\mu$ without knowledge of the exact mass of the lepton and the form of the weak couplings. If eμ universality is valid as expected both for sequential and ortholeptons /172/ (lepton number of the τ^- = lepton number of the e^- or μ^-) then $B_e \equiv B_\mu$ and the B_e value can be directly determined. However, if the τ is a paraelectron or a paramuon (lepton number τ^+ = lepton number of e^- or μ^-) then B_μ/B_e = 1/2 or 2.

The lepton assignment can in principle be determined from a measurement of the final states

$$e^+e^- \to \tau^-\bar{\tau} \to e^+e^- + \text{missing energy}$$
$$\to \tau^-\bar{\tau} \to \mu^+\mu^- + \text{missing energy}.$$

However, unlike the eμ channel which is rather clean, e^+e^- and $\mu^+\mu^-$ final state events are contaminated with electromagnetic events.

Information on semihadronic decays of the heavy lepton can be obtained from a measurement of inclusive electron (muon) events resulting from

$$e^+e^- \to \tau^-\bar{\tau} \to (\nu_\tau \ell \bar{\nu}_e) (\bar{\nu}_\tau \cdot \text{hadrons}).$$

The rate for the inclusive events is given by

$$N_{\ell,h} = 2\sigma_{\tau\bar{\tau}} \cdot L \cdot A_{\ell} \cdot A_h \cdot B_{\ell} \cdot B_h$$

with the nomenclature defined above.

As discussed above, τ production can be strongly enhanced relative to charm production either by selecting events with low multiplicity or by demanding the decay lepton to have momenta above 1.0 GeV/c. It was shown in Sect. 11.3 that only a small fraction of the semileptonic charm decays satisfies these conditions. Most experiments reported have therefore measured the lepton two-prong cross section

$$N_{\ell,1p} = 2\sigma_{\tau\bar{\tau}} \cdot L \cdot A_{\ell} \cdot A_{1p} \cdot B_{\ell} \cdot B_{1p}$$

with $B_{1p} = B(\tau \rightarrow \nu_{\tau} + 1\text{ charged} + \geq 0\text{ photons})$. Electrons are sometimes excluded from the one-prong modes.

12.2 Leptonic Decays of the τ

Table 12.2 summarizes the experiments that searched for heavy leptons in e^+e^- annihilation. Two early experiments done at Frascati at c.m. energies up to 3 GeV showed that besides e and μ no leptons exist with a mass below 1.15 GeV /173, 174/. Below we discuss some of the major experiments in detail.

12.2.1 eμ Events

We first discuss the experiments studying events of the type

$$e^+e^- \rightarrow e^{\mp}\mu^{\pm} + \text{nothing}$$

above c.m. energies of 3.5 GeV. In the experiment of SLAC-LBL the following criteria were used to select the events:

1) Only two charged tracks with opposite charges and no photons.
2) Each track should have a momentum greater than 0.65 GeV/c.
3) One prong is identified as an electron by the pulse height in the shower counter, the other as a muon by range. The probability of misidentifying a hadron as an electron or as a muon is, respectively, 18 % and 20 %.
4) The coplanarity angle between the planes defined by the electron and the beam direction and the muon and the beam direction must be greater than 20°.

Table 12.2. Summary of e^+e^- experiments searching for a heavy lepton

Experiment	c.m. energy [GeV]	Final state	p_μ^{min} [GeV/c]	p_e^{min} [GeV/c]	Events Observed	Background	Lepton identification
BERNARDINI et al. /173/	1.2 – 3.0	eμ	–	–	no signal above background $M_L > 1.0$ GeV		e,μ: shower and range chambers
ORITO et al. /174/	2.6 – 3.0	eμ	–	~ 0.30	no event found $M_L > 1.15$ GeV		e: shower chambers; μ: range ⌐ 290 g/cm of
SLAC-LBL /167, 175/	3.8 – 7.8	eμ	0.65	0.65	190	46	μ: by range
		ee	0.65	0.65			
		μμ	0.65	0.65			
		μX	0.91	–	230	94	e: lead-scintillation counters
Maryland, Princeton Pavia /176, 177/	7	μX	1.0	–	13	4	μ: by range
PLUTO /178, 179/	3.6 – 5.0	eμ	1.0	0.30	23	1.9	μ: by range
		μX	1.0	–	273	62	e: lead proportional chambers
LBL-SLAC /180/	3.7 – 7.4	eμ	0.65	0.40	21	0.4	e: lead-glass counters
		eX	0.65	0.40	31	12.1	
DASP /169, 181/	4.0 – 5.2	eμ	0.70	0.20	13	1.2	μ: by range
		eX	–	0.20	89	16	e: Cerenkov counters and lead scintillation counters
		μX	1.0	–	25	4.0	
Ironball /182/	7	μμX	1.3	–	16	5	μ: by range
DESY-Heidelberg /183/	3.6 – 4.4	eX+ee	–	0.5	182	≈ 11	e: NaI and lead glass counters
		μX+μμ	0.6	–	94	≈ 2	
		μe	0.6	0.5	23	–	μ: by range
DELCO /184/	3.77-7.84	eX	–	0.20	692	< 42	e: Čerenkov and shower counters

X contains in general one nonshowering track only.

A total of 190 events /175/ with a background of 46 events was observed in the c.m. range from 3.8 to 7.8 GeV. If these events result from charm then the leptons would be accompanied by undetected hadrons. The SLAC-LBL group made a careful evaluation of this possibility and they conclude with 90 % confidence that no more than 39 % of the eμ events can contain additional charged particles or photons. They conclude that the eμ events result from pair production of a new lepton,

$$e^+e^- \rightarrow \tau\bar{\tau} \rightarrow (\bar{\nu}_\tau e \bar{\nu})(\bar{\nu}_\tau \mu \bar{\nu}_\mu) \rightarrow e\mu + \text{missing energy.}$$

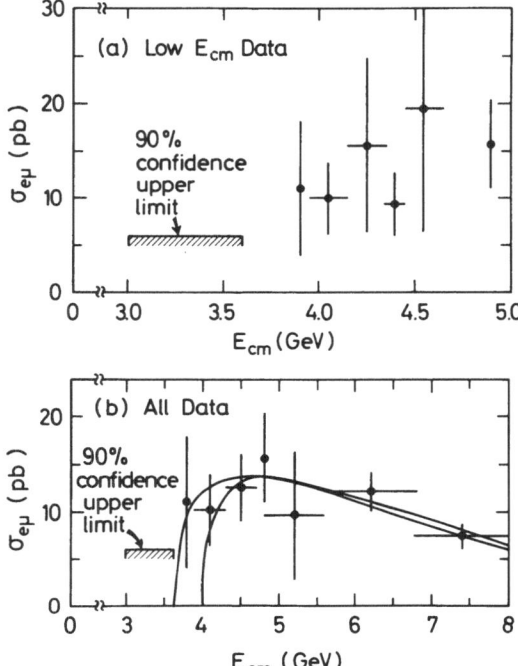

Fig. 12.2. The observed eμ production cross section plotted versus c.m. energy. The data are from SLAC-LBL /175/; only statistical errors are indicated and the horizontal lines show the energy range covered by each point. The 90 % confidence upper limit on the background based on data below 3.6 GeV is shown. The solid lines are predictions for a lepton of mass 1.8 GeV (2.0 GeV) assuming a V-A current

The observed cross section is plotted in Figure 12.2 versus the c.m. energy. The cross section varies smoothly with energy and agrees well with the energy dependence of pair production of a heavy lepton plotted as the solid lines in Figure 12.2. Both curves evaluated for m_τ = 1.8 and 2.0 GeV fit the data well. The momentum

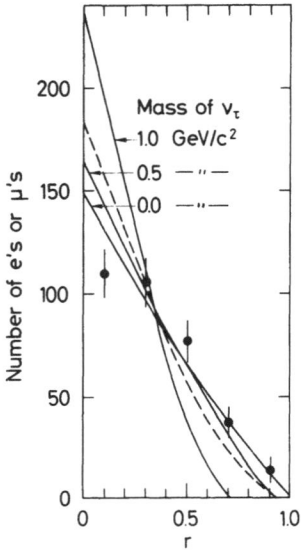

Fig. 12.3. Number of electrons and muons observed in the SLAC-LBL experiment /175/ plotted versus

$$r = \frac{P - 0.65 \text{ GeV/c}}{P_{max} - 0.65 \text{ GeV/c}} \; .$$

Solid curves are for m_τ = 1.9 GeV and V-A coupling with the mass of the neutrino as a parameter. The dashed curve is for V+A coupling with m_τ = 1.91 GeV and a massless τ neutrino

distribution is shown in Figure 12.3. Plotted are the number of electrons and muons versus the normalized momentum variable

$$r = \frac{P - P_0}{P_{max} - P_0}, \; P_0 = 0.65 \text{ GeV/c.}$$

P_{max} is defined as the maximum momentum of the electron (muon) which is consistent with the leptonic decay of a heavy lepton of mass 1.9 GeV with a massless neutrino. The solid curves are distributions predicted for various values of the neutrino mass assuming a V-A leptonic coupling. The dotted curve is for a V+A leptonic coupling and a massless neutrino. For m_τ = 1.9 GeV and a massless τ neutrino they find that the χ^2 probability is about 0.1 % for V+A compared to 50 % for V-A.

The 95 % confidence upper limit on the neutrino mass is m_{ν_τ} < 0.6 GeV for m_τ = 1.90 GeV and V-A coupling. The mass of the τ was determined to (1.90 ± 0.10) GeV. This is an average over various methods and assumes V-A and a massless neutrino. The leptonic branching ratios were determined from the observed eμ cross section (Figure 12.2) to

$$B_e \equiv B_\mu \equiv 0.186 \pm 0.01 \pm 0.028.$$

The first error is statistical and the second systematic. This value assumes M_{ν_τ} = 0, V-A, and eμ universality.

The PLUTO collaboration reported /178/ data on $e^+e^- \rightarrow e\mu$ + missing energy with a very small contamination of other events. Electrons above 0.3 GeV/c are identified by showers in two lead converters (0.44 and 1.7 radiation lengths thick) in 55 % of the full solid angle. Muons are required to penetrate the iron yoke and produce a spark in one of the proportional tube chambers. The μ acceptance is 43 % of 4π. The probability for misidentifying a hadron as either electron or muon is small: P_{he} = 3.5 ± 0.7 % and $P_{h\mu}$ = 2.8 ± 0.7 %.

Above 4 GeV $e^{\pm}\mu^{\mp}$ + nothing events were observed. No events with like sign leptons were found. The background due to misidentified hadrons is determined to 2 ± 0.5 events. No such events were found at 3.6 GeV. The number of eμ events accompanied by photons or additional charged particles is small and consistent with coming from background. The scarcity of such events precludes the possibility that the eμ + nothing events come from charmed particle decays: otherwise there should have been three times as many events of the type eμ + γ or eμ + $\geq$ 1 charged particle as there are eμ + nothing events. The observed number of events is at least an order of magnitude smaller. The total fraction of eμ events that could have originated from charm production is less than 9 %. The energy behavior of the cross section

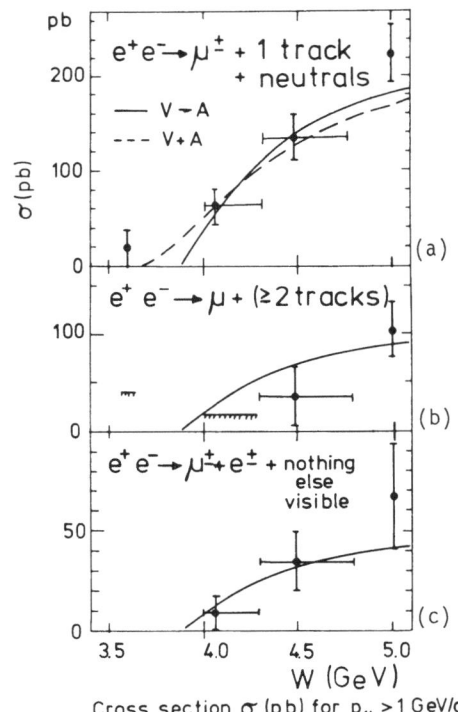

Cross section σ (pb) for p_μ >1 GeV/c

Fig. 12.4. c.m. energy dependence of muon cross sections measured by PLUTO /178, 179/. (a) Inclusive two prongs [muon + charged track ($T^{\pm}$) + neutrals]. (b) Inclusive multiprongs, and (c) exclusive μe events. Cross sections are given for muon momenta > 1 GeV/c and have been corrected for trigger and detector acceptance, and for hadron punchthrough. (a) has also been corrected for QED, (c) for electron detection efficiency. The solid curves are calculated for V-A decay with $M(\tau)$ = 1.91 GeV, $M(\nu_\tau)$ = 0

for eμ events is given in Figure 12.4. The shape and magnitude of the measured cross section are in agreement with the prediction for heavy lepton production (solid curve).

The leptonic branching ratios and the τ mass are determined from the eμ events and the data /179/ on the inclusive (μ + 1 prong) cross section discussed below. For V-A couplings they find $B(\tau \rightarrow \tau_\nu \; \mu\bar{\nu}_\mu) = 0.14 \pm 0.034$, $B(\tau \rightarrow \nu_\tau \; e\bar{\nu}_e) = 0.16 \pm 0.06$, and $m_\tau = 1.93 \pm 0.05$ GeV. A V+A coupling yields $B(\tau \rightarrow \nu_\tau \; \mu\bar{\nu}_\mu) = 0.17 \pm 0.05$, $B(\tau \rightarrow \nu_\tau \; e\bar{\nu}_e) = 0.13 \pm 0.05$, and $m_\tau = 1.82 \pm 0.08$ GeV. The V-A fit is slightly favored over the V+A fit.

The DASP group has measured /169/ $e^+e^- \rightarrow$ eμ + missing energy in the c.m. range of 4.0 to 5.2 GeV. Muons of momenta greater than 0.7 GeV/c were identified in the outer detector ($P_{h\mu} = 4.2 \pm 0.8$ %) by range, electrons with momenta above 0.2 GeV/c either in the inner detector ($P_{he} = 2 \pm 0.9$ %) or the outer ($P_{he} = 4 \times 10^{-4}$). A total of 13 eμ events with an estimated background of 1.2 ± 0.4 events was found. Using the known production cross section and assuming eμ universality yield $B_e = B_\mu = 0.182 \pm 0.028 \pm 0.014$ for a V-A current and $B_e = B_\mu = 0.206 \pm 0.033 \pm 0.015$ for a V+A current. The first error is the statistical, the second the systematic one.

The LBL-SLAC group has recently reported /180/ new results on $e^+e^- \rightarrow e^\pm$ + a nonshowering particle + ≥ 0 photons.

Data are collected for c.m. energies between 4.1 and 7.4 GeV. The electron ($P_e \geq 0.4$ GeV/c) is identified in the lead glass wall. The nonshowering particle must have a momentum larger than 0.65 GeV/c and it is classified as muon or hadron by range. The acoplanarity angle with respect to the beam axis should be at least 20° and the mass squared recoiling against the pair must be greater than 0.8 GeV^2 at lowest c.m. energy and 1.5 GeV^2 at the highest. They find a total of 21 eμ events with an estimated background of 0.4 events. Assuming V-A couplings $m_\tau = 1.9$ GeV, $m_{\nu_\tau} = 0$ and eμ universality they find $B_e = B_\mu = 0.224 \pm 0.055$.

12.2.2   ee and μμ Events

τ pair production will also lead to events of the type

$$e^+e^- \rightarrow \tau\bar{\tau} \rightarrow e^+e^- + \text{missing energy}$$
$$e^+e^- \rightarrow \tau\bar{\tau} \rightarrow \mu^+\mu^- + \text{missing energy}.$$

The SLAC-LBL group /175/ has selected events of this type demanding $P_e > 0.65$ GeV/c, $P_\mu > 0.65$ GeV/c. The acoplanarity angle with respect to the beam axis should be greater than 20°. After large corrections for QED and misidentified hadron events they report

$$\frac{\text{Number of ee events}}{\text{Number of e}\mu \text{ events}} = 0.56 \pm 0.14 \begin{smallmatrix} +0.16 \\ -0.19 \end{smallmatrix}$$

$$\frac{\text{Number of } \mu\mu \text{ events}}{\text{Number of e}\mu \text{ events}} = 0.70 \pm 0.15 \pm 0.19.$$

A sequential lepton or an ortholepton yield 0.5 for both ratios. The authors /175/ quote that an electron-related paralepton will yield $N_{ee}/N_{e\mu}$ = 0.86, $N_{\mu\mu}/N_{e\mu}$ = 0.29 for the kinematical cuts used. A muon-related paralepton yields $N_{ee}/N_{e\mu}$ = 0.29 and $N_{\mu\mu}/N_{e\mu}$ = 0.86. The paralepton assignment is disfavored by the data.

The Colorado-Pennsylvania-Wisconsin group /182/ searched for

$$e^+e^- \rightarrow \mu^+\mu^- + \text{missing energy}$$

using the iron ball detector at SPEAR. The detector is a solid magnetized iron spectrometer with an azimuthal field of 15 kGauss and a radius of about 1 m. It identifies muons above 1.3 GeV/c in nearly 90 % of 4π. They select muon pairs with an acoplanarity angle with respect to the beam axis greater than 10° and demand that the system recoiling against the pair should have a mass squared greater than 4 GeV^2. A total of 25 $\mu^+\mu^-$ events satisfies the criteria; the estimated background is 14 events. The remaining 11 events yield a branching ratio $B_\mu = 0.20 \begin{smallmatrix} +0.10 \\ -0.08 \end{smallmatrix}$.

This is evaluated assuming m_τ = 1.9 GeV, V-A coupling and a massless neutrino.

12.2.3 Two-Prong Inclusive Electron and Muon Final States

The first evidence for anomalous muon production in $e^+e^- \rightarrow \mu^\pm$ + charged track + ≥ 0 photons was reported by the Maryland, Pavia, and Princeton collaboration /176, 177/. They identify muons above 1.0 GeV/c by range in a single-arm magnetic spectrometer. The statistics are poor, but the data are consistent with the heavy lepton hypothesis based on the eμ events.

The SLAC-LBL group has measured /185/ the cross section for $e^+e^- \rightarrow \mu^\pm$ + charged particle + ≥ 0 photons for c.m. energies between 3.9 and 7.8 GeV. Muons with momenta above 0.91 GeV/c are defined by range (equivalent of 65 cm of iron) in the "muon tower". A hadron has approximately a 3 % chance to fake a muon. Events should be acoplanar with respect to the beam axis by at least 20° and the mass squared recoiling against the two prongs should be greater than 1.5 GeV^2. After the background subtraction a significant signal of 103 ± 18 events remained for c.m. energies between 5.8 and 7.8 GeV. The momentum spectrum of the muons is plotted in Figure 12.5a and it can be compared to the spectrum of muons associated with multiprong events plotted in Figure 12.5b. The latter spectrum which mainly results from charm is much softer than the muon spectrum of the two-prong sample. A fit to the data assuming V-A couplings, $B(\tau \rightarrow \nu_\tau + 1 \text{ charged} + \geq 0 \text{ photons})$ = 0.85, m_{ν_τ} = 0, and m_τ = 1.9 GeV shown as the solid line in Figure 12.5a yields B_μ = 0.175 ± 0.027 ± 0.03. The second error shows the systematic uncertainties. Data from PLUTO /179/,

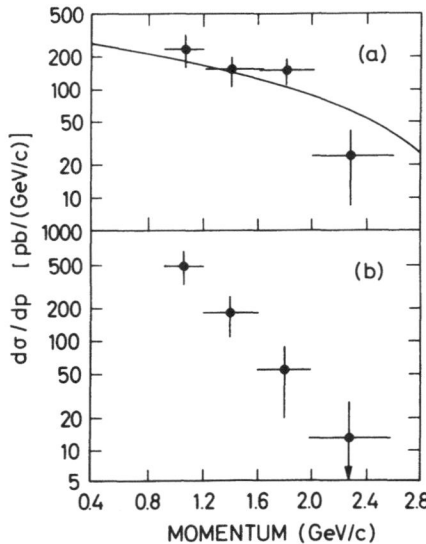

Fig. 12.5. (a) The momentum distribution of anomalous muon two-prong events. The data were obtained by SLAC-LBL /185/ in the c.m. energy range of 5.8 to 7.8 GeV. The solid curve represents the cross section expected from heavy lepton production with the parameters listed in the text, (b) the momentum distribution of inclusive muon multiprong events

DASP /169/, and DELCO /184/ indicate that the one-prong branching ratio might be smaller than the value assumed. Decreasing the one-prong branching ratio will increase B_μ.

PLUTO /179/ measured inclusive muon production for μ momenta above 1 GeV/c. The events had to contain at least one extra charged particle with p > 0.2 GeV/c and $|\cos\theta|$ < 0.87. The event sample was divided into two prongs (one extra track + any number of photons) and multiprongs. The two classes contain different contributions from conventional processes.

The main conventional sources of two-prong events are the QED processes 1) $e^+e^- \rightarrow \mu^+\mu^-$, 2) $e^+e^- \rightarrow \mu^+\mu^-\gamma$, and 3) $e^+e^- \rightarrow \mu^+\mu^-\gamma\gamma$. Reactions 1) and part of 2) were removed by requiring an acoplanarity angle of more than 10°. The contribution of 2) was further reduced by a cut on the missing mass squared. Because of the changing kinematical resolution this cut varied between 1.4 GeV^2 at $\sqrt{s}$ = 3.6 GeV and 2.7 GeV^2 at $\sqrt{s}$ = 5 GeV. The efficiency of this cut was checked with a 60 % subsample of type 2) events in which the photon converted in the detector. Reaction 3) cannot be separated by kinematical cuts. The contribution from 3) was calculated /186/ and subtracted from the data. It amounted to less than 7 % of the remaining muon signal. The background due to misidentified hadrons was typically 15 % and was also subtracted.

For multiprong events misidentified hadrons constitute the main source of background. Contributions from $e^+e^- \rightarrow e^+e^-\mu^+\mu^-$ were found to be negligible /186/.

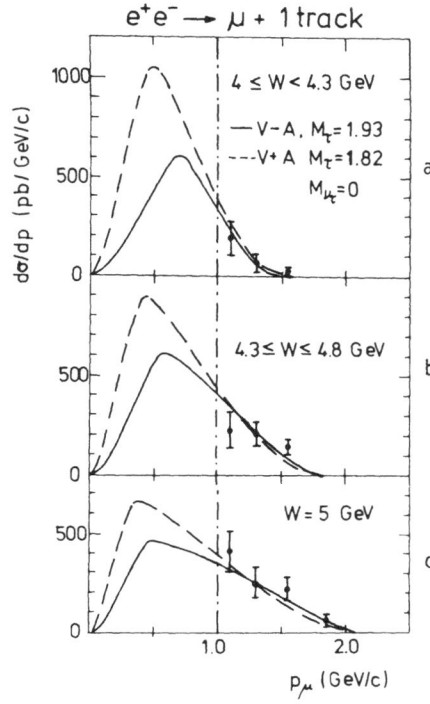

$e^+e^- \rightarrow \mu + 1$ track

$d\sigma/dp$ (pb/ GeV/c)

1000

$4 \le W < 4.3$ GeV

— V-A, $M_\tau = 1.93$
--- V+A $M_\tau = 1.82$
$M_{\nu_\tau} = 0$

500

a

0

$4.3 \le W \le 4.8$ GeV

500

b

0

$W = 5$ GeV

500

c

0

1.0   2.0

p_μ (GeV/c)

Fig. 12.6. Inclusive muon data for P_μ 1.0 GeV/c measured by PLUTO /179/. The data are corrected for QED production and other sources of background. (a,b,c) The muon momentum distribution of two-prong events for three different c.m. energies. The curves are fits to the data assuming pair production of a heavy lepton τ, a massless τ neutrino, and V-A (solid line) or V+A (dotted line) leptonic weak couplings

Inclusive J/ψ production with subsequent J/$\psi \rightarrow \mu^+\mu^-$ decay was eliminated by a cut in the invariant two-particle mass. The momentum distribution of the observed events are plotted in Figure 12.6a-c. The spectra show the triangular upper end characteristic of the three-body decay of a moving object.

The two-prong cross section is plotted in Figure 12.4a as a function of energy. The data indicate a threshold in the 3.6 to 4.0 GeV region. The data are fitted to the heavy lepton hypothesis for V+A (dotted line) and V-A (solid line) assuming a massless neutrino. The results, plotted in Figure 12.6, are in agreement with the heavy lepton hypothesis and yield 1.93 ± 0.05 GeV(V-A) and 1.82 ± 0.08 GeV (V+A) for the value of the mass. The size of the cross section near threshold excludes the possibility that the muons result from the pair production and weak decays of a spin 0 particle: it leads to unphysical values for the leptonic branching ratios.

The branching ratio for the τ to decay into multiprong and one-prong events is determined by assuming that all events in the multiprong sample with muon momenta above 1 GeV/c result from τ production. They find B(1 prong) = 0.70 ± 0.10 and B($\ge$ 3 prong) = 0.30 ± 0.10.

From an analysis of the vertex distribution the PLUTO group determined /187/ an upper limit of $9 \cdot 10^{-12}$ s for the τ lifetime. This is an order of magnitude larger than the value of $2.8 \cdot 10^{-13}$ s expected from conventional theory.

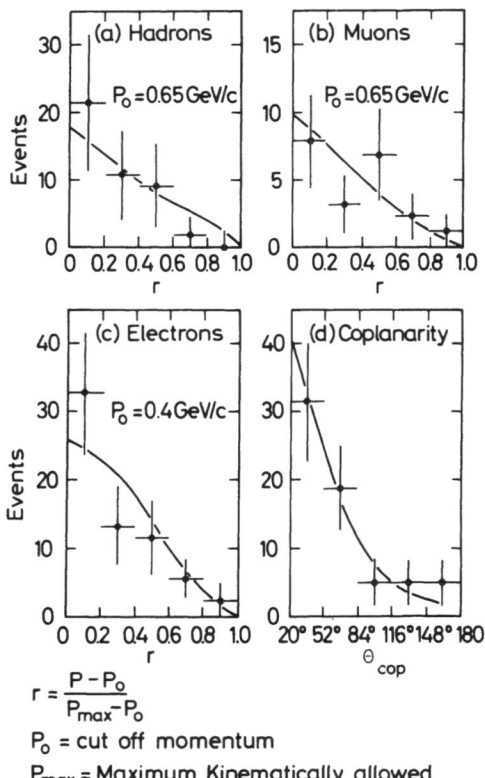

Fig. 12.7. The ratio $r = (P-P_0)/$ $(P_{max} - P_0)$ for a) hadrons from eh events, b) muons from eµ events, c) electrons from eh and eµ events, d) coplanarity distribution for eh and eµ events. The curves show the predictions for heavy leptons with m_τ = 1.9 GeV, m_{ν_τ} = 0 and V-A coupling

$r = \dfrac{P - P_0}{P_{max} - P_0}$

P_0 = cut off momentum

P_{max} = Maximum Kinematically allowed in τ decay

The LBL-SLAC group has measured /180/ $e^+e^- \rightarrow e^\pm$ + charged + $\geq$ 0 photons for c.m. energies between 4.1 and 7.4 GeV. A total of 31 events was observed with an estimated background of 12.1 events. The data are plotted in Figure 12.7 versus r. From a fit to these events and the eµ events discussed earlier they find for the branching ratio B_{1h} for $\tau \rightarrow \nu_\tau$ + 1 charged hadron + any number of photons B_{1h} = 0.45 ± 0.19. The fit was made assuming V-A couplings, m_{ν_τ} = 0, m_τ = 1.9 GeV. The solid line shows the results of the fit.

The DASP collaboration has measured /169/

1) $e^+e^- \rightarrow e^\pm$ + nonshowering track + $\geq$ 0 photons
2) $e^+e^- \rightarrow \mu^\pm$ + nonshowering track + $\geq$ 0 photons

for c.m. energies between 3.6 and 5.2 GeV.

The selection criteria for the inclusive electron sample have been discussed earlier It is important to note that electrons are measured down to 0.2 GeV/c momenta and hence the cross sections extracted from these data are nearly independent whether V-A or V+A couplings are assumed.

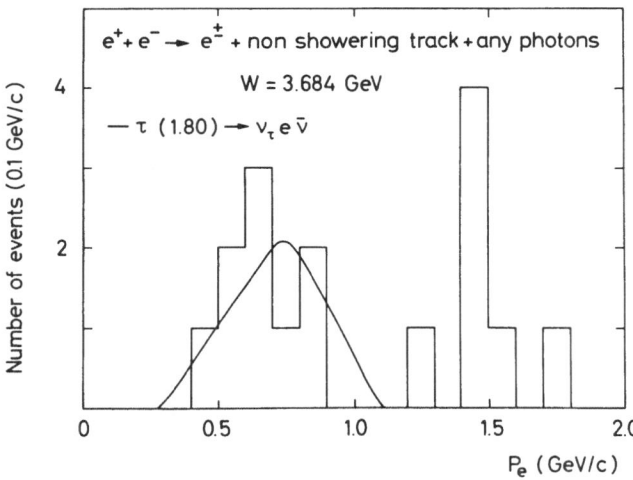

Fig. 12.8. Raw electron momentum distribution observed by DASP /169/ at 3.684 GeV. The events here are identified as electron, one nonshowering particle, and any number of photons

A total of 80 events were found at c.m. energies above 3.9 GeV, 17 events at the ψ' resonance and 1 event at 3.6 GeV.

The electron momentum spectrum measured at the ψ' and plotted in Figure 12.8 shows two clear clusters of events, one centered around 1.5 GeV/c and the other with momenta between 0.4 and 0.9 GeV/c. The first cluster can be associated with the cascade decay $\psi' \to J/\psi \, X \to e^+e^-X$. A careful evaluation shows that the second cluster of events cannot arise from hadrons mislabeled as electrons or higher order electromagnetic processes like $e^+e^- \to e^+e^- \mu^+\mu^-$ with one electron and one muon detected. These processes contribute, respectively, 0.84 ± 0.02 and 0.6 ± 0.2 events. Photons converting in the beam pipe and Dalitz decays of π^0 and η give 0.2 ± 0.1 events. Beam-gas interactions are estimated to contribute less than 0.1 event. The second cluster of events is therefore attributed to τ pair production. The data are indeed well fit by the solid line in Figure 12.8, which shows the spectrum evaluated for

$$\tau \to \nu_\tau \, e \, \nu_e$$

assuming a massless τ neutrino and $m_\tau = 1.80$ GeV. This conclusion is reinforced by the observed photon multiplicity listed in Table 12.3. The distributions are strikingly different. The electron events are accompanied by a few tracks as expected for τ decay whereas the hadron events have a large multiplicity.

Table 12.3. Photon multiplicity distributions

Number of photons	0	1	2	3	4	5	6	7
$e^{\pm}$ + nonshowering track at the ψ' ($0.4 < P_e < 0.9$ GeV/c)	4	3	1	1	0	0	0	0
$h^{\pm}$ + nonshowering track at the ψ' ($P_h > 0.4$ GeV/c)	270	370	440	428	312	199	99	32
$e^{\pm}$ + nonshowering track $\sqrt{s}$: 4-5.2 GeV ($P_e > 0.2$ GeV/c)	49	17	10	1	2	0	1	0

The authors of /169/ therefore conclude that they have observed an anomalous electron signal at a c.m. energy of 3.684 GeV. This shows conclusively that the τ particle is not associated with charm which has a threshold of 3.73 GeV.

We turn now to the DASP data on lepton two prongs at higher energies. A fraction of the two-prong electron events observed above 3.9 GeV might result from associated production and semileptonic decays of charmed particles. An upper limit can be obtained by assuming that all inclusive electron events with more than two prongs are due to charm production. From the measured multiplicity distributions of these events (Figure 11.15) and the known detection efficiency a total of (5 ± 2) events is expected from this source. The direct decay of a pair of charmed hadrons into a final state with one electron and one nonshowering track is expected to contribute less than one event. The background from all other sources has been estimated to (9 ± 3) events, in agreement with (7 ± 7) events extrapolated from the 3.6 GeV data.

The quantity $2\sigma_{\tau\bar{\tau}}B_e \cdot B_{ns}$ is plotted in Figure 12.9 as a function of c.m. energy. Radiative corrections were applied and the data were corrected for the enhancement at the ψ' due to vacuum polarization. Note the rapid rise near threshold which is characteristic for s-wave production. The data shown in Figure 12.9 were used to determine the mass of the τ and its spin /169/. The cross section for $\tau\bar{\tau}$ production for a τ spin of 0, 1/2, and 1 reads as follows:

Spin 0:
$$\sigma_{\tau\bar{\tau}} = (1/4)\sigma_{\mu\mu} \beta_\tau^3 |F|^2 B_e \cdot B_{ns} \qquad (12.5)$$

where $\sigma_{\mu\mu} = \dfrac{4\pi\,\alpha}{3\,s}$ and F is the τ form factor.

Spin 1/2:
$$\sigma_{\tau\bar{\tau}} = \sigma_{\mu\mu} \beta_\tau \{1 + (1/2)(1-\beta_\tau^2)\} B_e \cdot B_{ns} \qquad (12.6)$$

Spin 1:
$$\sigma_{\tau\bar{\tau}} = \sigma_{\mu\mu} \beta_\tau^3 \left\{ \left(\frac{s}{4M_\tau^2}\right)^2 + 5\frac{s}{4M_\tau^2} + 3/4 \right\} \cdot B_e \cdot B_{ns} \qquad (12.7)$$

The τ is assumed to have the same electromagnetic properties as the W boson /188/.

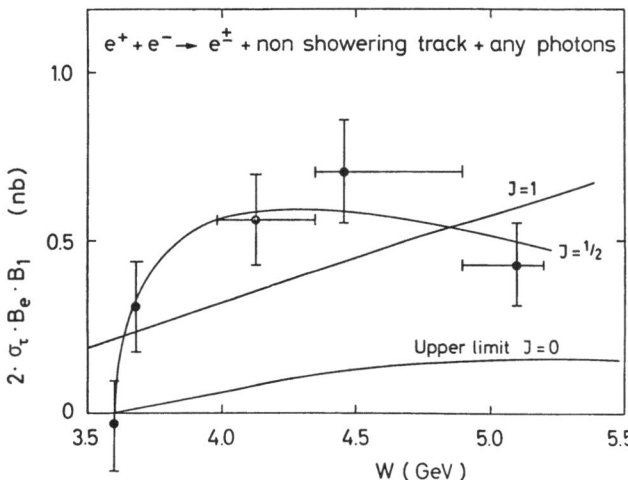

Fig. 12.9. Integrated inclusive cross section for events having an identified electron, a nonshowering particle, and any number of photons as a function of c.m. energy. The data are from DASP /169/. The solid curves show fits to the data assuming pair production of point particles with spin 0, 1/2, and 1

For spin 0 the upper limit on $2\sigma_{\tau\tau}\bar{}B_e B_{ns}$ was calculated with F = 1 and the conservative assumption that the τ has only leptonic decays and $B_e = B_\mu$. This upper limit is plotted in Figure 12.9 and is seen to be lower than the data by an order of magnitude. For spin 1/2 and 1 a fit was made treating the τ mass and the products of the branching ratios $B_e \cdot B_{ns}$ as free parameters. The spin 1 curve (see Figure 12.9) does not describe the data; including the data obtained at higher energies at SPEAR /175/ excludes spin 1. The data are well described by a pointlike fermion of spin 1/2. The fit yielded for the τ mass

$$m_\tau = 1.807 \pm 0.020 \text{ GeV.}$$

The fit used by DASP to evaluate the τ mass from the electron inclusive events yields $B_e \cdot B_{ns} = 0.086 \pm 0.012$. Using $B_e = 0.182 \pm 0.028$ the DASP group derived the branching ratio for $\tau \to \nu_\tau$ + nonshowering particle + ≥ 0 photons, $B_{ns} = 0.47 \pm 0.10$. The branching ratio B_{1h} for $\tau \to \nu_\tau$ + hadron + ≥ 0 photons is given by $B_{1h} = B_{ns} - B_\mu = (0.29 \pm 0.11)$. The systematic errors are small compared to the statistical error. The average number of photons associated with $\tau \to \nu_\tau$ + hadron + ≥ 0 photons can be obtained after making background corrections. Averaging the observed photon mulitplicity over all two-prong events in the higher energy data and correcting for the photon detection efficiency the decay of the type $\tau \to \nu_\tau$ + charged hadron + any number of photons was found to yield on the average 2.8 ± 0.7 photons.

The branching ratio B_{3h} for the τ to decay into final states with at least three charged particles can be obtained from $B_{3h} = 1-B_e-B_{ns}$ (the number of electron events with $P_e > 1$ GeV/c and 5 or more charged tracks were found to be small). The result is $B_{3h} = 0.35 \pm 0.11$.

The DASP group studied also μ inclusive events /169/. Candidates for muon inclusive events had to have one muon track in the spectrometer, a second nonshowering track, and any number of photons observed either in the inner detector or in the spectrometer arms. A charged particle was called a muon if it had a momentum greater than 1.0 GeV/c, gave no signal in the threshold Cerenkov counter, suffered an energy loss consistent with that of a minimum ionizing particle in the shower counter and penetrated at least 60 cm of iron. A total of 25 events with a background of 3.8 events was found.

After all corrections DASP observed (21 ± 5) muon inclusive events and (18.5 ± 4.6) electron inclusive events with momenta above 1.0 GeV/c. The ratios of the leptonic widths evaluated directly from these data are independent of the form of the coupling. This yields $B_\mu/B_e = 0.92 \pm 0.32$ with a systematic uncertainty of 0.07. The result is consistent with eμ universality.

The DELCO group /184, 189/ has measured the cross section for $e^+e^- \to e^\pm$ + nonshowering track + ≥ 0 photons for c.m. energies between 3.7 and 7.4 GeV. The particles were identified using threshold Cerenkov counters and shower counters. A track was called an electron if it had a momentum greater than 0.2 GeV/c and fired the proper Cerenkov counter. A nonshowering track must have a momentum greater than 0.3 GeV/c, not fire the Cerenkov counter, and give a low pulse height in the shower counter. In addition the tracks were required to have an acoplanarity angle with respect to the beam axis greater than 20°. A total of 692 events (459 with no

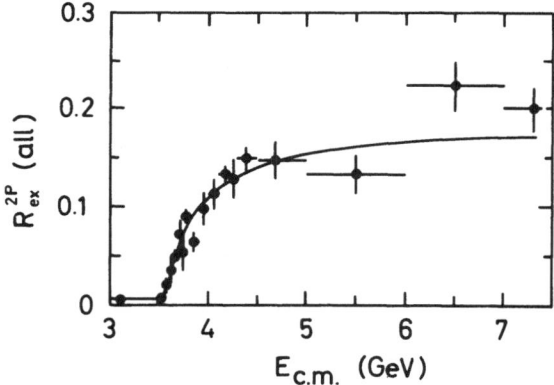

<u>Fig. 12.10.</u> The ratio $(e^+e^- \to e^\pm X^\mp, X \neq e^-)/\sigma_{\mu\mu}$ as measured /184/ by DELCO. The solid line is a fit to the data assuming lepton pair production

photon) were found with an estimated background from particle misidentification of less than 28 events. The contribution from higher order electromagnetic processes was found to be less than (2 ± 0.5) %. In particular they observe 70 events with an estimated background of 5 events below charm threshold.

In Figure 12.10 the ratio

$$R_{ex}^{2p}(\text{all}) = \frac{\sigma(e^+e^- \to e^{\pm} + X^{\mp} + \geq 0 \text{ photons})}{\sigma_{\mu\mu}}$$

is plotted as a function of energy between 3.5 and 7.4 GeV in c.m. Figure 12.11

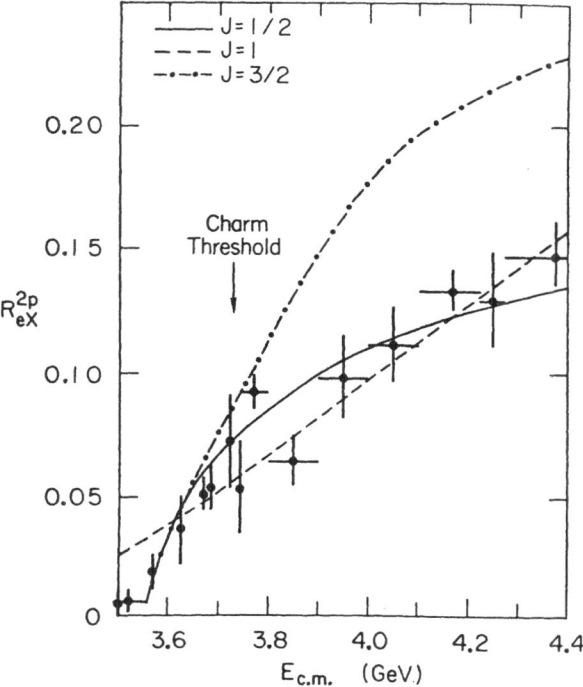

Fig. 12.11. The ratio $R_{ex}^{2p} = \sigma(e^+e^- \to e^{\pm}X^{\mp}, X \neq e^-)/\sigma_{\mu\mu}$ as measured by DELCO /184/

depicts the same ratio in the threshold region together with the theoretical cross sections for $J = 1/2$, 1 and 3/2. Only pair production of a spin 1/2 particle gives a good fit to the data. The resulting mass is $1.782 \, {}^{+\, 0.002}_{-\, 0.007}$ GeV in agreement with the DASP data. The topological branching ratios are determined as follows. The solid curve shown in Figure 12.10 yields $B_e \cdot B_{ns} = 0.084 \pm 0.004$ in excellent agreement with the DASP data /169/. A fit to the events without a photon yields $B_e B_{ns}^{0\gamma} = 0.0525 \pm 0.0035$. They find, using the theoretical values for one-prong decays including $\tau^\pm \to \nu\rho^\pm$ with both photons escaping detection, $B_{ns}^{0\gamma} = 2B_e$. This yields

$$B_e = 0.160 \pm 0.13 \quad \text{and} \quad B_{3h} = 0.32 \pm 0.05.$$

The DELCO group /189/ also obtained an upper limit on the τ liefetime of $\tau_0 < 2.3 \times 10^{-12}$ s compared to a predicted lifetime of $2.8 \cdot 10^{-13}$ s.

The DESY-Heidelberg group /183/ has measured the reaction $e^+e^- \to \frac{e^\pm}{\mu^\pm}$ + one charged track + 0 photon for c.m. energies between 3.6 and 4.40 GeV. They also obtain a clear signal below charm threshold. A fit to the cross section as a function of energy yields

$$m_\tau = 1.787 \begin{array}{c} + \ 0.010 \\ - \ 0.018 \end{array} \text{GeV}$$

in agreement with DASP and DELCO.

12.2.4 Space-Time Structure of the Weak Current in τ Decay

The lepton spectrum permits a study of the space-time structure of the weak current mediating the decay of the τ. If only V and A type couplings are considered the Hamiltonian for $\tau \to \nu_\tau \ell \bar{\nu}_e$ is of the form /190/

$$H_{int} = \frac{G_F}{\sqrt{2}} \left[\bar{\psi}_{\nu_e} \gamma_\mu (1-\gamma_5)\psi_e \right] \left[\bar{\psi}_\tau \{ g_+ \gamma_\mu(1+\gamma_5) \right. \tag{12.8}$$
$$\left. + g_- \gamma_\mu(1-\gamma_5) \} \ \bar{\psi}_{\nu_\tau} \right],$$

where $g_\pm$ are the coupling strengths for $V \pm A$ couplings. In the τ rest system the shape of the lepton spectrum can be expressed in terms of the lepton momentum p, energy E, maximum energy

$$E_{max} = (m_\tau^2 - m_e^2 - m_\nu^2)/2m_\tau \quad \text{and} \quad x = E/E_{max},$$

$$\frac{dN(x)}{dx} = \text{const } x^2 \{3(1 - x) + 2\rho \ (\tfrac{4}{3} x-1)\},$$

where terms of order m_e/m_τ have been neglected. The shape of the spectrum is then determined by the Michel parameter ρ /190/ where ρ is defined as

$$\rho = \frac{3}{4} \frac{g_-^2}{g_+^2 + g_-^2} .$$

Special cases are

pure V-A: $g_+ = 0$ $\rho = 3/4$
   V+A: $g_- = 0$ $\rho = 0$
   V : $g_+ = g_-$ $\rho = 3/8$
   A : $g_+ = -g_-$ $\rho = 3/8$.

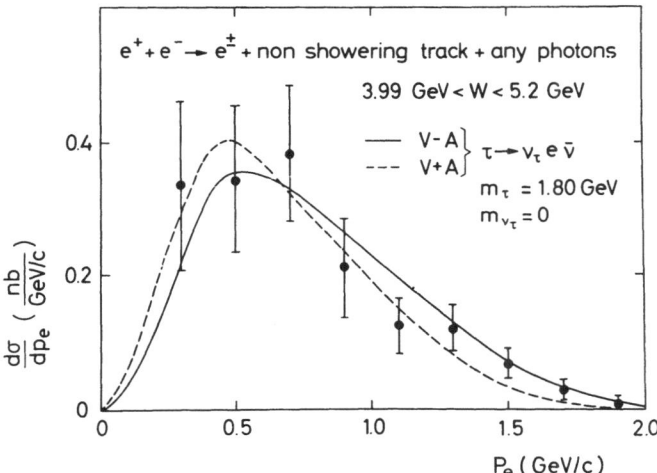

<u>Fig. 12.12.</u> Corrected electron momentum distribution measured by DASP /169/ for events having an identified electron, a nonshowering particle, and any number of photons. (Above 1 GeV/c data having a muon instead of an electron are combined with the electron data to form a weighted mean)

The lepton momentum spectrum, obtained by the DASP group /169/ by combining the electron and the muon data, is plotted in Figure 12.12 for c.m. energies between 4.0 GeV and 5.2 GeV. This spectrum extends to much higher momenta than the electron spectrum observed in the semileptonic decays of the charmed hadrons, reflecting the pointlike structure of the τ and the low mass of its neutrino.

The solid line shows a fit to the data assuming $m_\tau = 1.80$ GeV, a massless neutrino, and a (V-A) structure of the current. The dashed line is a fit keeping the masses constant but changing the left-handed V-A current into a right-handed V+A current. Both fits are clearly acceptable.

Fits were also made varying the mass of the τ neutrino. The 90 % confidence upper limits on the neutrino mass are $m_{\tau_\nu} < 0.74$ GeV for V-A and $m_{\nu_\tau} < 0.54$ GeV for V+A.

178

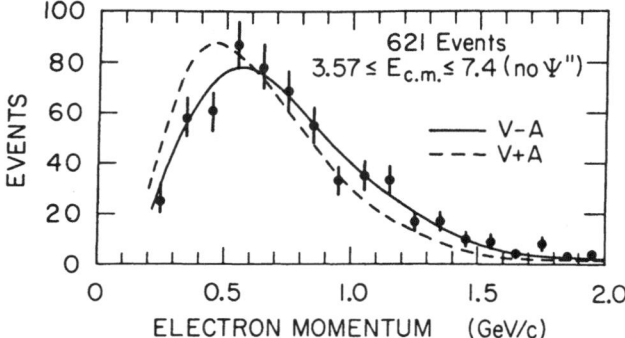

Fig. 12.13. Electron momentum spectrum for events with electron + one charged track as measured by DELCO /189/. The solid and dashed curves represent the spectra expected for V-A and V+A coupling at the $\tau\nu_\tau$ vertex, respectively

From their high statistics data (see Figure 12.13) the DELCO group /189/ finds

$\rho = 0.72 \pm 0.15.$

This result agrees well with V-A and excludes V+A. A pure V or pure A coupling appears to be unlikely. Tne analysis assumed a massless τ neutrino. If the ν_τ is massive the theoretical spectrum will be pushed towards smaller x values and hence V+A currents are ruled out for all values of x. An upper limit on the mass of the τ neutrino of $m_{\tau_\nu} < 250$ MeV was obtained assuming a V-A current.

12.3 Semihadronic Decays of the τ

A characteristic feature of the standard weak interaction is that decays involving strange particles are suppressed relative to decays involving nonstrange final states by $tg^2\theta_c \approx 0.05$. The DASP group determined /160/ the ratio of strange to nonstrange particles in semihadronic τ decays from a measurement of

$$\frac{\sigma(e^+e^- \rightarrow e^\pm + K^\pm + \geq 0 \text{ photons})}{\sigma(e^+e^- \rightarrow e^\mp + \pi^\pm + \geq 0 \text{ photons})}$$

It was shown that the two-prong cross section including one electron predominantly results from $e^+e^- \rightarrow \tau\bar{\tau} \rightarrow (\nu_\tau e\bar{\nu}_e)(\nu_\tau + \text{hadrons} + \geq 0 \text{ photons})$ with only a small contamination from charm decays. The hadrons were identified in either the inner or the outer detector.

The DASP group found

$$\frac{\sigma(e^{\mp} + K^{\pm} + \geq 0 \text{ photons})}{\sigma(e^{\mp} + \pi^{\pm} + \geq 0 \text{ photons})} = 0.07 \pm 0.06.$$

Therefore on the average only 7 % of all semihadronic τ decays yield a strange particle in accordance with theory. This should be compared to multiprong events where DASP found /160/

$$\frac{\sigma(e + K^{\pm} + \geq 1 \text{ prong} + \geq 0 \text{ photons})}{\sigma(e + \pi^{\pm} + > 1 \text{ prong} + \geq 0 \text{ photons})} = 0.24 \pm 0.05.$$

Since the charged multiplicity is on the order of 4 this is equivalent to (0.9 ± 0.18) charged kaons per multiprong event (see discussion in Sect. 3).

12.3.1 $\tau \to \pi\nu$

The DASP group /181/ searched for the $\tau \to \pi\nu$ decay studying the process

$$e^{+}e^{-} \to \tau\bar{\tau} \to (e\nu\nu)(\pi\nu)$$

leading to the final state $e\pi$ 0γ. No positive evidence was found for the existence of the decay into $\pi\nu$. The branching ratio deduced from the data, $B_{\pi} = 0.02 \, ^{+\ 0.03}_{-\ 0.02}$, was about 2.5 s.d. below its theoretical values, $B_{\mu} = 0.10$.

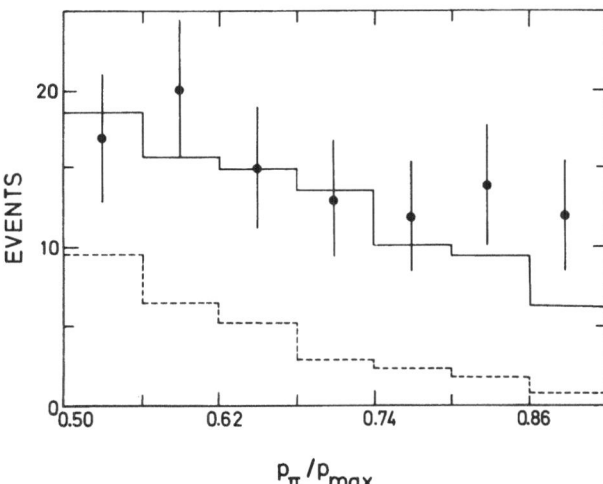

Fig. 12.14. The pion momentum distribution for events of the type $e^{+}e^{-} \to \pi^{\pm} + 1$ charged measured at 4.8 - 7.4 GeV by the SLAC-LBL group /191/

The first clear evidence for the $\pi\nu$ decay was presented by the SLAC-LBL group /191/ from an analysis of

$$e^+e^- \rightarrow \pi^{\pm}X^{\mp}\,0\gamma$$

for c.m. energies between 4.8 and 7.4 GeV. Figure 12.14 shows the pion momentum spectrum in terms of $x = p/p_{max}$ for x above 0.5. The x spectrum is seen to be almost constant in marked difference to the x region below 0.5 where hadronic channels dominate and produce an exponentially falling x distribution (see Sect. 14). The most probable mechanism for producing a constant x distribution at high x is the two-body decay into π + a of a particle with fixed momentum.

The solid line in Figure 12.14 shows the expected x distribution from $\tau \rightarrow \pi\nu$ plus background; the dashed line shows the background (mainly from $\tau^{\pm} \rightarrow \rho^{\pm}\nu \rightarrow \pi^{\pm}\pi^0\nu$) alone. The shape of the spectrum as well as the magnitude of the observed cross section agrees with the assumption that most of these high x events come from $\tau \rightarrow \pi\nu$ decay.

Further evidence for the $\pi\nu$ decay comes from experiments by PLUTO /192/, DELCO /184/, and SLAC-LBL (MARK) /193/. The measured branching ratios are summarized in Table 12.4 /194/.

Table 12.4. Results on the decay $\tau \rightarrow \pi\nu$

Experiment	Final state	Events	Background	$B(\tau \rightarrow \pi\nu)$ [%]
DASP /181/	$\pi^{\pm}$ + e + 0γ	2	1	2^{+3}_{-2}
SLAC-LBL (MARKI) /191/	$\pi^{\pm}$ + ch + 0γ	~200	~70	9.3 ± 3.9
PLUTO /192/	$\pi^{\pm}$ + ch + 0γ	32	9	$9.0 \pm 2.9 \pm 2.5$
DELCO /184/	$\pi^{\pm}$ + e + 0γ	15	5.6	$6.0 \pm 1.6 ^{+1.9}_{-1.2}$
SLAC-LBL(MARKII) /193/	$\pi^{\pm}$ + ch + 0γ	142	46	$8.0 \pm 1.1 \pm 1.5$
	$\pi^{\pm}$ + e + 0γ	27	10	$8.2 \pm 2.0 \pm 1.5$

12.3.2 $\tau \rightarrow K\nu$

The DASP group has also searched for events of the type

$$e^+e^- \rightarrow \tau\bar{\tau} \rightarrow \{(K\nu)[(e\nu\nu) + (\mu\nu\nu) + (\pi\nu)]\}$$
$$= K^{\pm} + \text{charged track} + \text{missing energy.}$$

Only one event with $P_K > 1.0$ GeV/c was found /181/. This yields a 90 % confidence upper limit of $B_K < 0.016$.

12.3.3 $\tau \to \rho\nu$

DASP measured /181/ the decay $\tau \to \rho\nu_\tau$ by selecting final states with $\pi^\pm$ + charged track + two photons. Events in which the two photons are compatible with resulting from a π^0 decay are retained provided that both computed photon energies are above

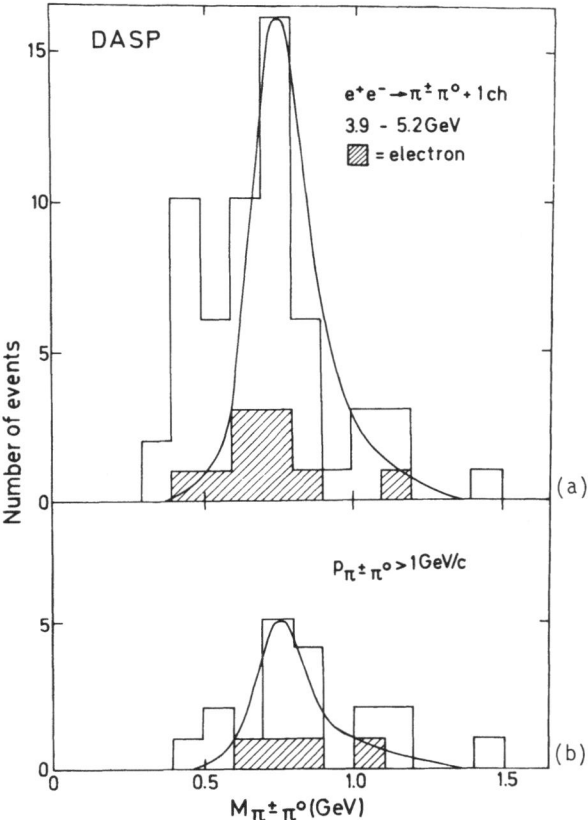

Fig. 12.15. (a) The distribution of $M(\pi^\pm\pi^0)$ observed by DASP /181/ for events with the topology $e^+e^- \to \pi^\pm\pi^0$ + charged track, (b) the $M(\pi^\pm\pi^0)$ distribution for events of the same topology as above but with $P_{\pi^\pm\pi^0} > 1$ GeV/c

50 MeV. The remaining events are plotted versus $M(\pi^\pm\pi^0)$ in Figure 12.15a. Events with an identified electron are hatched. Events within the ρ band [0.5 GeV < $M(\pi^\pm\pi^0)$ < 1.0 GeV] are plotted versus the momenta of the $\pi^\pm\pi^0$-system in Figure 12.16. The

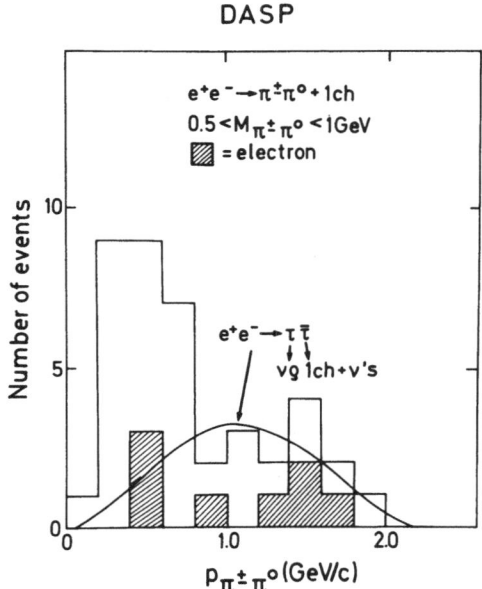

Fig. 12.16. The momentum distribution of the $(\pi^{\pm}\pi^0)$ system observed by DASP /181/ in $e^+e^- \to \pi^{\pm}\pi^0 T^{\mp}$. The mass of the $\pi^{\pm}\pi^0$ system was between 0.5 and 1.0 GeV

momentum distribution expected from the decay $\tau \to \rho\nu_\tau$ is shown as the solid line. Note the flat distribution above 0.9 GeV/c which is characteristic for a two-body decay of a moving object. The enhancement at low momenta is presumably due to multi-hadron events. To reduce the background, only events with a $(\pi^{\pm}\pi^0)$ momentum above 1.0 GeV/c are considered. The $\pi^{\pm}\pi^0$ mass distribution for these events is plotted in Figure 12.15b. These events yield as a preliminary value

$$B_\rho = 0.24 \pm 0.09$$

in good agreement with the theoretical predictions.

12.3.4 $\tau \to \rho\pi\nu$

The PLUTO group has measured /195/ the final state $e^+e^- \to e^{\mp}\pi^+\pi^-\pi^+$. They found, after background corrections, that these events result from

$$e^+e^- \to \tau\bar\tau \to (e\nu_\tau \bar\nu_e)(\rho^0\pi^-\bar\nu_\tau).$$

They observe a total of 39 events of the type e + 3 hadrons and assume that the hadrons are all pions. From the sample of 39 events those events are selected for

which the 3π system has an invariant mass and momentum consistent with pair production and subsequent decay of a particle of mass ≤ 1.9 GeV; 6 events are rejected.
The remaining sample of 33 events contains an estimated background of 13 ± 4 events
from hadrons mislabeled as electrons and at most 3 events which had unobserved
photons.

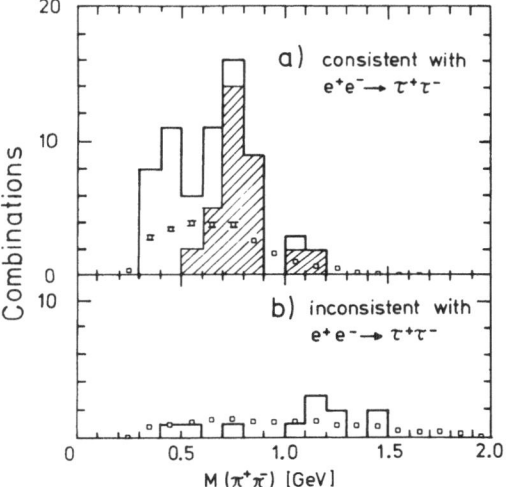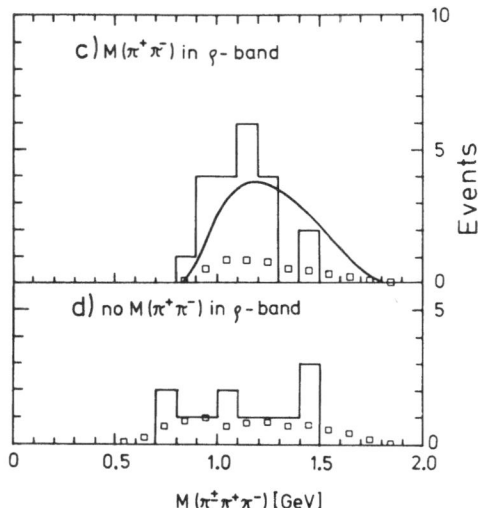

Fig. 12.17. Properties of the two- and three-pion mass distribution in the $e\pi\pi\pi$
event sample observed by PLUTO /195/. The squares represent the expected distributions for background from identified events. The solid line in (c) represents the
sum of the background and the $\tau \to \rho^0\pi$ distribution normalized to the data. (a)
$M(\pi^+\pi^-)$ for events kinematically consistent with τ pair production. There are two
combinations per event and the shaded area represents the distribution of the higher
$M(\pi^+\pi^-)$ combination, (b) $M(\pi^+\pi^-)$ for events kinematically inconsistent with τ pair
production, (c) $M(\pi^\pm\pi^+\pi^-)$ distribution for events consistent with pair production
and having at least $\pi^+\pi^-$ mass in the ρ band

The $\pi^+\pi^-$ mass distribution for the 33 events is plotted in Figure 12.17a. The
background is shown by the squares. Each event contributes two mass values and the
distribution can be understood as arising mainly from a ρ signal in one combination
and a broad distribution in the other. Plotting only the highest mass combination
shows the ρ more clearly. Twenty-one events have a $M(\pi^+\pi^-)$ between 0.70 and 0.84
GeV and are retained. The momentum distribution for the electrons in the remaining
sample is plotted in Figure 12.18b.

The electron spectrum is hard and consistent with $\tau \to e\nu_\tau\bar{\nu}_e$ shown as the solid
line but inconsistent with the charm spectrum shown as the dotted line. If the (3π)

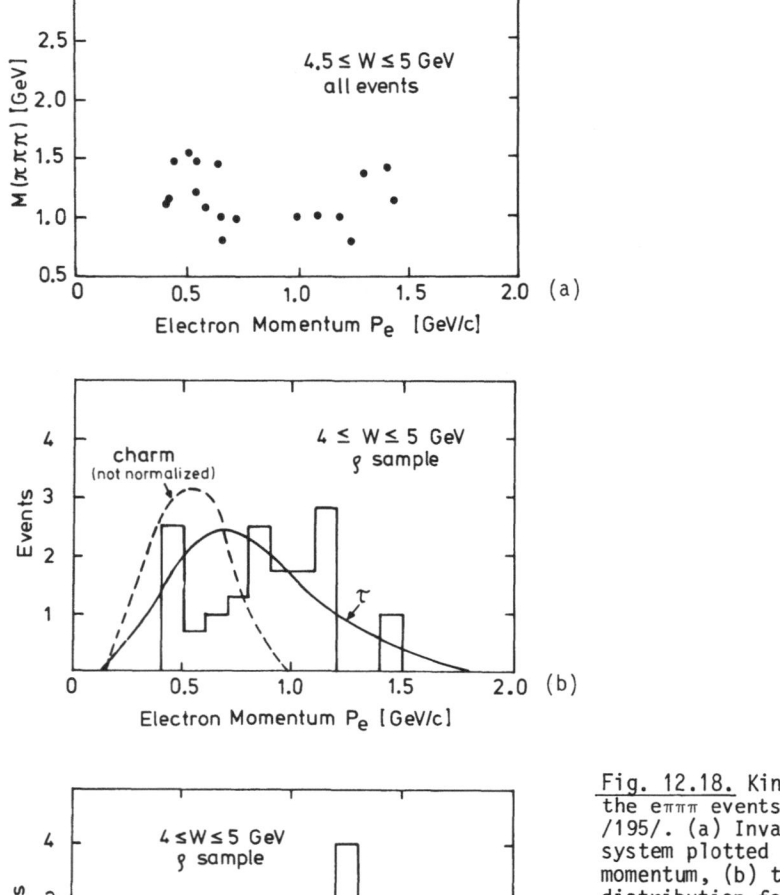

Fig. 12.18. Kinematic properties of the $e\pi\pi\pi$ events observed by PLUTO /195/. (a) Invariant mass of the 3π system plotted against the electron momentum, (b) the electron-momentum distribution for events from the ρ sample (see text). The solid curve is the prediction for τ decay normalized to the data with $p_e > 0.4$ GeV/c. The dotted curve is the electron spectrum from charm decays measured by DASP. The curves are corrected for the detection efficiency, (c) distribution of the normalized 3π momentum (ρ sample), compared with the expected $\tau \to \nu_\tau \pi\pi\pi$ decay distribution

system results from the two-body decay of a moving object it should uniformly populate momenta between P_{max} and P_{min}. These limits depend on the c.m. energy and the mass of the 3π system. However, the quantity $U = (P_{3\pi} - P_{min})/(P_{max} - P_{min})$ is uniformly populated for any set of c.m. energies and mass values. The events are plotted in Figure 12.18c versus U. The solid line indicates the event distribution predicted including all the experimental corrections, for a two-body decay $\tau \to (3\pi)\nu_\tau$ of a moving object.

The events can therefore safely be assigned to τ pair production. The 3π mass distribution of the 21 events with at least one $\pi^+\pi^-$ combination in the ρ band is shown in Figure 12.17c. The background expected from misidentified electrons is indicated by the squares. From a spin parity analysis the PLUTO group concluded that the $\rho\pi$ system is preferentially in a $J^P = 1^+$ state /196/. The mass histrogram shows a clustering at low $\rho\pi$ mass as expected for an A_1 signal. The nonresonant mass distribution $\tau \to \rho\pi\nu_\tau$ (solid line) was computed assuming (V-A) couplings, $m_\tau = 1.8$ GeV, and a massless τ neutrino. The χ^2 probability for this fit is 5 % compared to 80 % for an A_1 with mass 1.1 GeV and width of 0.25 GeV. The resonance interpretation is thus favored. Using $B_e = 0.16$ they find $B(\tau^+ \to \rho^0\pi^+\nu_\tau) = 0.05 \pm 0.03$.

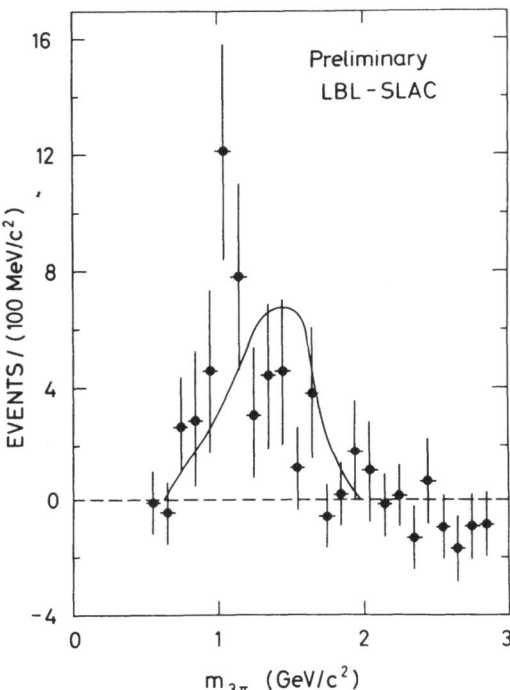

Fig. 12.19. Three-particle invariant mass distribution opposite muons corrected for hadron misidentification as measured by SLAC-LBL /191/. The solid curve is the phase space prediction for $\tau \to \nu\pi\rho$, corrected for acceptance effects. The prediction is normalized to events with a three-particle effective mass between 0.7 and 1.8 GeV

The SLAC-LBL group has also reported /191/ evidence for $\tau \to A_1 \nu_\tau \to \rho\pi\nu_\tau$. The mass spectrum for the 3π system is plotted in Figure 12.19. The peak observed is too narrow to result from a phase spacelike decay $\tau \rightarrow \rho\pi\nu$, plotted as the solid line. A resonance fit to the mass spectrum yields $m_{A_1} \cong 1.1$ GeV, $\Gamma \approx 0.2$ GeV in agreement with the PLUTO result.

12.4 Summary of the τ Properties

Table 12.5 summarizes the experimental information on mass, leptonic, and semihadronic decay modes of the τ.

The observation of the τ at the ψ' below charm threshold conclusively demonstrates that the τ signal has nothing to do with charm. The shape and magnitude of the $\tau\bar{\tau}$ production cross section exclude spin 0, 1, and 3/2 for the τ: they strongly favor the assignment as a pointlike, spin 1/2 fermion. The best value for the τ mass at present is $1.782 \, {}^{+\,0.002}_{-\,0.007}$ GeV. The analysis of the decay spectra puts a limit of 0.25 GeV on the mass of the τ neutrino. The lifetime of the τ is less than $2.3 \cdot 10^{-12}$ s. The lepton momentum spectrum strongly favors a V-A type coupling of the weak current to the $\tau\bar{\nu}_\tau$ system. The measured leptonic decay rates are consistent with e/μ universality. The leptonic and semihadronic branching ratios agree with those expected from theory for a heavy lepton of mass 1.8 GeV.

The consistency with e/μ universality classifies the τ as either a ortholepton or a sequential lepton. In the first case the τ has the same lepton number as e or μ. In the second case the τ carries a new lepton number. Recent neutrino experiments rule out that the τ is μ-like /198/.

Table 12.5. Summary of the experimental results

Quantity	Experimental results	Predicted by theory[a] (sequential lepton)	Experiment	Comments
Mass [GeV]	1.807 ± 0.020	–	DASP /169/	earlier experiments did not have data close to threshold
Mass [GeV]	$1.782 \, {}^{+0.002}_{-0.007}$	–	DELCO /184/	
Mass [GeV]	$1.787 \, {}^{+0.010}_{-0.018}$	–	D-HD /183/	
Spin	$1/2$	$1/2$	–	Spin 0, 1, and 3/2 are excluded
Lifetime [s]	$< 9 \cdot 10^{-12}$ $< 2.3 \cdot 10^{-12}$	$2.8 \cdot 10^{-13}$	PLUTO /187/ DELCO /189/	
m_{ν_τ} [GeV]	<0.25 GeV	0	DELCO /189/	
Structure	V-A $(\rho = 0.72 \pm 0.15)$	V-A $(\rho = 0.75)$	DELCO /189/	SLAC-LBL finds /175/ $\chi^2 = 50$ % for V-A compared to $\chi^2 = 0.1$ % for V+A
B_μ/B_e	0.92 ± 0.32	0.98	DASP /169/	
$N_{ee}/N_{\mu e}$	$0.56 \pm 0.14 \, {}^{+0.16}_{-0.19}$	0.5	SLAC-LBL /175/	ortholepton = 0.5 paraelectron $N_{ee}/N_{\mu e} = 0.86$, $N_{\mu\mu}/N_{\mu e} = 0.29$
$N_{\mu\mu}/N_{\mu e}$	$0.70 \pm 0.15 \pm 0.19$	0.5	SLAC-LBL /175/	ortholepton = 0.5 paramuon $N_{ee}/N_{\mu e} = 0.29$, $N_{\mu\mu}/N_{\mu e} = 0.86$
B_e	$0.186 \pm 0.010 \pm 0.028$	0.18	SLAC-LBL /175/	from eμ; assume $B_e = B_\mu$ and V-A
	0.16 ± 0.06		PLUTO /179/	
	$0.224 \pm 0.032 \pm 0.044$		LBL-SLAC /180/	from eμ; assume $B_e = B_\mu$
	$0.182 \pm 0.028 \pm 0.014$		DASP /169/	from eμ; assume $B_e = B_\mu$ and V-A
	0.16 ± 0.04		DELCO /184/	from eX; assume $B(0\gamma)^{exp} = 2B_e$ 1 pion

Table 12.5. continued

Quantity	Experimental results	Predicted by theory[a] (sequential lepton)	Experiment	Comments
	$0.175 \pm 0.072 \pm 0.030$	0.18	SLAC-LBL /185/	from μX; assume $B_{1\text{prong}}(\geq 0\gamma) = 0.85$ and V-A
	0.14 ± 0.034		PLUTO /179/	from μX; assume V-A
	$0.22 \begin{smallmatrix} + 0.07 \\ - 0.08 \end{smallmatrix}$		Iron ball /182/	from $\mu\mu$
	0.20 ± 0.10		Maryland-Princeton-Pavia /176, 177/	from μX
$B(\tau \to e\gamma)$	<0.026		LBL-SLAC /197/	
$B(\tau \to \mu\gamma)$	<0.013	0	LBL-SLAC /197/	
$B(\tau \to 3$ charged leptons)	<0.006 <0.01		SLAC-LBL /197/ PLUTO /168/	
$B(\tau \to 1$ charged + any photons)	0.70 ± 0.10		PLUTO /179/	
	0.90 ± 0.10		LBL-SLAC /180/	
	0.65 ± 0.12		DASP /169/	
	0.685 ± 0.03		DELCO /184/	3 charged particles

$B(\tau \to 1$ charged hadron + any photons)	0.40 ± 0.15		PLUTO /179/	
	0.45 ± 0.19		LBL-SLAC /180/	
	0.29 ± 0.11		DASP /169/	
$B(\tau \to \geq 3$ charged hadrons + any photons)	0.35 ± 0.11		DASP /169/	
	0.32 ± 0.05		DELCO /184/	
$B(\tau \to \pi\nu)$	0.077 ± 0.013	0.10	SLAC-LBL /191/, PLUTO /192/, DELCO /184/, and /194/	for details see Table 4.4
$B(\tau \to K\nu)$	<0.016	0.005	DASP /181/	
$B(\tau \to \rho\nu)$	0.24 ± 0.09	0.22	DASP /181/	
$B(\tau \to \rho\,\pi^{\pm}\nu)$	0.050 ± 0.015	~ 0.1	PLUTO /195/	spectrum consistent with A_1 decay. Including neutral decay modes would give $B(\tau \to A_1\nu) = 0.10 \pm 0.03$

[a] assuming a sequential lepton of mass 1.782.

13. The Γ Family

The observation of the Τ gave the impression of a déjà vu phenomenon in more than one respect. It was discovered by a Columbia-FNAL Stonybrook collaboration /199/ studying th $\mu^+\mu^-$ mass spectrum produced via

$$400 \text{ GeV } p + N \rightarrow \mu^+\mu^- + X$$

where Be, Cu, and Pt were used as targets.

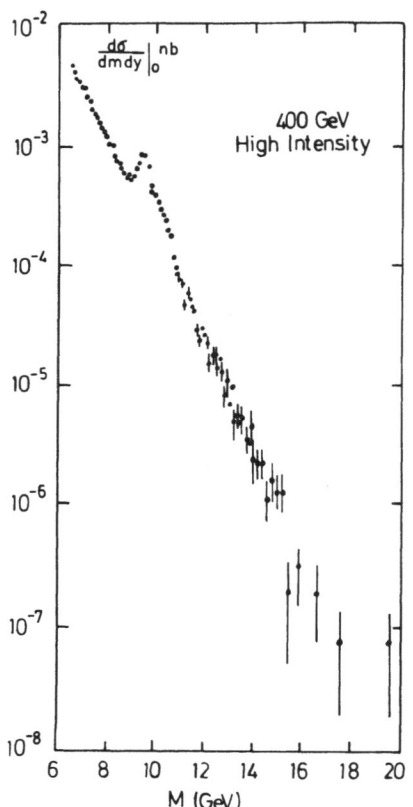

Fig. 13.1. Dimuon production at 400 GeV via pN → μ+μ⁻X as measured by the Columbia-FNAL-Stonybrook collaboration /199/. Plotted is the cross section dσ/dmdy at y = 0 as a function of the dimuon mass m

Figure 13.1 shows the latest data from this experiment /199/. Plotted is the $\mu^+\mu^-$ mass spectrum in terms of the differential cross section $B\frac{d^2\sigma}{dmdy}\Big|_{y=0}$ per nucleon with m = $\mu^+\mu^-$ mass, y = c.m. rapidity of the $\mu^+\mu^-$ system, and B = branching ratio for the $\mu^+\mu^-$ channel. A broad enhancement is seen around 9.5 GeV riding on a falling background. No other structure is observed between 6 and 13 GeV. The background

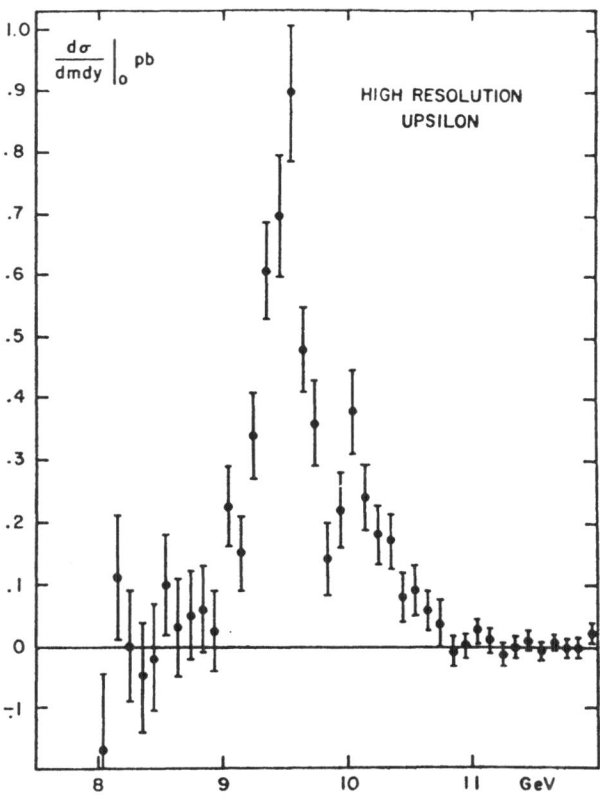

Fig. 13.2. Dimuon production via pN → $\mu^+\mu^-$X as measured by the Columbia-FNAL-Stony-brook collaboration /199/. Plotted is the dimuon mass spectrum obtained after subtracting a smooth background

subtracted signal is plotted in Figure 13.2. One observes a 500 MeV wide (FWHM) peak around 9.4 GeV followed by a second bump near 10 GeV. The rms mass resolution in this experiment was roughly 500 MeV.

The authors have fitted the mass spectra to a two-peak and a three-peak hypothesis. The latter was found to produce a lower chi square per degree of freedom.

In both cases the resonances were assumed to be narrow which implies that the mass shapes are determined by the experimental resolution. The results from these fits are listed in Table 13.1.

Table 13.1. Resonance fits to the background subtracted $\mu^+\mu^-$ mass spectrum /199/

		2 peaks	3 peaks	
T	$B\dfrac{d\sigma}{dy}\Big|_{y=0}$	9.41 ± 0.013	9.40 ± 0.013	GeV
		0.18 ± 0.01	0.18 ± 0.01	pb
T'	$B\dfrac{d\sigma}{dy}\Big|_{y=0}$	10.06 ± 0.03	10.01 ± 0.04	GeV
		0.069 ± 0.006	0.065 ± 0.007	pb
T''	$B\dfrac{d\sigma}{dy}\Big|_{y=0}$		10.40 ± 0.12	GeV
			0.011 ± 0.007	pb

Meanwhile further experiments have observed the T in hadron-initiated reactions. They are summarized in /199/.

From the experience with the J/ψ it seemed natural to associate the T with a new heavy quark Q. An important piece of information was missing, however: the widths of the T states. Are those states indeed narrow like the J/ψ,ψ'? This question can best be studied by producing the T in an e^+e^- storage ring. DORIS, which at the time could reach a total c.m. energy of 2 x 4.3 = 8.6 GeV, was converted to a single ring machine in the fall of 1977 and subsequently upgraded to 2 x 5 = 10 GeV.

The efforts of the DORIS machine crew led to the observation of the T by the PLUTO /200/ and the DASP2 groups /201/ in June of 1978. In a subsequent running period at DORIS the T was also seen by a DESY-Hamburg-Heidelberg-Munich collaboration /202/. Figure 13.3 shows the total cross section as measured by these experiments near the T mass region. Only the PLUTO data are corrected for acceptance. The T shows up as a narrow (<10 MeV wide) signal at 9.46 GeV. As in the case of the J/ψ and ψ' the observed width (σ = 7.8 ± 0.9 MeV) is determined by the energy spread of the beams which for m = 9.5 GeV was calculated to give an rms resolution of σ_m = 8 MeV.

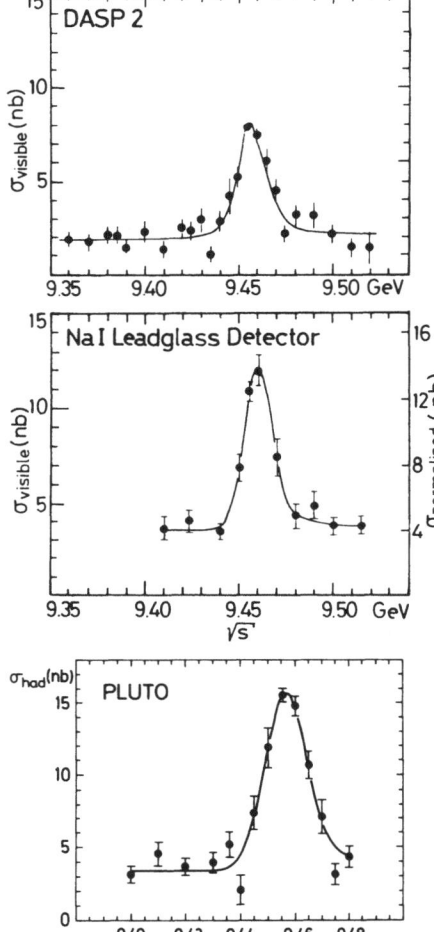

Fig. 13.3. The total cross section for hadron production in the Υ region. The top two measurements by DASP2 /201/ and DESY-Hamburg-Heidelberg-Munich /202/ give visible cross sections, the bottom measurement by the PLUTO group /200/ shows the corrected cross section

Integration of the resonance signal relates the hadronic cross section to the total, the hadronic, and the leptonic widths Γ_{tot}, Γ_{had}, and Γ_{ee}

$$\frac{M^2}{6\pi^2} \int \sigma_{had} \, dm = \frac{\Gamma_{ee}\Gamma_{had}}{\Gamma_{tot}}$$

For Γ_{tot} we can write

$$\Gamma_{tot} = \Gamma_{had} + \Gamma_{ee} + \Gamma_{\mu\mu} + \Gamma_{\tau\tau} = \Gamma_{had} + 3\Gamma_{ee},$$

where e,μ,τ universality was assumed. In order to determine Γ_{had} and Γ_{ee} separately, a measurement of a leptonic decay mode, say, $\tau \to \mu^+\mu^-$ is required. This yields

$$\frac{M^2}{6\pi^2} \int \sigma_{\mu\mu} \, dm = \frac{\Gamma_{ee}^2}{\Gamma_{tot}} \, .$$

In addition to the observation of the τ in the hadronic final states a first attempt was made to measure $B_{\mu\mu} = \Gamma_{\mu\mu}/\Gamma_{tot}$ /203-205/ and to deduce Γ_{ee} and Γ_{tot}. The mass and width parameters as obtained by the three experiments are listed in Table 13.2.

Table 13.2. Mass and width parameters of the τ as measured at DORIS

		PLUTO /203/	DASP2 /204/	D-H-HD-M /205
M_γ	[GeV]	9.46 ± 0.01	9.46 ± 0.01	9.46 ± 0.01
Γ_{ee}	[keV]	1.33 ± 0.14	1.5 ± 0.4	1.1 ± 0.3
$B_{\mu\mu}$	[%]	2.2 ± 2.0	2.5 ± 2.1	1.0 $^{+\,3.4}_{-\,1.0}$
Γ_{tot}	[keV]	>23 (2 s.d.)	>20 (2 s.d.)	>15 (2 s.d.)

The mean values are /206/

M_γ = 9.46 ± 0.01 GeV (The error reflects the 1 $^o/_{oo}$ uncertainty in the energy calibration of DORIS)

Γ_{ee} = 1.3 ± 0.2 keV

$B_{\mu\mu}$ = 2.6 ± 1.4 %

Γ_{tot} > 25 keV (95 % C.L.)
< 18 MeV

best value: Γ_{tot} = 50 keV

The DASP2 /204/ and DESY-Hamburg-Heidelberg-München /202/ collaboration also searched for the τ'. The visible cross sections measured by these experiments around 10 GeV is displayed in Figure 13.4. Both experiments observe a peak at 10.02 GeV which is consistent with the τ' mass value quoted by the Columbia-FNAL-Stonybrook collaboration /199/. The τ' parameters deduced by the storage ring experiments are listed in Table 13.3.

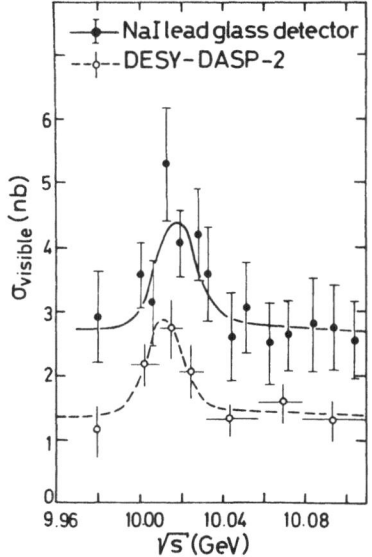

Fig. 13.4. Evidence for the T' as measured by the DASP2 /204/ and DESY-Hamburg-Heidelberg-Munich group /202/

Table 13.3. Mass and width parameters of the T' as measured at DORIS

	DASP2 /204/	D-H-HD-M /202/
$M_{T'}$ [GeV]	10.012 ± 0.01	10.02 ± 0.02
$M_{T'} - M_T$ [MeV]	555 ± 11	560 ± 10
Γ_{ee}	0.35 ± 0.14	0.32 ± 0.13
$\Gamma_{ee}(T)/\Gamma_{ee}(T')$	≈ 3	3.3 ± 0.9

Charge of the New Quark

In the nonrelativistic quark model the leptonic decay width for the vector states is directly proportional to the quark charge, Q,

$$\Gamma_{vee} = \frac{16\pi\alpha^2 Q^2}{M^2} \; |^3S_1(0)|^2 . \tag{13.1}$$

It is an empirical fact that the leptonic decay width divided by the average quark charge is approximately the same for all vector ground states, viz., ρ, ω, ϕ, and

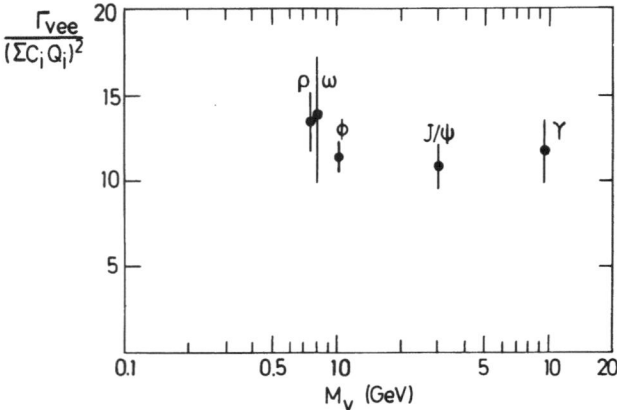

Fig. 13.5. The leptonic decay widths of vector mesons divided by the square of the average quark charge plotted as a function of the vector meson mass

J/ψ: $\Gamma_{vee}/\Sigma|c_i Q_i|^2 \approx 11$. This is shown by Figure 13.5. This rule applied to the Υ yields $Q = 0.34 \pm 0.04$ or $Q = 1/3$.

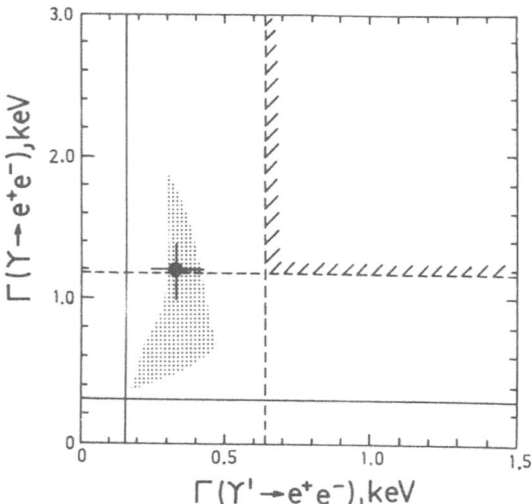

Fig. 13.6. Predictions of the quarkonium model for $\Gamma(\Upsilon \to ee)$ and $\Gamma(\Upsilon' \to ee)$ as computed by QUIGG and ROSNER /207/. The solid lines show the lower limits for charge $Q = 1/3$ of the b quark, the dashed lines for charge $Q = 2/3$. The hatched area shows the most likely region for $Q = 1/3$. The data point indicates the average value of the DORIS experiments

The leptonic width depends on the wave function which is a function of the quark potential. In QCD the quark potential is flavor independent. Therefore the same potential which describes the J/ψ should also be applicable to the Υ provided that effects due to the difference in mass can be neglected. Their influence on the extraction of the quark charge can be reduced by comparing the leptonic widths of Υ and Υ' in relation to those of J/ψ and ψ'.

The theoretical prediction /207/ for $\Gamma(\Upsilon \rightarrow ee)$ and $\Gamma(\Upsilon' \rightarrow ee)$ is shown in Figure 13.6. The solid lines indicate the lower limits for $Q = 1/3$, the dashed lines for $Q = 2/3$. The shaded region indicates the most probable values for the leptonic widths if $Q = 1/3$. They were determined from twenty different potentials which reproduce Γ_{ee} for J/ψ and ψ'. The cross shows the position of the average value measured by the DORIS experiments. The data disagree with the $Q = 2/3$ prediction but are consistent with $Q = 1/3$. The experiments do not fix the sign of the charge. The comparison with the d and s quarks suggests $Q = -1/3$, however. It has become customary to name the new quark b for bottom (or beauty).

The Υ - Υ' Mass Difference

Despite the large difference in mass between J/ψ and Υ the Υ - Υ' mass difference is almost the same as between J/ψ and ψ'. In a nonrelativistic quark model the mass difference depends on the type of the potential. If Υ and Υ' are assigned to the 1^3S_1 and 2^3S_1 states of the $b\bar{b}$ system, respectively, the model predicts the mass difference shown in Table 13.4.

Table 13.4. The Υ - Υ' mass difference in the nonrelativistic quark model. m_Q = mass of quark.

a) Experiment: $m(\psi' - J/\psi) = 591 \pm 1$ MeV, $m(\Upsilon' - \Upsilon) = 558 \pm 10$ MeV.

b) Theory:

Type of potential	$\Delta m(2^3S_1 - 1^3S_1)$	$\Delta m(\Upsilon' - \Upsilon)^a$
Coulomb	$\sim m_Q$	1800 MeV
Linear	$\sim m_Q^{-1/3}$	400 MeV
Logarithmic	const	591 MeV

[a] computed from $\Delta m(\psi' - J/\psi)$.

A pure logarithmic potential which predicts $\Delta m(T' - T)$ and $\Delta m(\psi' - J/\psi)$ to be the same is at variance with experiment while a superposition of a coulomb and a linear term could give the correct description.

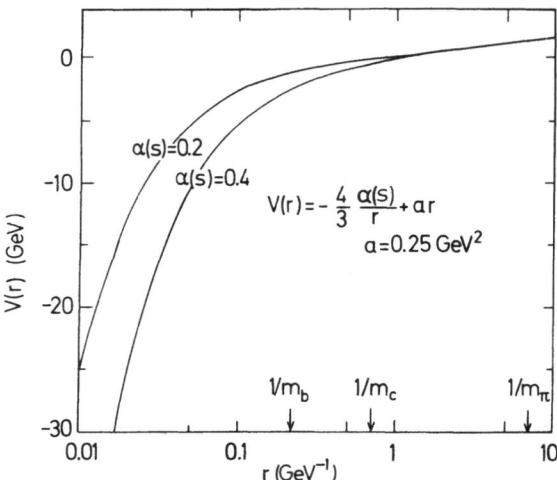

Fig. 13.7. The radius dependence of the quarkonium potential

Figure 13.7 shows the potential

$$V = -\frac{4}{3}\frac{\alpha_s}{r} + ar$$

with α_s = 0.2, 0.4 and a = 0.25 GeV as a function of the radius. Indicated are also the Compton wavelengths of the pion, the charm quark, and the b quark. It seems reasonable to expect the $b\bar{b}$ spectroscopy to be slightly more Coulomb-like than the $c\bar{c}$ spectroscopy.

T Final States

Apart from the μ pair decay no other exclusive decay channel of the T has yet been identified. The analysis of the overall features of the hadronic final states, however, produced very interesting results. Maybe the most remarkable is the difference in jet formation on and off the T. These results will be discussed in Sect. 15.

The average number of charged particles resulting from direct T decays is larger by roughly one unit than the charge multiplicity observed at nearby energies. This feature is observed by all three storage ring experiments (see Table 13.5).

Table 13.5. Average charged particle multiplicities at the Υ and off resonance. The values quoted are raw numbers and have not been corrected for acceptance.

	DASP2 /204/	D-H-HD-M /202/	PLUTO /206/
$\langle n_{ch} \rangle$ off resonance	5.2 ± 0.1	6.4 ± 0.2	4.9 ± 0.1
$\langle n_{ch} \rangle$ on resonance	5.7 ± 0.1	6.9 ± 0.2	5.4 ± 0.1
$\langle n_{ch} \rangle$ Υ direct decay	6.2 ± 0.1	7.3 ± 0.2	5.9 ± 0.1

The PLUTO group has determined the cross section for inclusive K_S^0 production. The ratio

$$\frac{\sigma_{on}(K_S^0)}{\sigma_{off}(K_S^0)} = 4.0 \pm 1.7$$

should be compared to the ratio of the total cross section

$$\frac{\sigma_{on}^{tot}}{\sigma_{off}^{tot}} \cong 2.5.$$

In both cases an average was taken over the resonance region. No significant change is seen in K^0 production on and off the resonance.

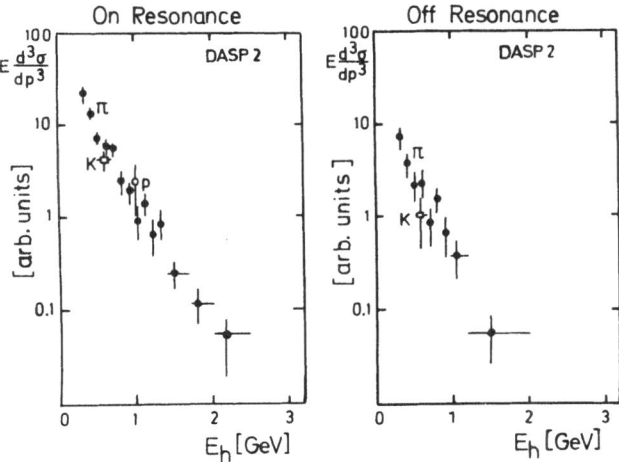

Fig. 13.8. The invariant cross section $E \, d^3\sigma/dp^3$ for the sum of $(\pi^+ + \pi^-)$, $(K^+ + K^-)$ and $2 \cdot \bar{p}$ production on and off the Υ resonance /204/

A similar conclusion was drawn by the DASP2 group from a measurement /204/ of inclusive charged hadron production. The data are shown in terms of $\frac{s}{\beta} \frac{d\sigma}{dx}$ in Figure 13.8. Within the large errors the same yield is found for $K^{\pm}$ and $\bar{p}$ relative to $\pi^{\pm}$ production.

14. Inclusive Hadron Production

14.1 Basic Formulae

One of the basic properties of electron hadron scattering is the almost perfect scale invariance exhibited by the structure functions in the deep inelastic region. In-clusive hadron production in e^+e^- annihilation is expected to possess similar proper-ties. First measurements on this subject were carried out by the SLAC-LBL collabora-tion who observed approximate scaling of the cross section obtained by summing all charged hadrons produced. In the DASP experiment the $\pi^\pm$, $K^\pm$, and $\bar{p}$ spectra were determined separately which allowed testing of scaling for each particle species.

We start with a brief description of the formalism /208/ for inclusive hadron production

$$e^+e^- \rightarrow h\,X \tag{14.1}$$

depicted by the following diagram:

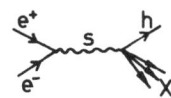 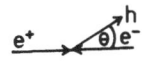

Define: $q = p_+ + p_-$ four momentum vector of the virtual photon

$\quad\quad q^2 = s$

$\quad\quad p = (\vec{p}, E)$ four momentum vector of the hadron h

$\quad\quad \Theta$ = production angle of h with respect to the e^+ direction

$\quad\quad x = \dfrac{2q\cdot p}{s} = \dfrac{2E}{\sqrt{s}}$ fractional energy of h

$\quad\quad \nu = \dfrac{q\cdot p}{m} = \dfrac{E}{m}\sqrt{s}$ energy of γ in h rest system

Note that $\dfrac{2m\nu}{s} = x$.

The virtual photon, as seen in the rest system of h, has transverse (T) and longitu-dinal (L) components. As a consequence the process (14.1) is described by two in-dependent structure functions, e.g., $\bar{W}_T\,(s,\nu)$ and $\bar{W}_L\,(s,\nu)$. The differential cross section has the form

$$\frac{d^2\sigma}{dxd\Omega} = \frac{\alpha^2}{s} \frac{|\vec{p}|}{\sqrt{s}} m \{\bar{W}_T (1 + \cos^2\theta) + \bar{W}_L (1 - \cos^2\theta)\}. \tag{14.2}$$

Special cases of (14.1) are pair production of fermions (e.g., $e^+e^- \to \mu^+\mu^-$) where $\bar{W}_L = 0$ and of scalar or pseudoscalar mesons (e.g., $e^+e^- \to \pi^+\pi^-$) where $\bar{W}_T = 0$. It is customary to use instead of $\bar{W}_L$, $\bar{W}_T$ the structure functions $\bar{W}_1$ and $\bar{W}_2$ which are defined in terms of the tensor $\bar{W}_{\mu\nu}$

$$\overline{W_{\mu\nu}} \equiv 4\pi^2 \frac{E}{m} \sum_n <0|J_\mu(0)|p,n> <n,p|J_\nu(0)|0> \quad . \quad (2\pi)^4 \delta^4(q - p - p_n)$$

$$= -(g_{\mu\nu} - \frac{q_\mu q_\nu}{s})\bar{W}_1(s,\nu) + \frac{1}{m^2} (P_\mu - \frac{p \cdot q}{s} q_\mu) (P_\nu - \frac{p \cdot q}{s} q_\nu) \cdot \bar{W}_2(s,\nu) \tag{14.3}$$

and

$$\bar{W}_1(s,\nu) = \bar{W}_T(s, \nu) \tag{14.4}$$

$$\bar{W}_2(s,\nu) = \frac{m^2}{|\vec{p}|^2} (\bar{W}_L(s,\nu) - \bar{W}_T(s,\nu)).$$

Note that in the pure transverse case ($\bar{W}_L = 0$) the equivalent of the Callan-Gross relation reads for $\beta = 1$

$$\nu\bar{W}_2 = -\frac{2m}{x} \bar{W}_1. \tag{14.5}$$

From (14.2) and (14.4) we find

$$\frac{d^2\sigma}{dxd\Omega} = \frac{\alpha^2}{s} \beta x \{m \bar{W}_1 + \frac{1}{4} \beta^2 x \bar{W}_2 \sin^2\theta)\} \tag{14.6}$$

where $\beta = |\vec{p}|/E$. Note that $m\bar{W}_1 \geq 0$ since the cross section has to be a positive quantity. After integrating over the angles and replacing $4\pi\alpha^2/3s$ by $\sigma_{\mu\mu}$ one has

$$\frac{d\sigma}{dx} = 3\sigma_{\mu\mu}\beta x \{m \bar{W}_1 + \frac{1}{6} \beta^2 x \nu\bar{W}_2\}. \tag{14.7}$$

For $E \gg m$ this simplifies to

$$\frac{d\sigma}{dx} = 3\sigma_{\mu\mu}x \{m \bar{W}_1 + \frac{1}{6} x \nu\bar{W}_2\}. \tag{14.8}$$

If the structure functions $\bar{W}_1$ and $\nu\bar{W}_2$ obey scaling they become functions of the ratio ν/s alone. Using $x = \frac{2m\nu}{s}$ as the scaling variable and substituting

$$-m\bar{W}_1(s,\nu) \equiv \bar{F}_1(x,s)$$

$$\nu\bar{W}_2(s,\nu) \equiv \bar{F}_2(x,s)$$

(14.9)

scale invariance is defined as

$$\lim_{\substack{\nu \to \infty \\ x = \text{const}}} -m\bar{W}_1(s,\nu) = \lim_{\substack{s \to \infty \\ x = \text{const}}} \bar{F}_1(x,s) \equiv \bar{F}_1(x)$$

(14.10)

and similarly

$$\lim \nu\bar{W}_2(s,\nu) = \bar{F}_2(x).$$

Scale invariance leads to the following expression for the inclusive cross section:

$$\frac{d\sigma}{dx} = 3\sigma_{\mu\mu}x \ \{-\bar{F}_1(x) + \frac{1}{6} x \ \bar{F}_2(x)\}.$$

(14.11)

If scale invariance is fulfilled the shape of the particle energy spectra, $d\sigma/dx$, is independent of s.

Inclusive production of hadrons in e^+e^- annihilation is closely related to electron hadron scattering. We consider electroproduction on protons,

$$ep \to e'X$$

(14.12)

Notation: $q = p_e - p_{e'}$ = virtual photon, $q^2 < 0$

$\quad\quad\quad$ p $\quad\quad\quad\quad\quad$ = incoming proton

$\quad\quad\quad$ Θ $\quad\quad\quad\quad\quad$ = scattering angle of e' with respect to e

$\quad\quad\quad$ $\nu \equiv \frac{p \cdot q}{m}$ $\quad\quad$ = photon energy in the rest system of the incoming proton (lab. system)

$\quad\quad\quad$ $\nu = E_e - E_{e'}$

$\quad\quad\quad$ $\omega = \frac{2p \cdot q}{-q^2} = \frac{2m\nu}{-q^2}$.

The electroproduction cross section reads

$$\frac{d^2\sigma}{dE_e'\ d\cos\theta} = \frac{8\pi\alpha^2}{(q^2)^2}\ E'^2\ \{W_2(q^2,\nu)\ \cos^2\frac{\theta}{2} + 2W_1(q^2,\nu)\ \sin^2\frac{\theta}{2}\}, \qquad (14.13)$$

where the structure functions W_1 and W_2 are defined through the tensor W_μ

$$W_{\mu\nu} = 4\pi^2\ \frac{E}{m}\ \Sigma\ <p|J_\mu(0)|n> <n|J_\nu(0)|p>\ (2\pi)^4\ \delta^4(q + p - p_n) \qquad (14.14)$$

$$= -(g_{\mu\nu} - \frac{q_\mu q_\nu}{q^2})W_1(q^2,\nu) + \frac{1}{M^2}(p_\mu - \frac{p\cdot q}{q^2}\ q_\mu)(p_\nu - \frac{p\cdot q}{q^2}\ q_\nu)W_2(q^2,\nu).$$

With the substitutions

$$mW_1(q^2,\nu) \equiv F_1(\omega,q^2)$$
$$\nu W_2(q^2,\nu) \equiv F_2(\omega,q^2). \qquad (14.15)$$

Scale invariance leads to

$$\lim_{\substack{\nu \to \infty \\ \omega\ fixed}} mW_1(q^2,\nu) = \lim_{\substack{-q^2 \to \infty \\ \omega\ fixed}} F_1(\omega,q^2) \equiv F_1(\omega) \qquad (14.16)$$

In field theory the tensors $W_{\mu\nu}$ and $\bar{W}_{\mu\nu}$ are related by crossing symmetry

$$\overline{W_{\mu\nu}}(q,p) = -\ W_{\mu\nu}(q,-p). \qquad (14.17)$$

[We consider here only the case where h is a fermion. If h would be a boson, $\bar{W}_{\mu\nu}(q,p)$ $(q,p) = W_{\mu\nu}(q,-p)$].

Consequently

$$m\bar{W}_1(q^2,\nu) = -\ mW_1(q^2,-\nu)$$
$$\nu\bar{W}_2(q^2,\nu) = (-\nu)W_2(q^2,-\nu). \qquad (14.18)$$

The kinematical regions for annihilation and the scattering processes are separated in the q^2, ν plane ($q^2 > 4\ m^2$ and $q^2 < 0$). The analytic continuation does not exist in general. In the scaling limit the two processes have the point $\omega = x = 1$ in common /209/

$$\bar{F}_1(x = 1) = -F_1(\omega = 1)$$
$$\bar{F}_2(x = 1) = F_2(\omega = 1).$$

(14.19)

In certain field theories GRIBOV and LIPATOV /210/ found for $e^+e^- \to pX$

$$\bar{F}_1(x) = -\frac{1}{x} F_1(\omega = \frac{1}{x})$$

and

(14.20)

$$\bar{F}_2(x) = -\frac{1}{x^3} F_2(\omega = \frac{1}{x}).$$

14.2 Hadron Multiplicity

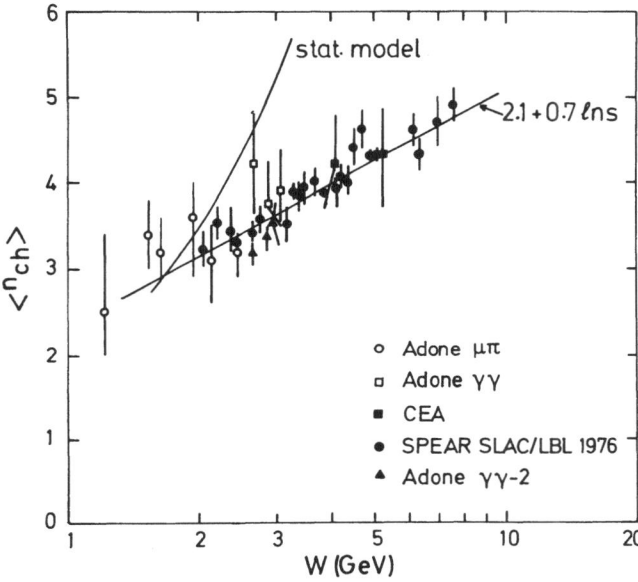

Fig. 14.1. The average multiplicity of charged hadrons produced in e^+e^- collisions /108/

Figure 14.1 shows the average number of charged hadrons, $\langle n_{ch} \rangle$, produced in e^+e^- annihilation /108/. The data are compatible with a logarithmic increase,

$$\langle n_{ch} \rangle = 2 + 0.7 \ln s.$$

(14.21)

206

In contrast to the data a phase spacelike production mechanism would lead to a linear increase of $<n_{ch}>$ with $\sqrt{s}$ /211/

$$<n_{ch}> \sim \frac{2}{3} \frac{\sqrt{s}}{3.375} \qquad \sqrt{s} \text{ in GeV.}$$

A logarithmically rising multiplicity can be obtained with an experimental cutoff of the momentum distribution /212/

$$\frac{d^3}{dp^3} \sim e^{-ap}. \qquad\qquad (14.22)$$

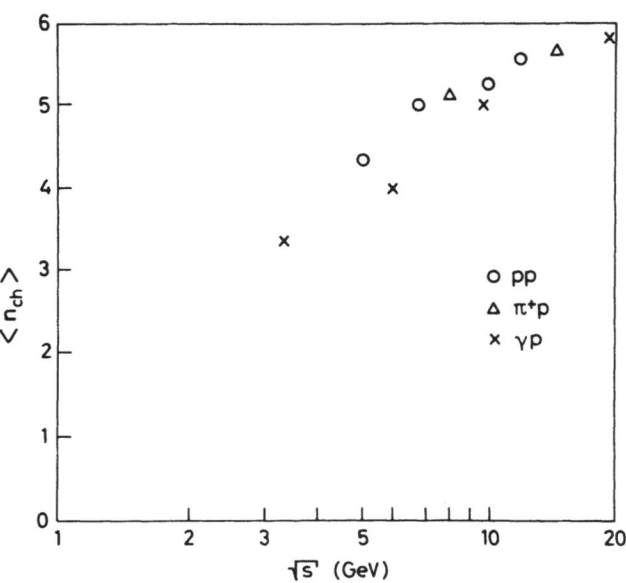

Fig. 14.2. The average multiplicity of charged hadrons produced in pp, π^+p, and γp collisions /213/

The observed behavior of $<n_{ch}>$ is surprisingly similar to the results for hadron-hadron, γN, eN, or νN collisions /213/ (see Figure 14.2). Obviously, the multiplicity depends only on the available energy and is not sensitive to the specific properties of the initial state.

14.3 Average Particle Energy

The SLAC-LBL collaboration has measured the average energy $<E_{track}>$ for charged hadrons /108/. Since the particle identification is not complete (π/K separation

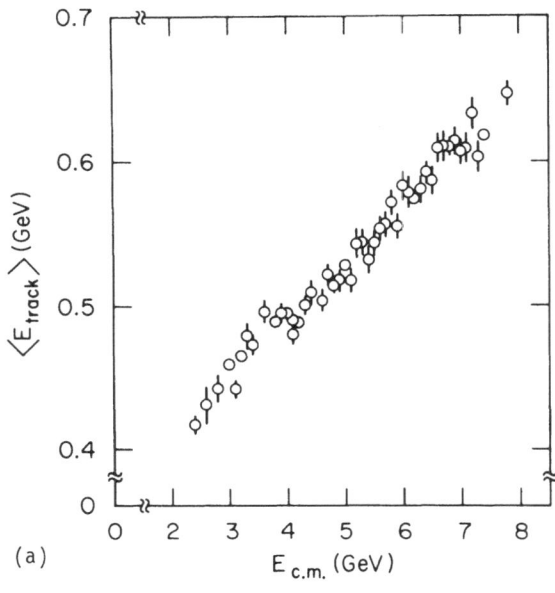

(a)

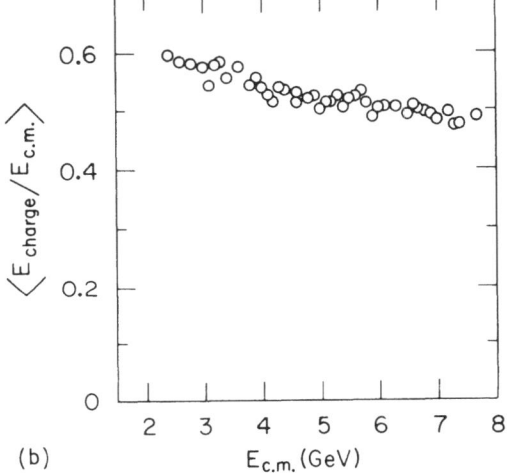

(b)

Fig. 14.3. (a) Mean energy per track (assuming pion mass) for e^+e^- events with three or more charged particles observed /108/, (b) average fraction of total c.m. energy appearing in charged particles. Pion masses are assumed /108/

for p ≲ 0.7 GeV/c, K/p for p < 1.2 GeV/c) the high momentum particles were assumed to be pions. The data show $\langle E_{track}\rangle$ to increase linearly with $\sqrt{s}$ (Figure 14.3a). At $\sqrt{s}$ = 3.7 GeV a step is observed. Beyond this step, $\langle E_{track}\rangle$ is ≈ 8 % smaller than a linear extrapolation would predict. This reduction is presumably caused by charmed particle (and perhaps heavy lepton pair) production where part of the energy is carried away by neutrinos.

The ratio of total charged particle energy to total energy ($\sqrt{s}$) is plotted /108/ in Figure 14.3b. If only π^+, π^- and π^0 are produced and if they are produced with equal strength, $\langle E_{ch}\rangle/\sqrt{s}$ = 2/3. Experimentally one observes that the fraction of the energy taken by charged particles decreases from 0.6 at 2 GeV to ~0.47 at 8 GeV. This may be caused by an increase of η, η', and K^0 production and by semileptonic decays of charmed particles produced above 4 GeV.

14.4 Momentum Spectra Without Particle Separation

We first discuss spectra of charged particles not separated by mass. In the SLAC-LBL experiment /214/ momentum spectra were determined for events with three or more charged particles registered in the detector. In the PLUTO experiment /215/ two charged particles were required per event while the DASP group employed a genuine inclusive trigger /216/. Figure 14.4 shows a plot of $s\frac{d\sigma}{dx}$ as measured by SLAC-LBL for $\sqrt{s}$ = 3.0, 4.8, and 7.4 GeV. The data were summed over all charged particle

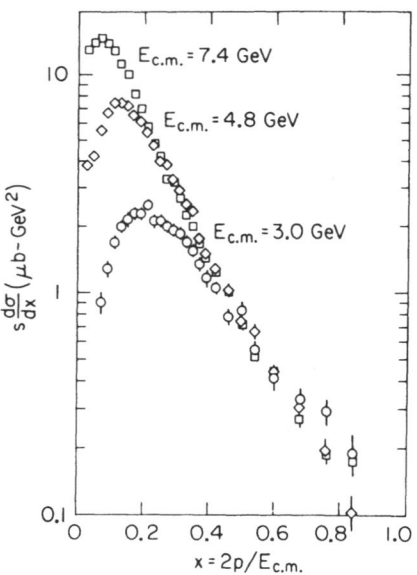

Fig. 14.4. $sd\sigma/dx$ for c.m.s. energies E_{cm} of 3.0, 4.8, and 7.4 GeV; $x \equiv 2p/E_{cm}$ /108/

species. Consequently, instead of the scaling variable $x_E = 2E_h/\sqrt{s}$ the variable $x = x_p = 2p_h/\sqrt{s}$ was used [note: $x_p \cong x_E - 2m_h^2/(x_E \cdot s)$]. The PLUTO data obtained

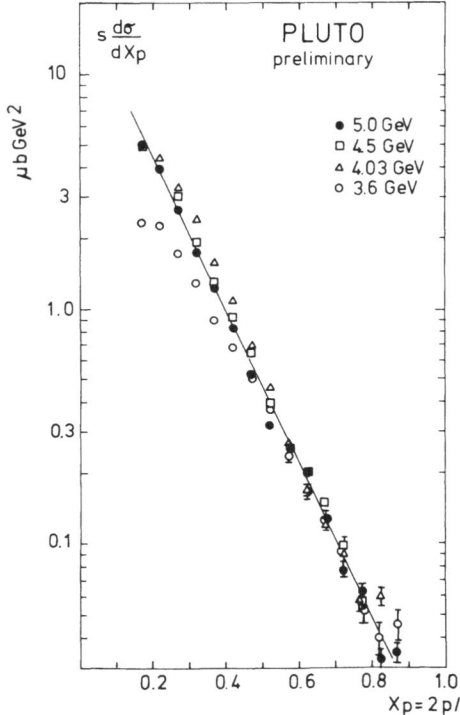

Fig. 14.5. Momentum spectra of charged particles /215/

at energies between 3.6 and 5.0 GeV are shown in Figure 14.5. For $x \lesssim 0.6$ the cross section points measured in the three experiments agree within their systematic uncertainties as shown in Figure 14.6. At larger values of x the SLAC-LBL data are above those measured by PLUTO. All three experiments, however, observe the same qualitative behavior: the x spectrum passes through a maximum at small x values which is followed by an exponential fall-off. The latter is reminiscent of hadron-hadron collisions.

As a function of s, $sd\sigma/dx$ shows the following features: a) For $x \geq 0.4$ (and $\sqrt{s} \geq 3$ GeV) the spectra are the same to within a factor of ~1.5, i.e., scale invariance is satisfied to within this accuracy. The region in x over which scaling is fulfilled is extended downwards to smaller values of x with increasing s: For $\sqrt{s} \geq 5$ GeV, $x_{scaling} \geq 0.2$. b) $\sigma^{tot}/\sigma_{\mu\mu}$ rises by a factor of 2 between 3.8 and 4.6 GeV. This additional contribution to σ^{tot} is seen to add only events at small $x(x \leq 0.4)$.

$$S\frac{d\sigma}{dx_p} \ (\mu b \ GeV^2)$$

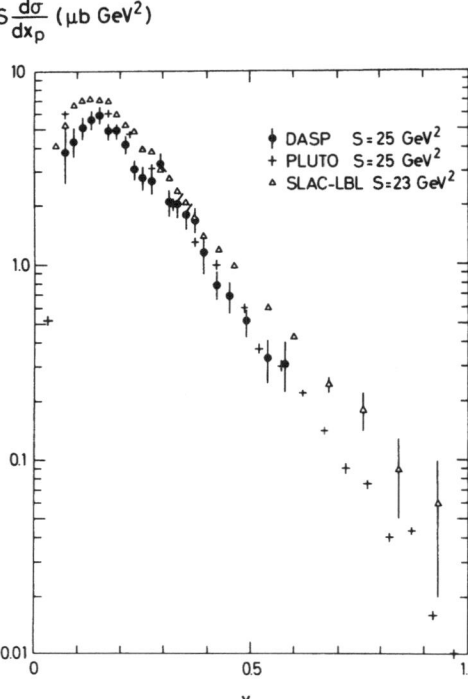

DASP   S = 25 GeV^2
PLUTO   S = 25 GeV^2
SLAC-LBL S = 23 GeV^2

x_p

Fig. 14.6. Momentum spectra of charged particles /216/

In Figure 14.7 scale invariance is tested by plotting $s\frac{d\sigma}{dx}$ as a function of s for fixed intervals of x. It leads to the same conclusions as above. The structure seen around 4 GeV corresponds to the bump observed in σ^{tot}.

14.5 Momentum Spectrum of $\pi^\pm$, $K^\pm$, and $\bar{p}$

In the DASP experiment /216/ a complete particle separation was possible for momenta up to 1.5 GeV/c and scale invariance was tested separately for each particle species. For c.m. energies between 3.6 and 5.2 GeV a total of 13 000 $\pi^\pm$, 890 $K^\pm$, and 130 $\bar{p}$ was used in the analysis. Since the majority of the protons was due to beam-gas interactions only antiprotons were considered. The proton yield was assumed to be the same as for antiprotons.

The differential cross section for inclusive production in general depends on the polar angle Θ [see, e.g., (14.6)]. In the DASP experiment the polar angular acceptance was $|\cos\Theta| < 0.55$. Within this range no statistically significant $\cos\Theta$ dependence was observed and a constant angular distribution was assumed in order to

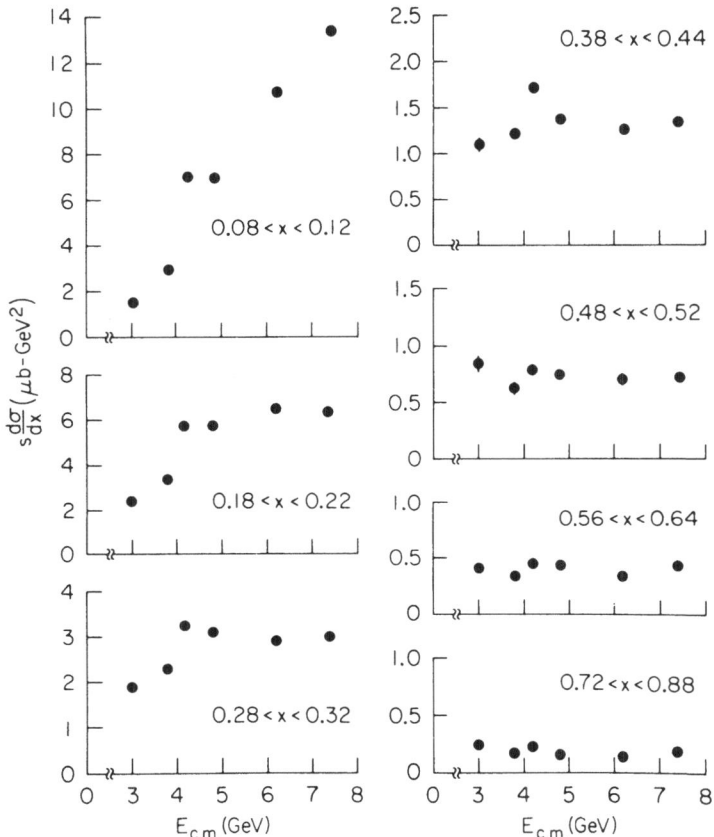

Fig. 14.7. sdσ/dx as a function of the c.m. energy $E_{c.m.}$ for various x intervals; x = $2p/E_{c.m.}$ /214/

integrate the cross section over cosθ. The possible error introduced by this procedure was estimated by considering the limit that only transverse photons contribute, W_L = 0. In this case the DASP cross sections given below would have to be increased by at most 24 %. With the angular dependence observed by SLAC-LBL /214/ the estimated increase is less than 5 % for x ≤ 0.5 and ≈13 % above.

The relative frequency of pions, kaons, and nucleons can be read off from Figure 14.8 which shows the cross section dσ/dp as a function of momentum averaged over c.m. energies from 4 to 5.2 GeV for the sum of $π^+$ and $π^-$, K^+ and K^-, and twice the antiproton yield. The dashed curves were obtained from an exponential extrapolation of the invariant cross sections $(E/4πp^2)$dσ/dp (see below). They indicate the expected momentum dependence at low momenta, where the particles are swept out of the spectrometer due to the magnetic field (pions), are lost by decay (kaons), or suffer too

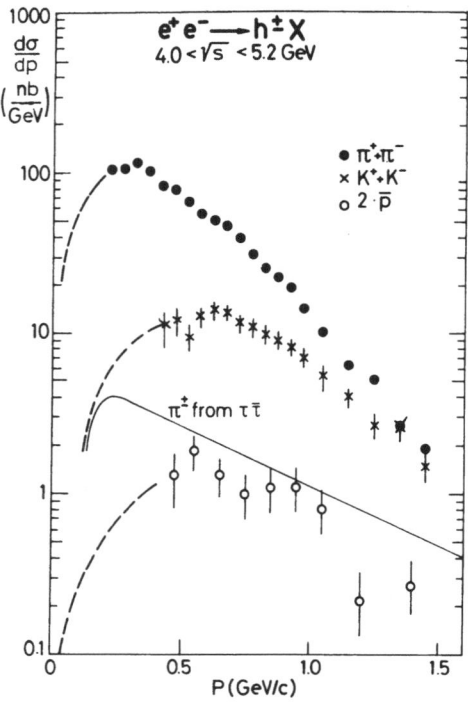

Fig. 14.8. Momentum spectrum of $\pi^{\pm}$, $K^{\pm}$, and $\bar{p}$ as measured by DASP /216/

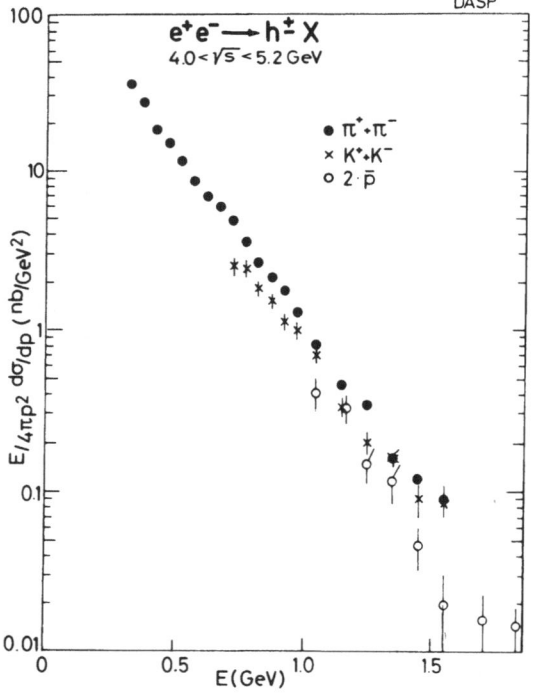

Fig. 14.9. The invariant cross section for $\pi^{\pm}$, $K^{\pm}$, and $\bar{p}$ as measured by DASP /216/

big an energy loss in the material in front of the magnet (antiprotons). At 0.5 GeV/c the $(\pi^+ + \pi^-)$: $(K^+ + K^-)$: $2\bar{p}$ yields are roughly in the ratio 100 : 10 : 1. With increasing momentum the differences between $\pi^\pm$ and $K^\pm$ yields become smaller: at 1.5 GeV/c they are of the same magnitude.

The inclusive cross section includes the contribution from τ pair production

$$e^+e^- \rightarrow \tau\bar{\tau} \rightarrow h^\pm X.$$

The pion yield from τ production was calculated by Monte Carlo using the decay branching ratio predicted by theory for a τ of 1.8 GeV mass. The pion yield is shown by the solid curve in Figure 14.8. It is suppressed below 0.6 GeV/c because of the selection criteria described above. The τ contribution accounts for ~4 % of all pions at p = 0.6 GeV/c and ~25 % at 1.5 GeV/c. The τ contribution to $K^\pm$ and $\bar{p}$ production are negligibly small. The data shown below have not been corrected for the contribution from τ production.

Figure 14.9 shows the same data plotted in terms of the invariant cross section $E d^3\sigma/dp^3 = (E/4\pi p^2)\, d\sigma/dp$. To within 20 or 30 % accuracy the $\pi^\pm$, $K^\pm$, and $2\bar{p}$ cross sections fall on the same curve, which is well approximated by an exponential,

$$(E/4\pi p^2)\, d\sigma/dp \sim \exp(-bE). \tag{14.23}$$

E_{cm} = 4 - 5.2 GeV b_π = 5.0 ± 0.1 GeV^{-1} b_K = 4.9 ± 0.2 GeV^{-1} $b_{\bar{p}}$ = 5.4 ± 0.5 GeV^{-1}.

Inclusive spectra in hadronic collisions behave in a similar manner when plotted as a function of the transverse energy $E_T = \sqrt{m^2 + p_T^2}$.

14.6 Particle Multiplicities

From the measured momentum distribution the DASP group /216/ calculated the average numbers of charged pions, kaons, and nucleons, e.g.,

$$\langle n_{\pi^\pm}\rangle = \frac{\displaystyle\int_0^{p_{max}} \frac{d\sigma}{dp}(e^+e^- \rightarrow \pi^\pm X)dp}{\sigma_{tot}}.$$

The extrapolation of $d\sigma/dp$ to zero momentum was done using the exponential fits described by (14.23). For σ_{tot} the data obtained in the same experiment /120/ were used. The fraction of the cross section determined by extrapolation amounted to

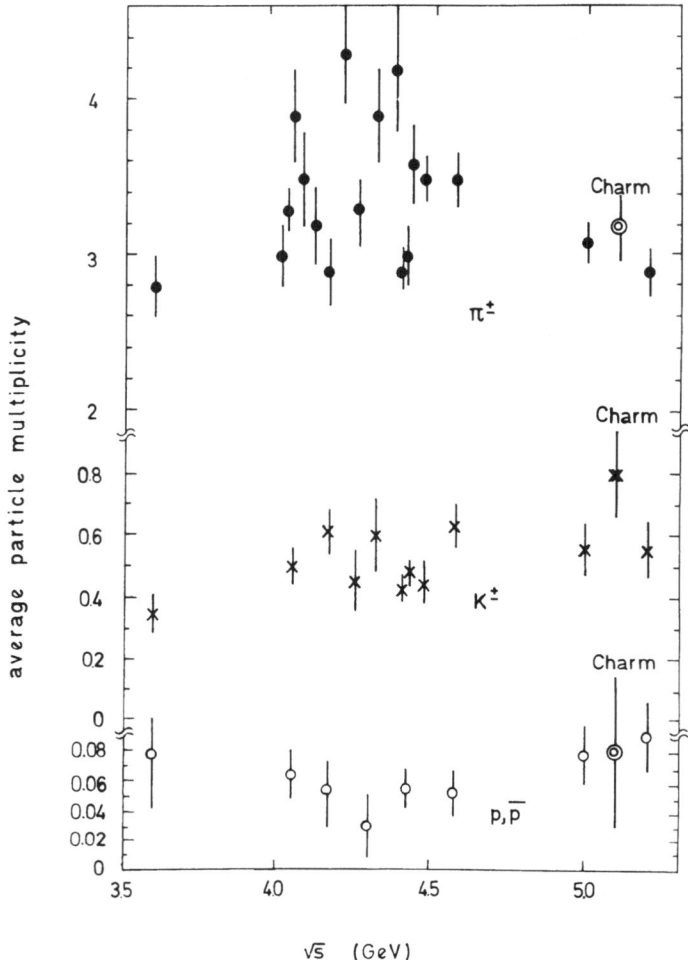

Fig. 14.10. Average $\pi^{\pm}$, $K^{\pm}$, and p,$\bar{p}$ multiplicities per event /216/

10 % for the majority of the pion data, to 20 % for kaons, and to 30 % for antipr
tons. The particle multiplicities are shown in Figure 14.10 as a function of the
c.m. energy. The errors given are purely statistical. The systematic errors of th
integrated inclusive cross sections and of σ_{tot} add a further uncertainty of 20 -
25 % of which ~15 % are due to normalization. Above charm threshold on the averag
3 - 4 $\pi^{\pm}$, 0.5 - 0.6 $K^{\pm}$, and 0.05 - 0.08 p and $\bar{p}$ are produced per event.

14.7   Test for Scaling

In general the structure functions $\bar{W}_1$ and $\nu\bar{W}_2$ depend on two variables, e.g., x an
s. If scale invariance is fulfilled $\bar{W}_1$ and $\bar{W}_2$ are functions of x alone and the cr

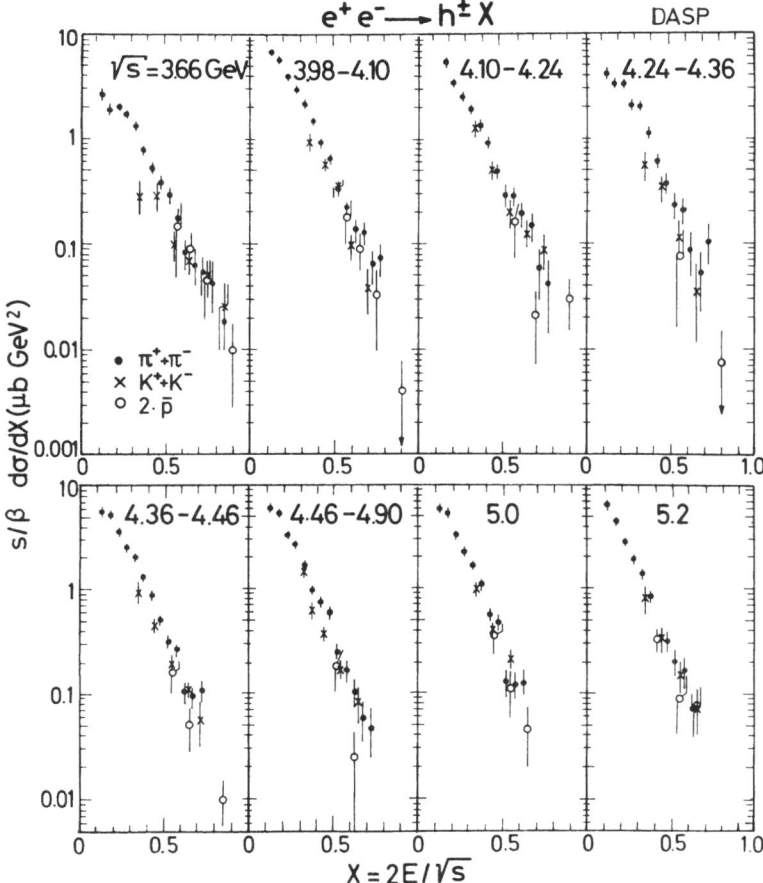

Fig. 14.11. The cross section (s/β) dσ/dx, x ≡ 2E/√s, versus x for the sum of π⁺ and
π⁻, K⁺ and K⁻ and twice the p̄ yield, as measured by DASP /216/

section (s/β)dσ/dx is almost the same for all values of s. [See (14.6)].

In Figure 14.11 the scaling cross sections (s/β)dσ/dx as measured by the DASP
group /216/ for π±, K±, and 2p̄ production are plotted as a function of x in eight
energy intervals between 3.6 and 5.2 GeV. Most spectacular is the similarity between the three types of particles. Within a factor of two their cross sections seem
to fall on a common curve decreasing exponentially for x ≳ 0.2. No peculiarities
are observed when comparing the shapes of the cross sections measured at the 4.04,
4.16, and 4.4 GeV resonances (second, third, and fifth intervals) and outside.

In order to test the pion data for scaling Figure 14.12 compares the π cross
sections outside the resonance region at energies of √s = 3.6 and 5.2 GeV. Below
x = 0.25 the cross section rises by a factor of 1.5 to 2 between √s = 3.6 and

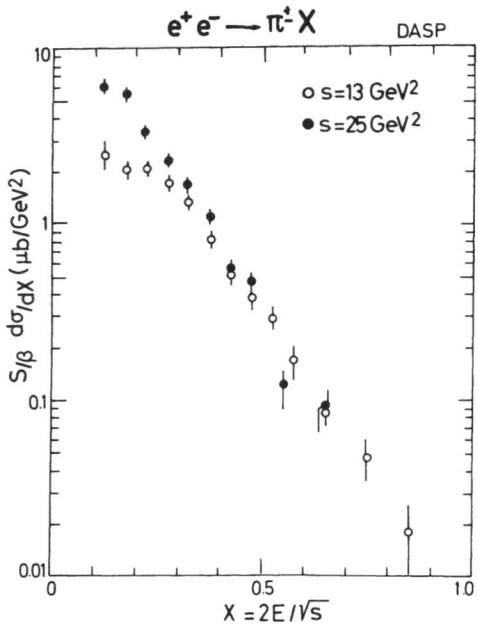

$$e^+e^- \longrightarrow \pi^\pm X \qquad \text{DASP}$$

Fig. 14.12. Comparison of the cross section $(s/\beta)d\sigma/dx$ for $\pi^\pm$ production at $s = 13$ and 25 GeV^2 /216/

5.2 GeV. At higher x values the two cross section sets agree within errors. This shows that the rise of R from a value of 2.3 at $\sqrt{s} = 3.6$ GeV to 4.5 above $\sqrt{s} = 5$ GeV is associated with <u>low</u> x pions.

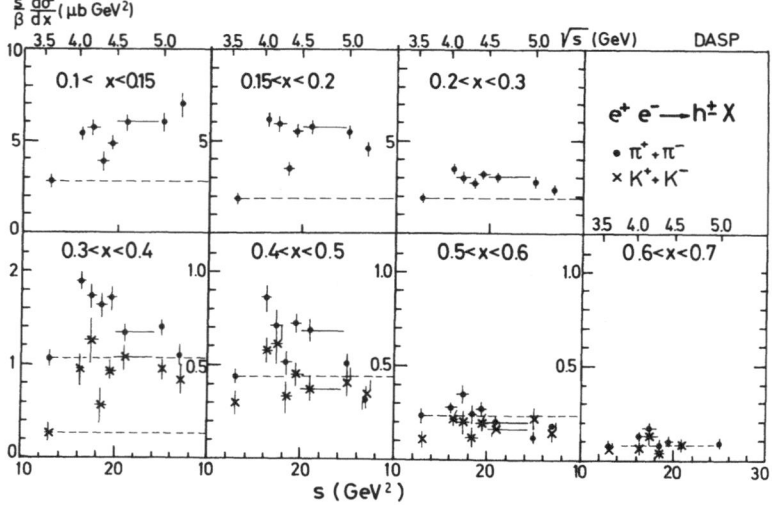

Fig. 14.13. The cross section (s/β) $d\sigma/dx$ versus s for fixed x for $\pi^\pm$ and $K^\pm$ production as measured by DASP /216/

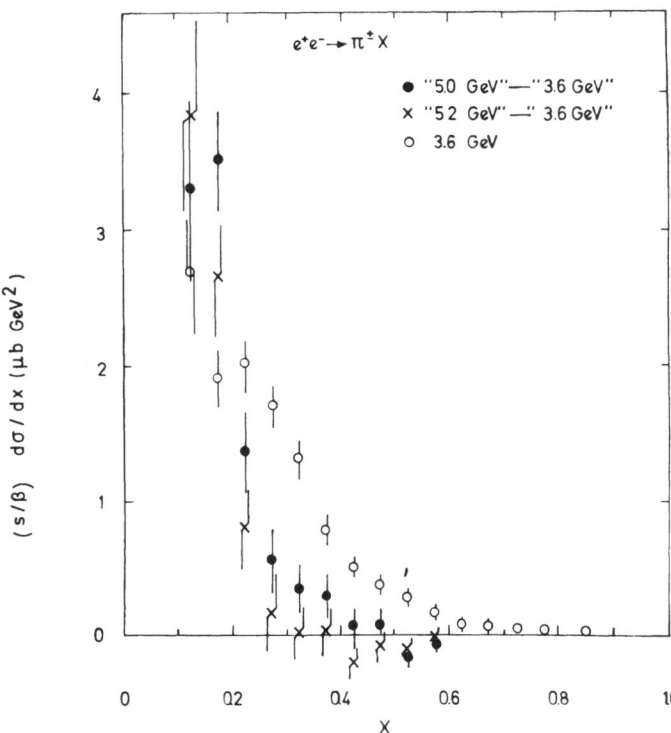

Fig. 14.14. The charm contribution to the scaling cross section of $\pi^{\pm}$ production determined as the difference between the 3.6 and 5.0 GeV data (♦) and as the difference between the 3.6 and 5.2 GeV data (✕). The open points show the $\pi^{\pm}$ scaling cross section measured below charm threshold at 3.6 GeV /216/

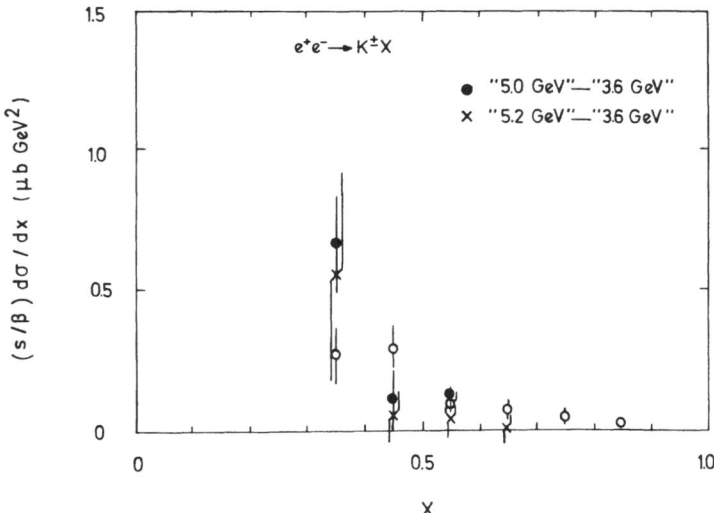

Fig. 14.15. The charm contribution to the scaling cross section of $K^{\pm}$ production (♦, ✕). The open points show the $K^{\pm}$ scaling cross section measured below charm threshold at 3.6 GeV /216/

Figure 14.13 shows the $\pi^\pm$ and $K^\pm$ data plotted as a function of s for fixed x. Both cross sections for x < 0.5 rise by a factor of two to three when crossing the charm threshold near $\sqrt{s}$ = 4 GeV. In the pion case for x > 0.2 this added contribution disappears above the resonance region. For x > 0.3 the scaling cross section at $\sqrt{s}$ = 5.2 GeV has reached its precharm level measured at $\sqrt{s}$ = 3.6 GeV. For kaons this occurs at slightly higher x values, x ≈ 0.4. We may therefore expect that at much higher energies ($\sqrt{s}$ >> 5 GeV) the charm contribution is confined to x values less than 0.3.

14.8 Charm Contribution to $\pi^\pm$ and $K^\pm$ Production

The DASP group /216/ determined the charm contribution to charged pion and kaon production as the difference in the cross sections for c.m. energies above charm threshold outside the resonances (5.0 and 5.2 GeV) and below charm threshold (3.6 GeV), viz.,

$$\frac{s}{\beta} \frac{d\sigma}{dx}^{charm} (\pi^\pm) = \frac{s}{\beta} \frac{d\sigma}{dx} (\pi^\pm, \sqrt{s} = 5 \text{ GeV}) - \frac{s}{\beta} \frac{d\sigma}{dx} (\pi^\pm, \sqrt{s} = 3.6 \text{ GeV}).$$

The charm contributions determined in this manner are plotted in Figures 14.14 and 14.15. Within errors the same result is found whether the 5.0 or 5.2 GeV data are used to define the postcharm-threshold data. For comparison the precharm scaling cross sections measured at 3.6 GeV are also shown. In the pion case the charm contribution is large for small x (x ≤ 0.2) and exceeds the 3.6 GeV values. It falls off rapidly towards higher x values. The descent is steeper than for the precharm data. For x values above 0.3 the charm contribution is close to zero. The data for kaon production behave in a similar manner although the conclusions are less firm because of the larger statistical errors. For x values between 0.3 and 0.4 the charm contribution is larger than the 3.6 GeV data. For x values above 0.4 the charm contribution is small.

14.9 e^+e^- Annihilation and Inelastic ep Scattering

Inclusive antiproton production by e^+e^- annihilation

$$e^+e^- \rightarrow \bar{p}X$$

and electron proton scattering,

$$ep \rightarrow eX$$

are related by crossing. If scaling holds the structure functions for one process can be calcualted from those of the other one for the elastic case, x = ω = 1, where ω = 2p·q/(-q^2) is the scaling variable for electron nucleon scattering [see (14.19)]. If the Gribov-Lipatov relations are assumed to hold the cross section for antiproton production can be calculated from the proton structure functions:

$$\frac{x}{\sigma_{\mu\mu}} \frac{d\sigma}{dx} (e^+e^- \rightarrow \bar{p}X) = 3\beta \{xF_1(\omega = \tfrac{1}{x}) - \tfrac{1}{6} \beta^2 F_2(\omega = \tfrac{1}{x})\}, \qquad (14.24)$$

where $\sigma_{\mu\mu}$ is the muon pair production cross section, $\sigma_{\mu\mu} = \frac{4\pi\alpha^2}{3s}$. Note that the first term in (14.24) is the dominant one.

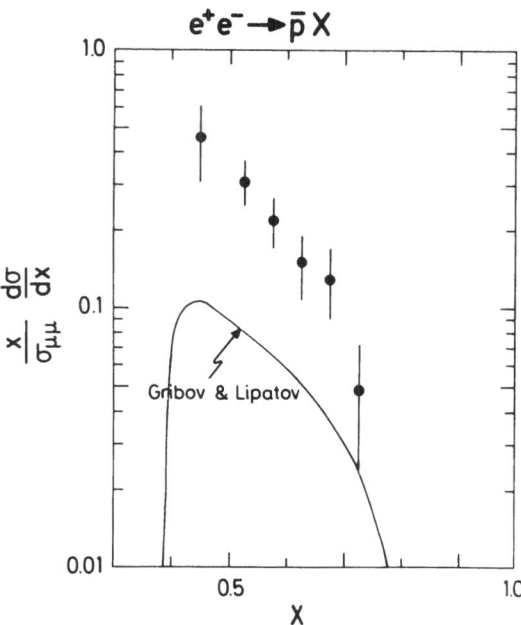

Fig. 14.16. The quantity $(x/\sigma_{\mu\mu}) d\sigma/dx$ versus x for $e^+e^- \rightarrow \bar{p}x$ averaged over c.m. energies from 4.0 to 4.5 GeV. The curve shows the prediction of Gribov and Lipatov /216/

Figure 14.16 shows $(x/\sigma_{\mu\mu})d\sigma/dx$ averaged over the c.m. energy region from 3.6 to 4.5 GeV which is below the charm threshold. The Gribov-Lipatov prediction (see curve in Figure 14.16) was computed from the values of the structure functions measured by /218/ imposing the same acceptance criteria as for the data. The Gribov-Lipatov prediction fails to describe the data quantitatively. The theoretical curve is always below the measured points. The discrepancy appears to increase towards smaller values of x reaching a factor of three at x = 0.5. Part of this failure — if not all — may have to be attributed to contributions of the type

$$e^+e^- \rightarrow h*X, \quad h* = \bar{\Lambda} \qquad \bar{\Sigma} \qquad \bar{N}* \qquad \text{etc.,}$$
$$\downarrow_{\rightarrow \bar{p}..,} \quad \downarrow_{\rightarrow \bar{p}..,} \quad \downarrow_{\rightarrow \bar{p}...}$$

which should be excluded from the e^+e^- data before a comparison is made /219/.

14.10 Inclusive Rho Production

The PLUTO group /220/ analyzed inclusive ρ^0 production. Identifying all charged particles as pions the PLUTO group observed a clear ρ^0 signal in the $\pi^+\pi^-$ effective
mass distribution (Figure 14.17).

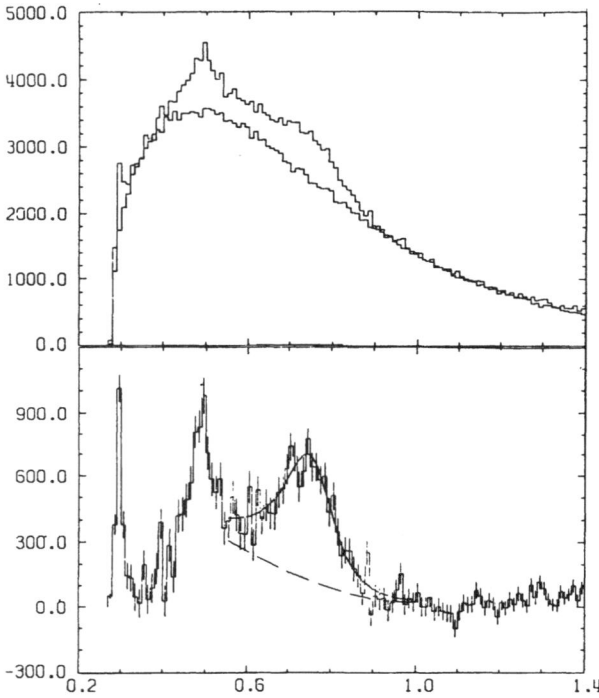

<u>Fig. 14.17.</u> $e^+e^- \rightarrow$ hadrons for c.m. energies between 4 and 5 GeV. The unsubtracted
$\pi^+\pi^-$ mass distribution (top) and after subtracting a smooth background (bottom)
/220/

Figure 14.18 shows the scaling cross section $(s/\beta)\, d\sigma/dx$ $(x \equiv 2E_\rho/\sqrt{s})$ versus x
averaged over c.m. energies between 4 and 5 GeV. The cross section lies above the
corresponding values for π^+ <u>or</u> π^- production as measured by DASP (dashed line).
Assuming that charged rho production is of similar magnitude the PLUTO group concludes that the majority of the pions oberserved in e^+e^- annihilation result from
rho production and decay.

221

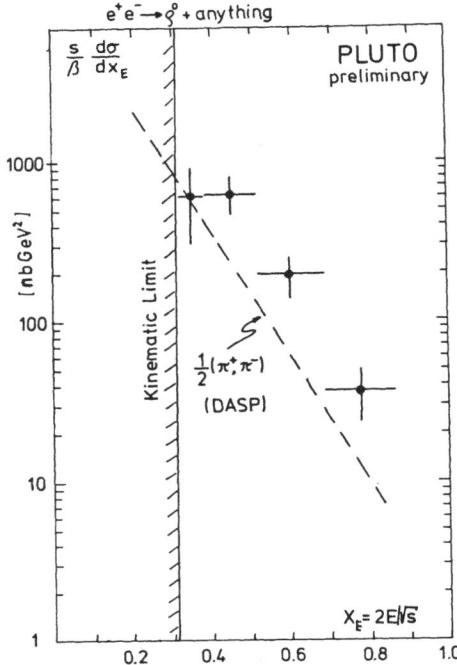

Fig. 14.18. The scaling cross section for inclusive ρ^0 production between 4 and 5 GeV /220/

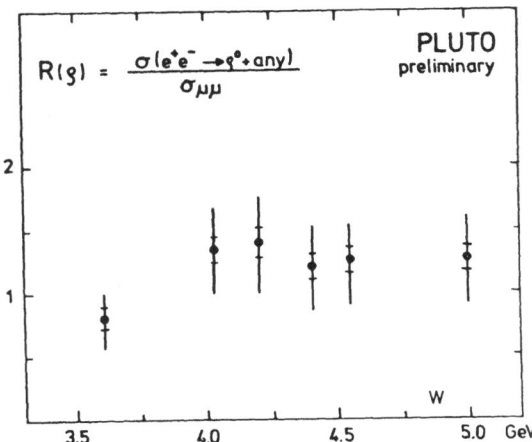

Fig. 14.19. The total cross section for inclusive ρ^0 production relative to $\sigma_{\mu\mu}$ /220/

In Figure 14.19 the total ρ^0 cross section is shown relative to $\sigma_{\mu\mu}$; R_ρ is seen to rise from a value around 0.8 below charm threshold to ≈ 1.3 above.

14.11 Inclusive D Production

The SLAC-LBL /221/ experiment measured the cross section for inclusive D production,

$e^+e^- \rightarrow Dx.$

The D mesons were identified by the $K^{\pm} \pi^{\mp}$ and $K^{\pm} \pi^{\mp} \pi^{\mp}$ decay modes.

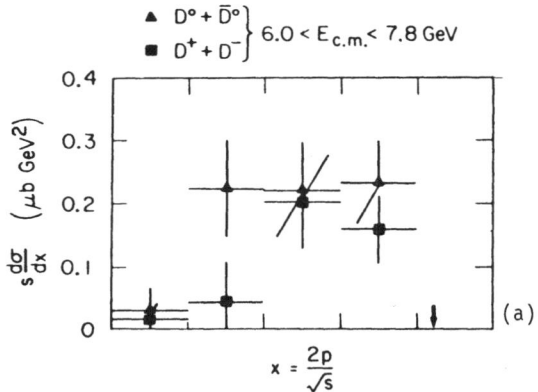

(a)

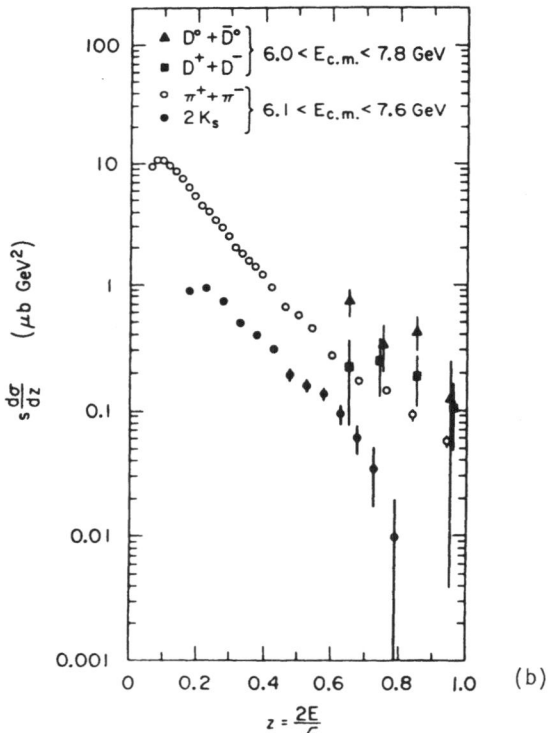

(b)

Fig. 14.20. Inclusive production spectra for D, charged pion, and K_S^0 mesons /221/

Figure 14.20a shows the cross sections $sd\sigma/dx_p$ versus the variable $x_p = 2p/\sqrt{s}$ for D^0, $\bar{D}^0$, and $D^\pm$ production at an average c.m. energy of 7 GeV. Within errors the two charge states have equal cross sections. In Figure 14.20b the cross sections $s\,d\sigma/dz$ are plotted versus the scaling variable $z = 2E/\sqrt{s}$ and compared to the measurements for $\pi^\pm$ and K_S^0 production. Note that $s\,d\sigma/dz$ is not exactly equal to the scaling cross section which is defined as $(s/\beta)d\sigma/dz$. The additional factor $1/\beta$ would increase the D cross section values at low z by roughly 30 %.

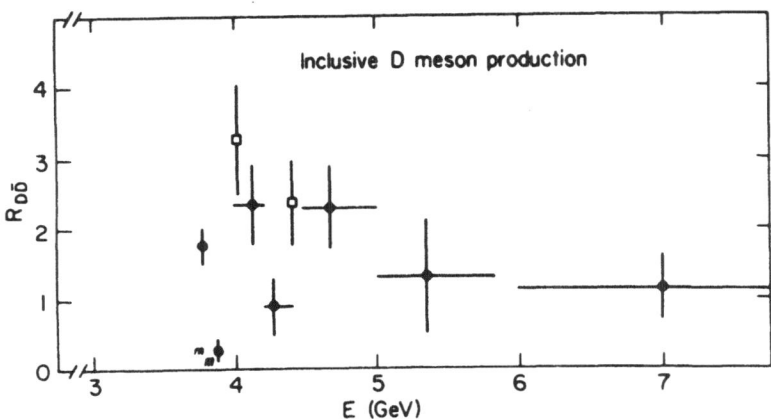

<u>Fig. 14.21.</u> The cross section for $D\bar{D}$ production, $e^+e^- \rightarrow D\bar{D}$ + anything in units of the μ pair cross section, $\sigma_{\mu\mu}$ /222/

Figure 14.21 presents the cross section for $D\bar{D}$ + anything as a function of the c.m. energy /222/; $R_{D\bar{D}}$ has a maximum near 4 - 4.5 GeV with a value around 2 - 3 followed by a decrease to a level of unity above 5 GeV.

15. Jet Formation

15.1 Angular Distributions

According to (14.2) the angular distribution of particles produced is of the form

$$\frac{d\sigma}{d\Omega} \sim \sigma_T + \sigma_L + (\sigma_T - \sigma_L)\cos^2\theta, \tag{15.1}$$

where σ_T and σ_L refer to the contributions from transverse and longitudinal photons. Figure 15.1 shows the angular distribution for $e^+e^- \to h\,X$ measured by the SLAC-LBL group at 4.8 and 7.4 GeV separately for low x (0.1 < x < 0.2) and higher x (x > 0.3)

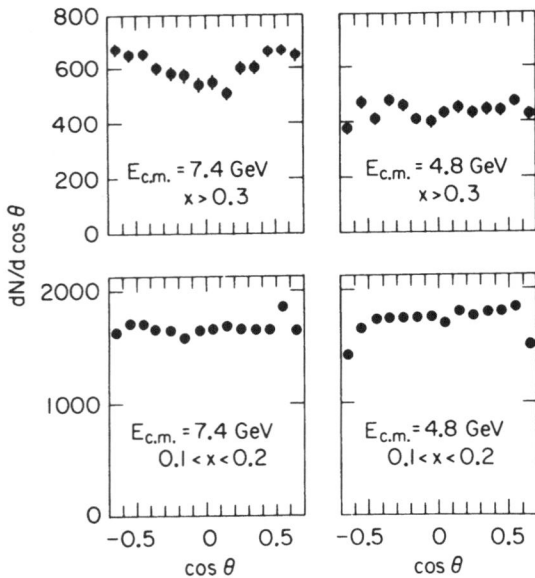

Fig. 15.1. $\cos\theta$ distribution for e^+e^- events with three or more charged hadrons at $E_{c.m.}$ = 4.8 and 7.4 GeV for different $x = x_p = 2p/\sqrt{s}$ intervals /223/

values /223/. For low x the cosΘ distributions are flat at both energies. At higher x values only the 7.4 GeV data show a cosΘ dependence. There, the data imply $\sigma_T > \sigma_L$. Since the acceptance is limited to $|\cos\Theta| < 0.6$ these data are not sensitive to small differences between σ_T and σ_L. The sensitivity is greatly improved by using transverse polarized e^+ and e^- beams. In this case

$$\frac{d\sigma}{d\Omega} \sim \sigma_T + \sigma_L + (\sigma_T - \sigma_L)\cos^2\Theta$$

$$+ P_+P_- (\sigma_T - \sigma_L) \sin^2\Theta \cos2\phi, \qquad (15.2)$$

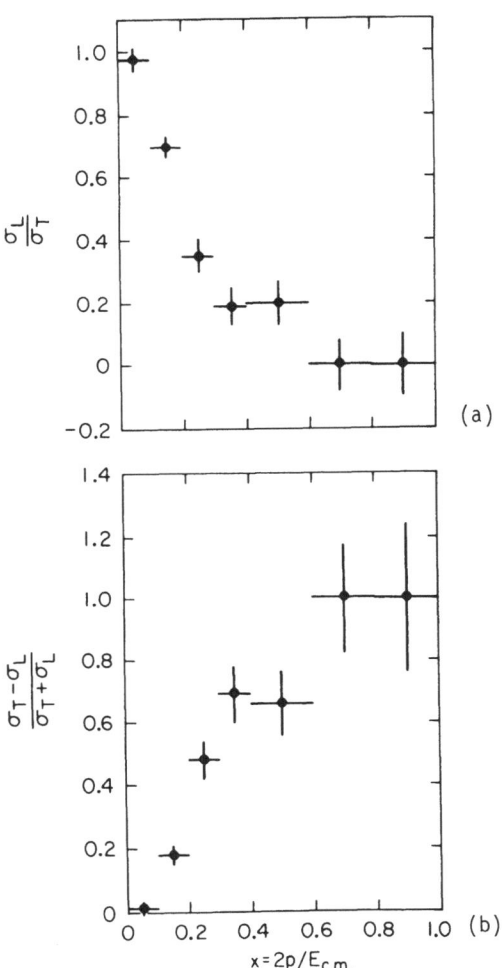

(a)

(b)

Fig. 15.2. (a) σ_L/σ_T versus $x = x_p = 2p/\sqrt{s}$ at $E_{c.m.} = 7.4$ GeV, (b) $(\sigma_T - \sigma_L)/(\sigma_T + \sigma_L)$ versus x_p at $E_{c.m.} = 7.4$ GeV /223/

where P_+, P_- measure the degree of polarization and $\phi = 0$ in the plane of the storage ring beams. For a measurement of this type a detector like that of SLAC-LBL with full azimuthal acceptance near $\cos\Theta = 0$ is optimal. The results of SLAC-LBL obtained at $\sqrt{s} = 7.4$ GeV are summarized in Figure 15.2 in terms of σ_L/σ_T and $(\sigma_T - \sigma_L)/(\sigma_T + \sigma_L)$ /223, 224/. As $x \to 0$ σ_L approaches σ_T. With increasing x the relative contribution of σ_L decreases and vanishes near $x = 1$ ($\sigma_L \ll \sigma_T$ at $x = 1$).

This behavior is predicted by the quark model. The primary process is the formation of a $q\bar{q}$ pair. Because the quark spin $s_q = 1/2$ the production angular distribution of the quark pair is $\sim 1 + \cos^2\Theta$, i.e., $\sigma_L = 0$ (quark spin $s_q = 0$ would lead to $\sigma_T = 0$).

In a second step physical particle states are formed. The exchange of quantum numbers occurs primarily along the original $q\bar{q}$ direction and the produced particles reflect the $\cos\Theta$ distribution of the primary $q\bar{q}$ pair. A particle with x close to 1 is emitted along the direction of the original quark momentum and therefore $\sigma_L \to 0$ as $x \to 1$. For x near zero the particle emission is practically independent of the original $q\bar{q}$ direction and therefore $\sigma_L \to \sigma_T$ as $x \to 0$. This leads qualitatively to the following picture:

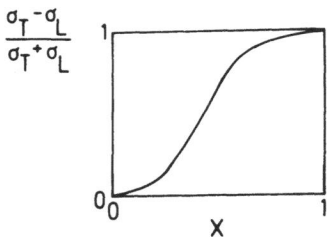

It is amusing to note that a similar behavior is obtained if the dominant processes are, e.g., of the type

$$e^+e^- \to \pi M,$$

where M is a meson or meson system of natural parity, $P_M = (-1)^{J_M}$ (e.g., $M = \rho, A_2$). KRAMER and WALSH showed that $\sigma_L = 0$ and $\frac{d\sigma}{d\Omega} \sim 1 + \cos^2\Theta$ for this case as well /225/.

15.2 Jet Structure

If the quark pair picture of hadron production is valid jet structure must be observed: the hadrons must be emitted in two narrow cones back to back. The axis of
the two jets is the direction of the primary $q\bar{q}$ pair. We expect, in analogy with
hadronic reactions, that the transverse momentum of a hadron with respect to the
jet axis is of the order of 300 MeV/c. Consequently, jet structure will produce
noticeable effects when the average particle energy is large compared to 0.3 GeV.

Jet structure was first observed by the SLAC-LBL group, studying e^+e^- collisions
up to 7.4 GeV c.m. energy. It was recently confirmed by experiments at DORIS which
extended the measurements up to 10 GeV. In the latter experiments the neutral component was analyzed as well. It was found that the neutral particles (mainly π^0's)
are also produced jetlike; furthermore, the neutral jet axis coincides with the
one determined from charged particles. We start with a discussion of the SLAC-LBL
data /223, 224, 226/.

The SLAC-LBL collaboration has tested their data for jet structure in the following way /227/.

1) For each event determine the axis which minimizes the square of the
   transverse momenta of the detected particles,

$$\sum_i p_\perp^2 = \text{min.} \tag{15.3}$$

2) Calculate for each event the sphericity,

$$S = \frac{3 \sum_i p_\perp^2}{2 \sum p^2}. \tag{15.4}$$

The sphericity is limited to values $0 \le S \le 1$. It is instructive to consider
two extreme cases:

Isotropic particle production; all particles have the same momentum:

$$S = 1$$

The end points of the momentum vectors lie on a sphere. If the particle multiplicity is large

$$S = \frac{3 \int p^2 \sin^2\theta \, d\cos\theta d\phi}{s \int p^2 d\cos\theta d\phi} = 1.$$

The other extreme is $p_\perp^2 \to 0$ for all particles. In this case $S \to 0$.

$$S \to 0$$

The sensivity of the sphericity to the formation of jets suffers because only the momenta of charged particles are measured. The axis determined from (15.3) will in general not be the axis which minimizes the transverse momenta of <u>all</u> particles. In order to interpret the measured sphericity distributions the authors have generated

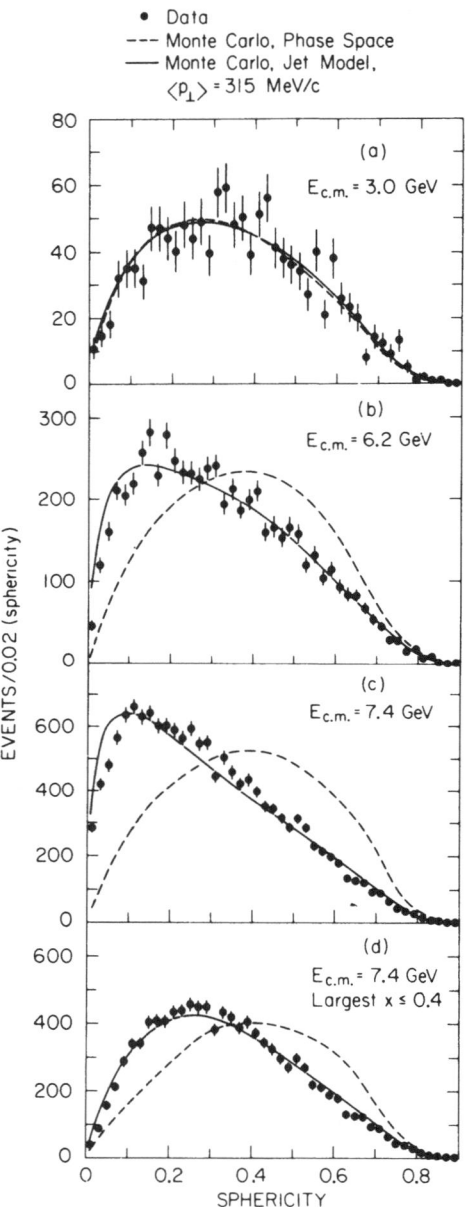

Fig. 15.3. Sphericity distributions for e⁺e⁻ events with three or more charged hadrons at various c.m. energies. The solid curves are the predictions of the jet model ($<p_\perp>$ = 315 MeV/c); the dashed curves show fits with invariant phase space. In (d), only those events where all particles have $x = x_p \leq 0.4$ were used /223/

Monte Carlo events according to two models:

1) phase space
2) jet formation with limited transverse momentum
$$(<p_\perp> = 0.3 \text{ GeV/c}).$$

The artificial events contained charged as well as neutral particles and were subjected to the same selection criteria as the real events. For the analysis only events with three or more charged particles were selected.

Figure 15.3 shows the S distributions at $\sqrt{s}$ = 3.0, 6.2, and 7.4 GeV and a comparison with these two models. At 3 GeV phase spacelike behavior and jet structure are indistinguishable. This was to be expected since $<E> \approx \sqrt{2} <p_\perp>$. For 6.2 and 7.4 GeV the measured sphericities are on the average much smaller than predicted by

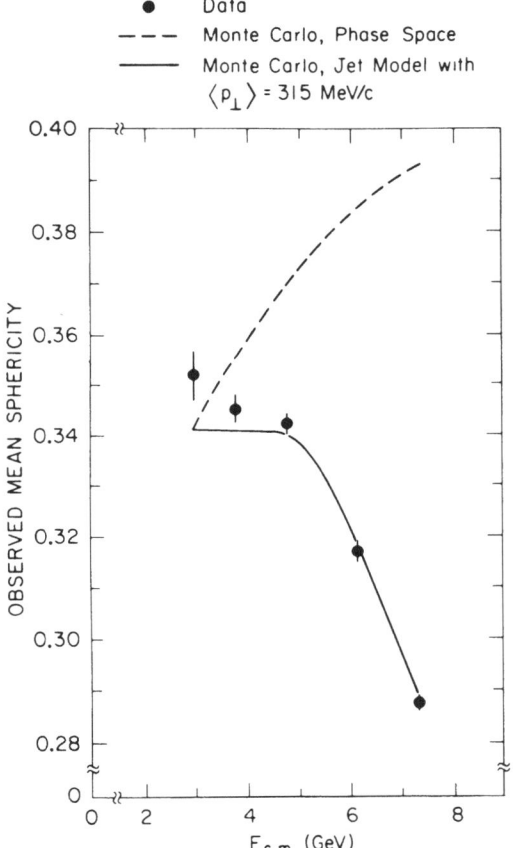

Fig. 15.4. Mean sphericity versus c.m. energy, $E_{c.m.}$. The solid curve is the result of the jet model with $<p_\perp>$ = 315 MeV/c. The dashed curve is the invariant phase space prediction /223/

230

phase space. This is also seen in Figure 15.4 where the mean value of S is given versus $\sqrt{s}$. Above $\sqrt{s}$ = 4 GeV the observed <S> values deviate markedly from a phase space behavior but agree with the jet picture.

In the quark model the jet axis is given by the direction of the primary $q\bar{q}$ pair. In this case, for transversely polarized beams the jet axis must show the same azimuthal anisotropy as the $q\bar{q}$ pair would, namely (see 15.2)

$$\frac{d\sigma}{d\Omega} \sim 1 + \alpha \cos^2\theta + \alpha P_+ P_- \sin^2\theta \cos2\phi \qquad\qquad (15.5)$$

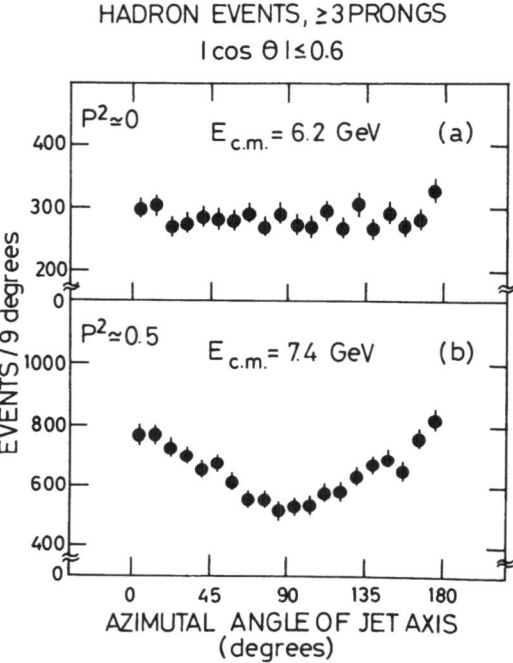

Fig. 15.5. Azimuthal distribution of the reconstructed jet axis; zero degree is in the ring plane: (a) for a c.m. energy of 6.2 GeV where the beam polarization is zero, (b) for a c.m. energy of 7.4 GeV where the product of the e$^+$ and e$^-$ beam polarizations is P^2 = 0.5 /223/

with α = 1. Figure 15.5 shows the ϕ distribution of the measured jet axis at $\sqrt{s}$ = 6 GeV where the beams were unpolarized, and at 7.4 GeV where $P_+ P_- \approx 0.5$ /223, 224/. The latter data exhibit a clear $\cos2\phi$ signal. A fit of the observed ϕ anisotropy as a function of x yielded α = 0.97 ± 0.1 again in agreement with the quark picture.

231

15.3 Particle Emission in the Jet Frame

The observed characteristics of jets as well as the energy dependence and the magnitude of the total hadronic cross section support strongly the view that hadron production in e^+e^- collisions is a two-step process where first a pair of quark-antiquarks is formed which then fragment into hadrons. The jet axis on the average measures the quark direction of flight.

If this view is adopted an analysis of hadron production with respect to the jet axis offers the possibility to determine the structure functions for quark fragmentation into hadrons. These may be compared to those deduced from hadron-hadron or lepton-hadron scattering. However, electron-positron annihilation has the advantage that effects due to the quark motion within the nucleon which, e.g., may smear the $p_\perp$ distributions are absent.

The SLAC-LBL group studied the hadron emission in the jet frame /226/. As mentioned before neutral particles were omitted in the analysis. This led to biases in particular for the $p_\perp$ distributions. The biases were corrected for by means of Monte Carlo simulations. The essential input to the Monte Carlo was that only pions are produced and that the frequency of π^0 mesons is given by $N_{\pi^0} = (N_{\pi^+} + N_{\pi^-})/2$. The Lorentz invariant phase space was modified by the following matrix element

$$|M|^2 = \exp\{-(\Sigma p_{i\perp}^2)/2b^2\}. \tag{15.6}$$

Note that $b^2 = \langle p_T^2 \rangle \cdot (N-1)/N$ where N is the number of particles produced. The b value was chosen such that an average $p_\perp$ of 340 MeV/c was obtained in accordance with the bulk of data.

The data were analyzed in terms of the Feynman scaling variable $x_\parallel = 2p_\parallel/\sqrt{s}$, the transverse momentum $p_\perp$ and the rapidity $y = (1/2)\ln\{(p_\parallel + E)/(p_\parallel - E)\}$ where $p_\parallel$ and $p_\perp$ denote the particle momentum components parallel and perpendicular to the jet axis.

In order to have a sample of events with a well-defined jet axis, events with at least one high-momentum particle were selected. High momentum was defined as $x_{max} > 0.3$. Having found a particle with $x > x_{max}$ the opposite jet (not containing this particle) was studied. This minimizes the biases introduced by the x_{max} cut. The distributions were normalized to the total number of jets studied. The effect of the x_{max} cut is illustrated by Figure 15.6 which shows the particle density distribution $(1/\sigma)\,d\sigma/dx_\parallel$ versus $x_\parallel$ around 7.4 GeV for different values of x_{max}. Except for $x_{max} > 0.7$ the "opposite jet" distributions are independent of the particular value of x_{max}.

Figure 15.7 shows the density distributions versus x for all events (i.e., $x_{max} > 0$) at various c.m. energies between 3 and 7.8 GeV. Here σ is the total cross section. For $x > 0.1$ approximate scaling is observed.

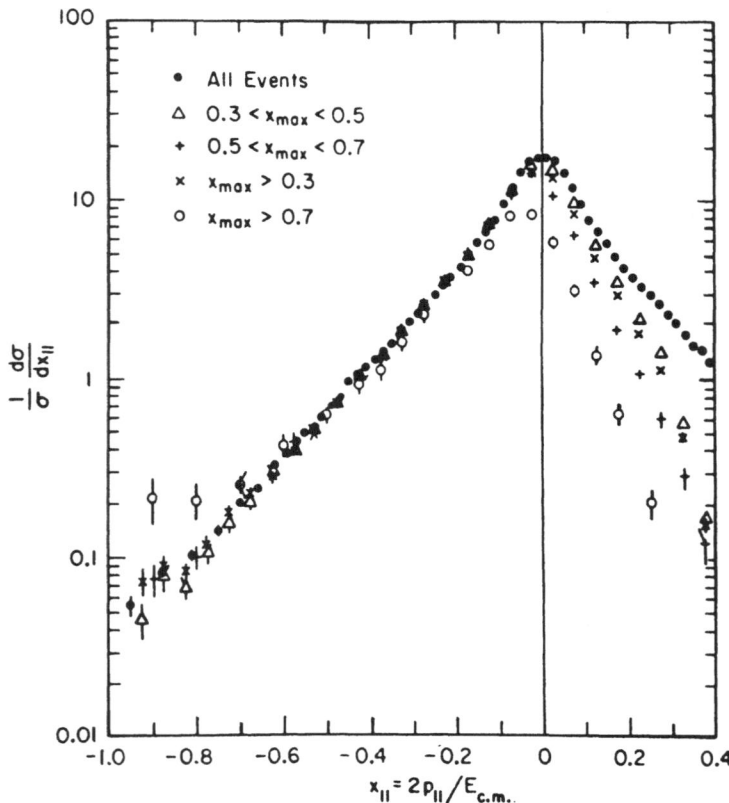

Fig. 15.6. Particle density distributions $(1/\sigma)\ d\sigma/dx_{\parallel}$ vs $x_{\parallel}$ for various x_{max} cuts for $7.0 < \sqrt{s} < 7.8$ GeV. x_{max} is the highest-x particle on one side of the event and is not plotted. The jet direction is oriented so that x_{max} is at positive $x_{\parallel}$ /226/

The rapidity distributions are presented in Figure 15.8. In this case $x_{max} > 0.3$ was required. The x_{max} cut distorts the y distributions for positive y. Two things are noteworthy: with increasing c.m. energy a plateau develops around y = 0. The rise of the distribution from y_{max} to the plateau occurs within two units in y. Both features are also common to hadron-hadron scattering.

The behavior with respect to the transverse momentum is shown in Figure 15.9 where $(1/\sigma)d\sigma/dp_{\perp}^2$ is plotted versus $p_{\perp}^2$. An x_{max} cut of 0.3 was applied. The area under these curves increases with increasing c.m. energy due to the rise in multiplicity. In Figure 15.10 the $p_{\perp}^2$ distribution near 7.4 GeV is compared to the Monte Carlo model. The model describes the data well up to $p_{\perp}^2 \approx 0.7$ (GeV/c)^2. At least part of the excess in events observed at higher transverse momenta is attributed by the authors to D^0 production and decay into $D^0 \rightarrow \pi^+ K^-$.

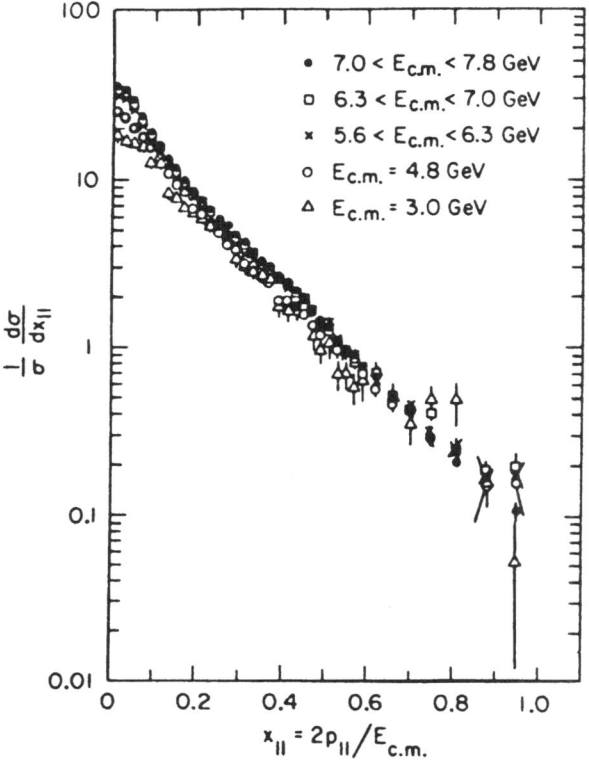

Fig. 15.7. The density distribution $(1/\sigma)\,d\sigma/dx$; σ is the total cross section /226/

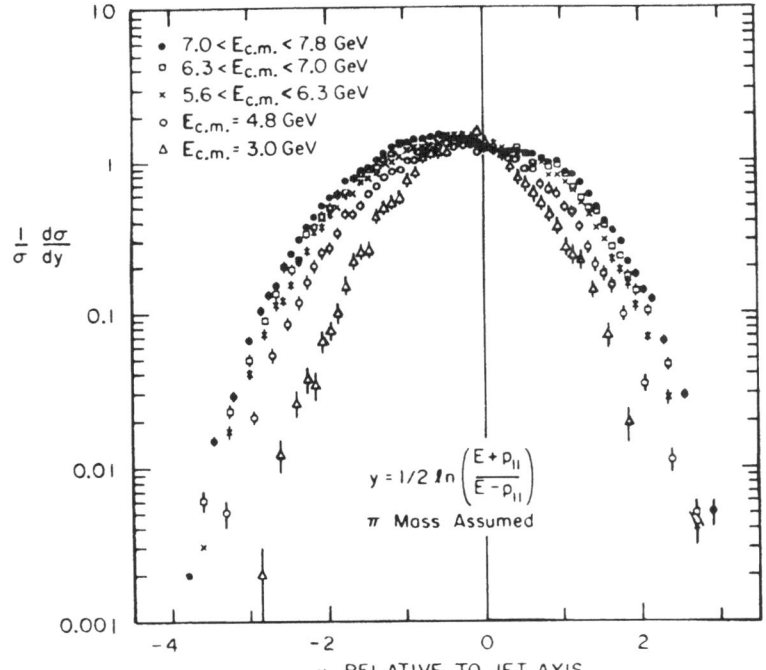

$$y = 1/2 \,\ell n\left(\frac{E + p_{||}}{E - p_{||}}\right)$$

π Mass Assumed

y RELATIVE TO JET AXIS

Fig. 15.8. The rapidity distribution /226/

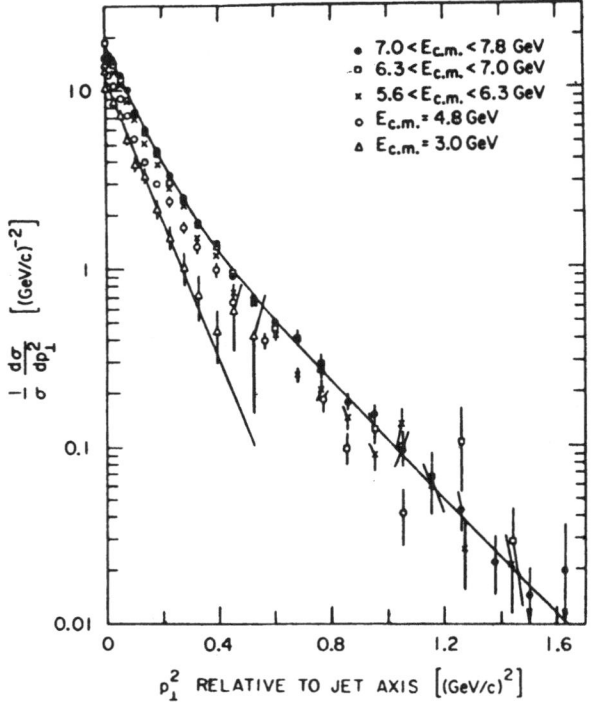

Fig. 15.9. The $p_\perp^2$ distribution /226/

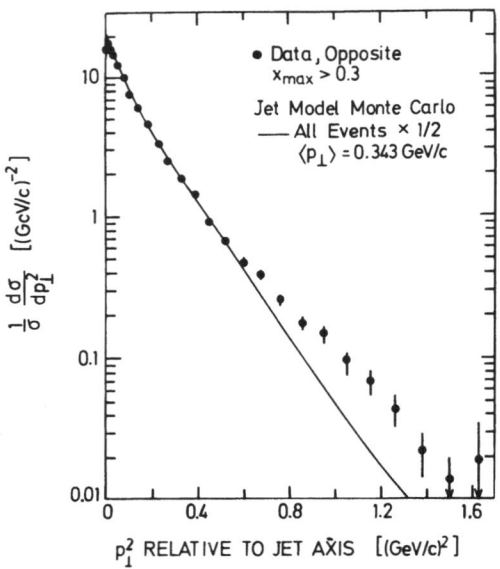

Fig. 15.10. The $p_\perp^2$ distribution at 7.4 GeV /226/

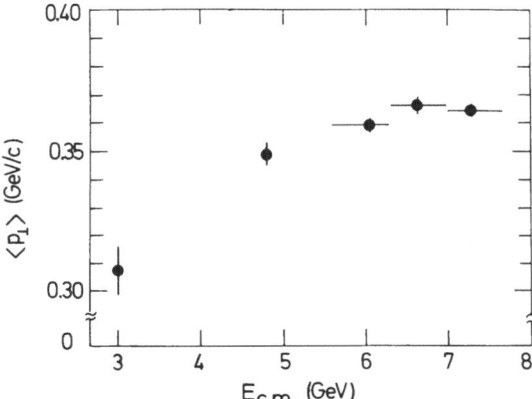

Fig. 15.11. Average p⊥ as a function of c.m. energy /226/

Figure 15.11 gives the c.m. energy dependence of the average transverse momentum. It is seen to rise from 0.31 to 0.35 GeV/c between 3 and 5 GeV and to become almost constant at higher energies. The rise can be understood as a phase space effect: at 3 GeV the average particle energy is only ~0.45 GeV.

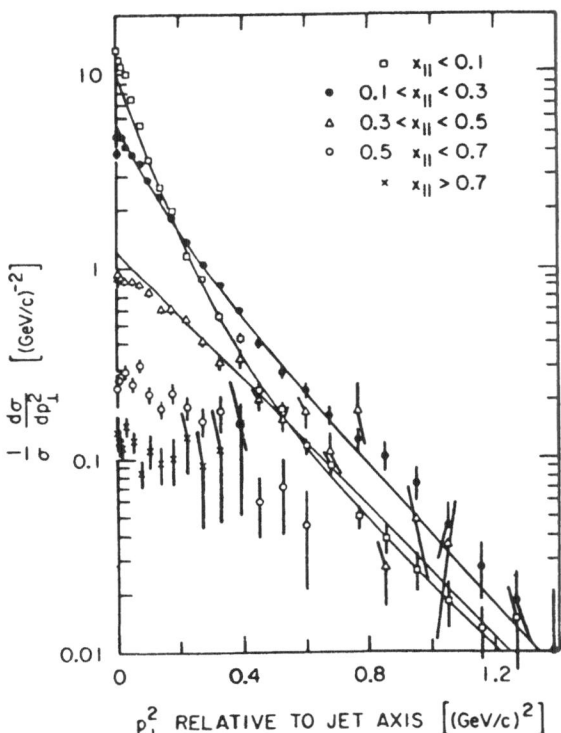

Fig. 15.12. The $p_\perp^2$ distribution for different $x_{\parallel}$ intervals /226/

236

An analysis of the $p_\perp^2$ distributions for different $x_{||}$ is shown in Figure 15.12 for $\sqrt{s}$ = 7.0 - 7.8 GeV. The transverse momentum distributions become wider with growing values of x .

The most simple correlation between two opposite jets that can be studied is the charge correlation. Consider, e.g., events originating from a u $\bar{u}$ quark pair. We expect at higher $x_{||}$ to find an excess of positive charges in the direction of the u quark and a corresponding excess of negative charges in the opposite direction.

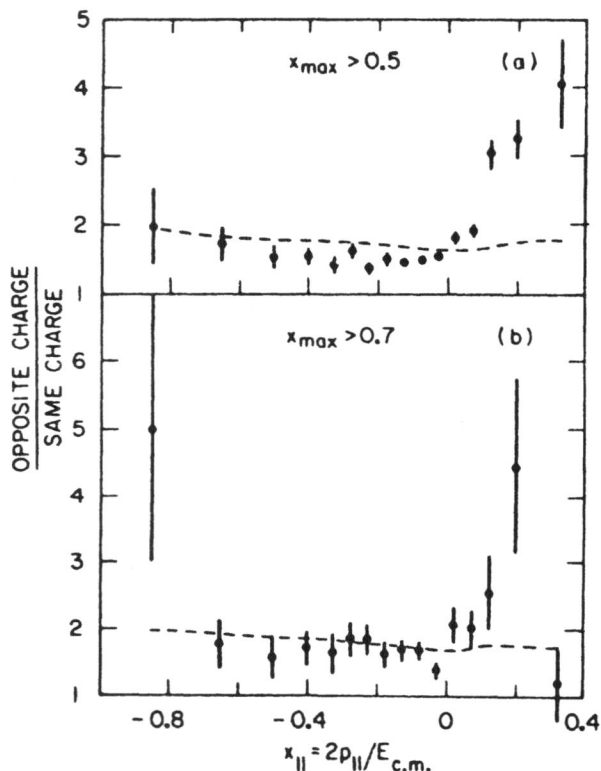

<u>Fig. 15.13.</u> Ratio of the number of particles with charge opposite to the charge of the x_{max} particle, to the number of particles with the same charge: (a) x_{max} > 0.5; (b) x_{max} > 0.7 /226/

The SLAC-LBL group looked for charge correlations in the following way. The x distributions were determined a) for all particles that had the same charge as the x_{max} particle; b) for those particles that had the charge opposite to the x_{max} particle.

Figure 15.13 shows the observed ratio (opposite charge)/(same charge) of these two distributions for two values of x_{max}, 0.5 and 0.7. The dashed line shows the expectation for no charge correlation. One observes a strong charge correlation on the same side and (within the statistical errors) no charge correlation between opposite jets. The same-side correlation most probably is due to low-mass pion resonance formation, such as the ρ^0 decaying into positive and negative particles (see the strong ρ^0 production observed by the PLUTO group, Sect. 14). The absence of the expected opposite-side charge correlation may have its cause in insufficient statistics and too low c.m. energies.

15.4 Jet Studies in the Υ Region

This section deals with the recent jet studies performed by three DORIS experiments for c.m. energies up to 10 GeV. Jet analyses off and on the Υ resonance are of great importance: they may provide a decisive test on QCD. In QCD the direct hadronic decays of the Υ proceed via a three-gluon intermediate state. As a consequence we expect the hadrons to emerge in three rather than two jets (as in nonresonant $q\bar{q}$ formation). The expected properties of the three-gluon jets have widely been discussed in recent papers /228-230/. We mention briefly some of the salient features /229/.

Consider a $Q\bar{Q}$ system of mass $M_{Q\bar{Q}}$ which decays into three gluons of energies E_i. Define the scaled energies

$$\xi_i = 2E_i/M_{Q\bar{Q}}. \tag{15.7}$$

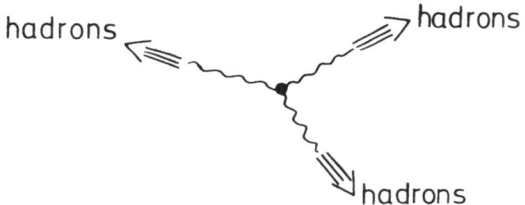

The energy distribution of the gluons is the same as for photons from orthopositronium decay:

$$\frac{1}{\sigma}\frac{d^2\sigma}{d\xi_1\,d\xi_2} = \frac{1}{\pi^2-9}\left\{\frac{1-\xi_3}{\xi_1\xi_2} + \frac{1-\xi_2}{\xi_1\xi_3} + \frac{1-\xi_1}{\xi_2\xi_3}\right\}. \tag{15.8}$$

238

By integrating over, e.g., ξ_2 one finds the energy distribution of one gluon (one jet),

$$\frac{1}{\sigma}\frac{d\sigma}{d\xi} = \frac{2}{\pi^2-9}\, F(\xi)$$

(15.9)

$$F(\xi) = \frac{\xi(1-\xi)}{(2-\xi)^2} + \frac{2-\xi}{\xi} + 2\left\{\frac{1-\xi}{\xi^2} - \frac{(1-\xi)^2}{(2-\xi)^3}\right\}\ln(1-\xi).$$

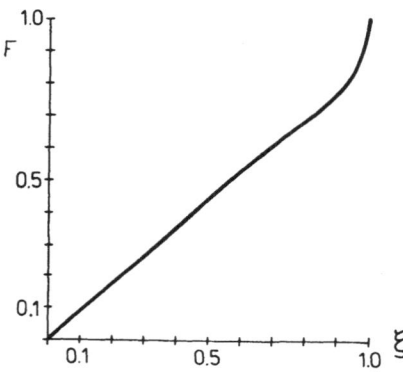

Fig. 15.14. Distribution of the gluon jet energy. (From /229/)

The function $F(\xi)$ is shown in Figure 15.14. The most probable configuration is one where two of the three gluons share basically all of the available energy. Of course, such a configuration will produce two jet events.

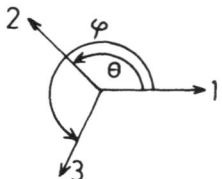

The geometrical structure of the three gluons can also be analyzed in terms of the angles Θ and ϕ between the gluons.

The scaled energies are related to Θ and ϕ in the following manner:

$$\xi_1 = \frac{2}{A} \sin(\Theta - \phi)$$
$$\xi_2 = \frac{2}{A} \sin\phi \tag{15.10}$$
$$\xi_3 = -\frac{2}{A} \sin\phi,$$

where $A = \sin\phi - \sin\Theta + \sin(\Theta - \phi)$.

We may ask for the probability of observing clean three-jet events. Theory predicts that in 38 % of the τ gluonic decays the three gluons are emitted within $\pm20°$ of the symmetric three-star directions.

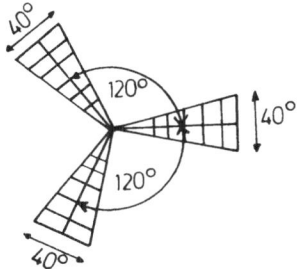

The orientation of the plane defined by the three gluons is a direct consequence of the vector nature of the gluons. If $\hat{n}$ denotes the normal to this plane and Θ_n the angle between $\hat{n}$ and the incoming e^+ (or e^-) then

$$\frac{1}{\sigma} \frac{d\sigma}{d\cos\Theta_n} = \frac{3}{16} (2 + \sin^2\Theta_n). \tag{15.11}$$

Another test for the vector nature of gluons is provided by the angular distribution between one of the gluon and the incoming e^+. In the limit $\xi_i \to 1$

$$\frac{1}{\sigma} \frac{d\sigma}{d\cos\Theta_i} \to \frac{3}{8} (1 + \cos^2\Theta_i). \tag{15.12}$$

According to present wisdom gluons do not become free but manifest themselves in hadron jets. Besides sphericity two other variables have been devised in the search for jets. These are

$$\underline{\text{thrust}} \,\, /231/ \quad T = 2 \frac{\tilde{\Sigma} \, p_{\parallel}^i}{\Sigma_i |p_i|}; \quad \frac{1}{2} \leq T \leq 1, \tag{15.13}$$

where the summation $\tilde{\Sigma}$ is to be extended over all particles in one hemisphere.

For practical applications the following definition of T was found
to be better suited /232/:

$$T = \frac{\sum_i |p_\parallel^i|}{\sum_i |p_i|} \; . \tag{15.13a}$$

For an event with no missing momentum the two definitions give the
same result. In both cases the jet axis is chosen such that T is
a maximum.

spherocity /233/

$$S_o = \left(\frac{4}{\pi}\right)^2 \left(\frac{\sum |p_\perp^i|}{\sum |p_i|}\right)^2 \; . \tag{15.14}$$

The jet axis is chosen such that S_o is a minimum. In contrast to sphericity, the momenta are summed linearly. The advantage of S_o over S is that S_o is infrared insensitive and is therefore better suited for QCD calculations.

The PLUTO group /232, 234/ for their jet analyses selected events with four or
more charged particles. The distributions presented below are the observed ones; no
corrections were applied for acceptance, cuts, or radiative effects. In comparing the

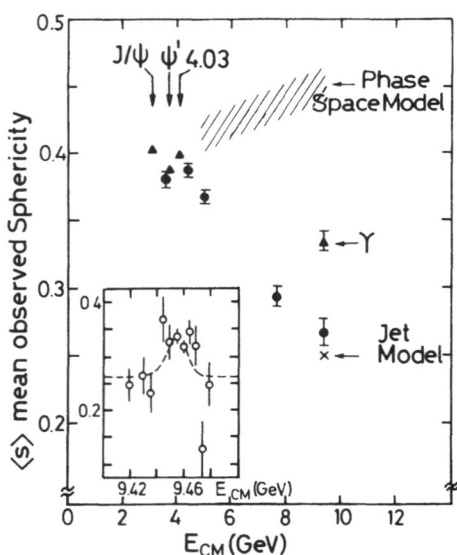

Fig. 15.15. Average observed sphericity
as a function of energy /232/

observed distributions with theory a Monte Carlo technique was used to impose the same acceptance criteria and cuts onto the theoretical events.

The mean sphericity <S> is plotted in Figure 15.15 as a function of c.m. energy between 3 and 10 GeV. In agreement with the SLAC-LBL data <S> is seen to decrease with growing energy. The points measured at the J/ψ and ψ' are significantly higher than the value at 3.6 GeV. Perhaps most interesting is the rise in <S> seen directly above charm threshold at 4.03 GeV. The c$\bar{c}$ events do not show a jet structure.

At the Υ the average sphericity is again larger than outside of the resonance at 9.4 GeV. The dashed band shows the expected <S> behavior for phase space distributed events. At higher energies (E$_{cm}$ ≳ 6 GeV) the observed sphericity values are well below the phase space prediction. The point measured at 9.4 GeV is in reasonable agreement with the jet model calculated according to FIELD and FEYNMAN /235/.

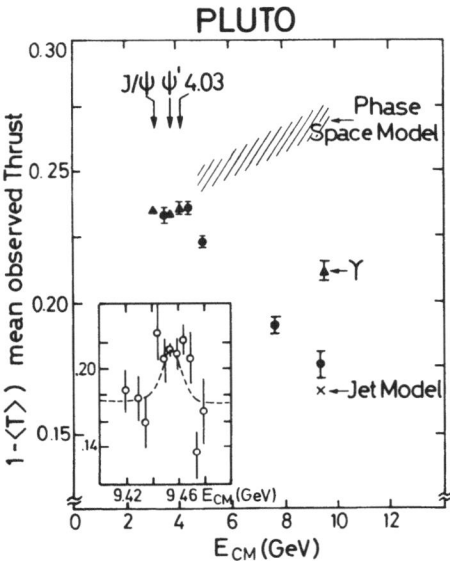

Fig. 15.16. The quantity 1 - <T> as a function of energy where <T> is the average observed thrust /232/

The analysis in terms of thrust leads to the same conclusions. Figure 15.16 shows a plot of 1 - <T> versus E$_{cm}$. The jet axes determined via S and T, respectively, are not too different. This is illustrated in Figure 15.17a where the angle between the two jet axes was plotted. For half of the events this angle is less than 15°. There is a strong correlation between the jet axis and the direction of the particle with the largest momentum. This is shown by Figures 15.17b and c where the distribution of the angle between the two directions was plotted. The correlation is stronger for the S jet axis.

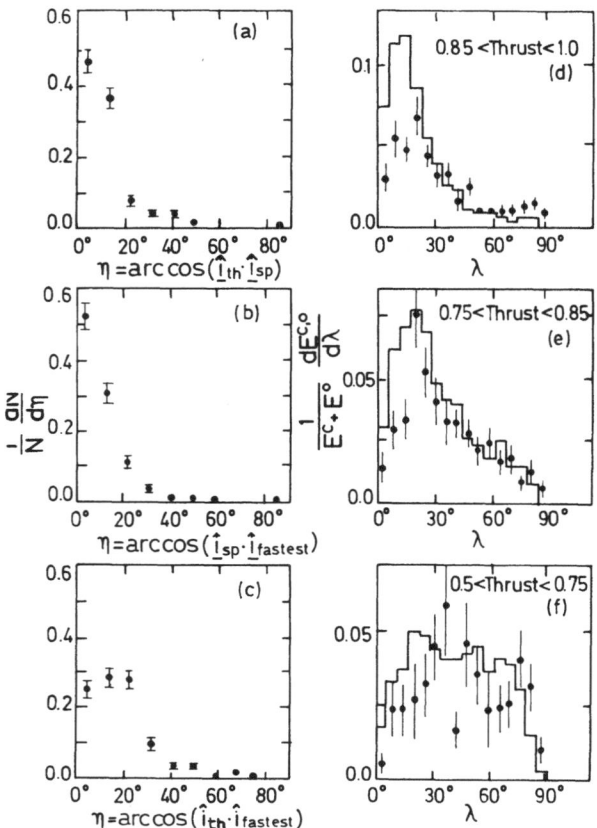

Fig. 15.17. (a) The distribution of the angle between the sphericity and the thrust axes at 9.4 GeV; (b,c) the distributions of the angle between the charged particle with the largest momentum and the sphericity and thrust axes, respectively, at 9.4 GeV; (d-f) the angular distributions $(1/E) dE^c/d\lambda$ of charged energy (histograms) and $(1/E) dE^o/d\lambda$ of neutral energy (data points) with respect to the thrust at 9.4 GeV for different T intervals /232/

    The angular distribution of the jet axis with respect to the beam direction is consistent with a $1 + \cos^2\theta$ behavior expected for $q\bar{q}$ production.

    The energy dependences of the average transverse and longitudinal momenta measured relative to the T jet axis are plotted in Figure 15.18a. The longitudinal component is seen to grow much more rapidly with energy than the transverse momentum. The average half-opening angle of jets computed via $\alpha = \arctan (<p_\perp>/<p_\parallel>)$ is 33° at 5 GeV and 28° at 9.4 GeV. We see that the jet cone shrinks with increasing energy but at a relatively slow rate.

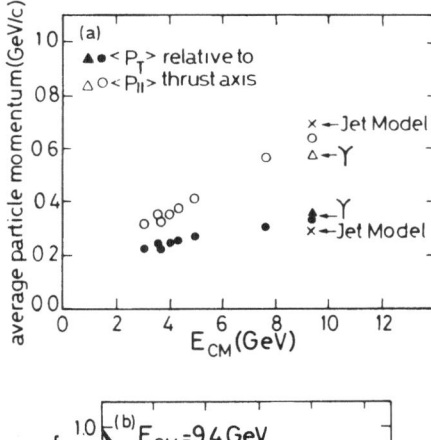

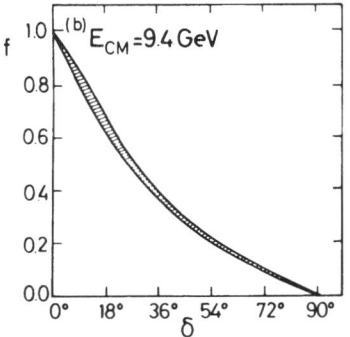

Fig. 15.18. (a) The energy dependence of the average transverse and longitudinal momenta; (b) the average fraction of visible energy (f) outside of a cone of half angle δ at 9.4 GeV /232/

Another measure of the jet opening angle is provided by the amount of energy emitted at an angle λ with respect to the jet axis.

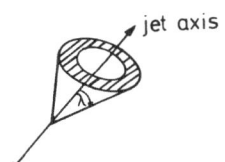

Figures 15.17d-f show at 9.4 GeV $(E^C + E^O)^{-1} dE^{C,O}/d\lambda$ versus λ for different thrust regions; E^C denotes the charged energy (the charged particles were assumed to be pions) and E^O the neutral energy (photons) as measured in the PLUTO shower counters. The histograms show the result for charged particles; the data points measure the neutral component. Note that the jet axis was defined by considering charged particles only. Figures 15.17d-f tell us that the neutral energy is also concentrated near the jet axis in much the same way as charged particles. The larger width of the E^O distributions may disappear when the neutrals are included in the determination of the jet axis.

Charged and detectable neutral particles carry on the average 85 % of the total available energy (visible energy). Figure 15.18b presents for 9.4 GeV the fraction of the visible energy observed outside a cone of half angle δ. For example 70 % of the visible energy are within a cone of δ = 43°.

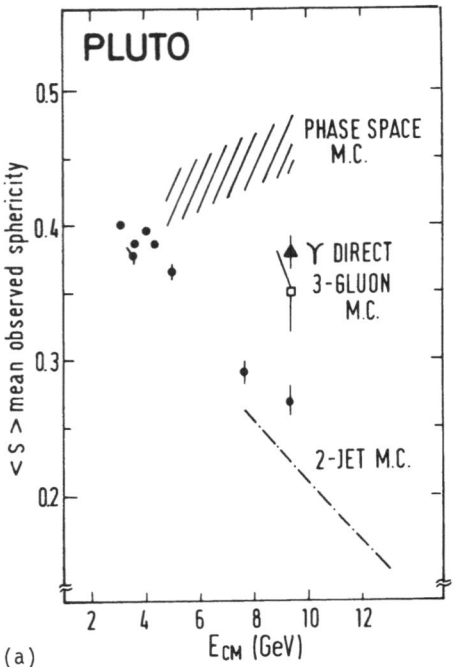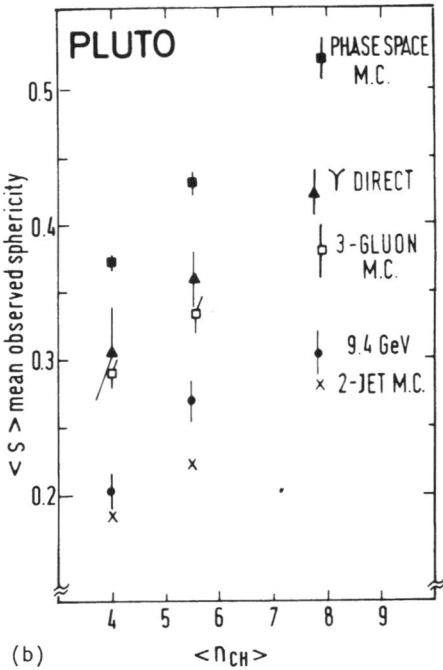

Fig. 15.19. (a) The average sphericity as a function of energy, (b) the average sphericity for 9.4 GeV from T decay for events with different number of charged particles /234/

We turn now to the PLUTO analysis of the T events /234/. Three processes can contribute to events in the T region, the direct hadronic decay, the decay through the one-photon channel, and the nonresonant continuum

$$\sigma = \sigma_{dir} + \sigma_{1\gamma} + \sigma_{cont}.$$ (15.15)

For the continuum contribution the data at 9.4 GeV were used. Neglecting possible interference effects the one-photon part leads to the same final states as the continuum. The size of $\sigma_{1\gamma}$ relative to σ_{cont} can be computed from the rate of μ pairs observed on and off the T

$$\sigma_{1\gamma} = \frac{\sigma_{\mu\mu}^{on} - \sigma_{\mu\mu}^{off}}{\sigma_{\mu\mu}^{off}} \cdot \sigma_{cont} = (0.24 \pm 0.22)\sigma_{cont}.$$

The total number of events available for the analysis was 1418 at the τ(9.45 - 9.47 GeV) and 420 in the continuum (9.30 - 9.44 GeV). The latter events were used to determine by proper subtraction the distribution for the direct decays of the τ.

Figure 15.19a shows the average sphericity plotted versus E_{cm}. The <S> value of τ direct decays is markedly larger than for the continuum. The increase in <S> is not caused by the larger multiplicity. This is demonstrated by Figure 15.19b where <S> is plotted for events with 4, 6, and 8 charged particles for 9.4 GeV and for the direct τ decays. The <S> values for the τ are significantly higher for every topology.

From the preceding discussion it is clear that the τ final states are less jet-like than those of the continuum. They are incompatible with the two-jet model used to describe the continuum but also with pure phase space (see Figure 15.19). The PLUTO group compared their data also with the three-gluon model formulated by KOLLER and WALSH /228/. The gluon jet was assumed to be similar to a quark jet observed in the continuum for the equivalent energy. It should be noted that the predicted hadron distributions depend decisively on this assumption. As for the other model calculations the effects of the detector acceptance and of the experimental cuts were included in the Monte Carlo simulation. Figure 15.19 shows that the sphericity observed for the τ as well as its dependence on the number of charged particles produced is well described by the three-gluon model.

The three-gluon decay predicts the hadrons to be concentrated in a plane. The PLUTO group tested the data for flatness in the following manner.

1) Compute for each event the tensor

$$T^{\alpha\beta} = \sum_i (p_i^2 \delta^{\alpha\beta} - p_i^\alpha p_i^\beta), \tag{15.16}$$

where α,β are the coordinate indices.

2) Solve for the eigenvalues λ_i, i = 1...3 of $T^{\alpha\beta}$ and order them according to $\lambda_1 \geq \lambda_2 \geq \lambda_3$. The sphericity is then given by

$$S = \frac{3\lambda_3}{\lambda_1 + \lambda_2 + \lambda_3} . \tag{15.17}$$

3) Compute the quantities

$$Q_K = 1 - \frac{2\lambda_K}{\lambda_1 + \lambda_2 + \lambda_3} = \frac{\sum_i (p_{\parallel i}^K)^2}{\sum_i (p_i)^2}, \tag{15.18}$$

where $p_\parallel^K$ is the momentum component along the axis associated with λ_K. Q_1 is a measure of the flatness of the event. $Q_1 = 0$ corresponds to a perfectly planar event, $Q_1 = 1/3$ to a spherical event; of course, an extreme two-jet event with all particle momenta along one axis will also lead to $Q_1 = 0$.

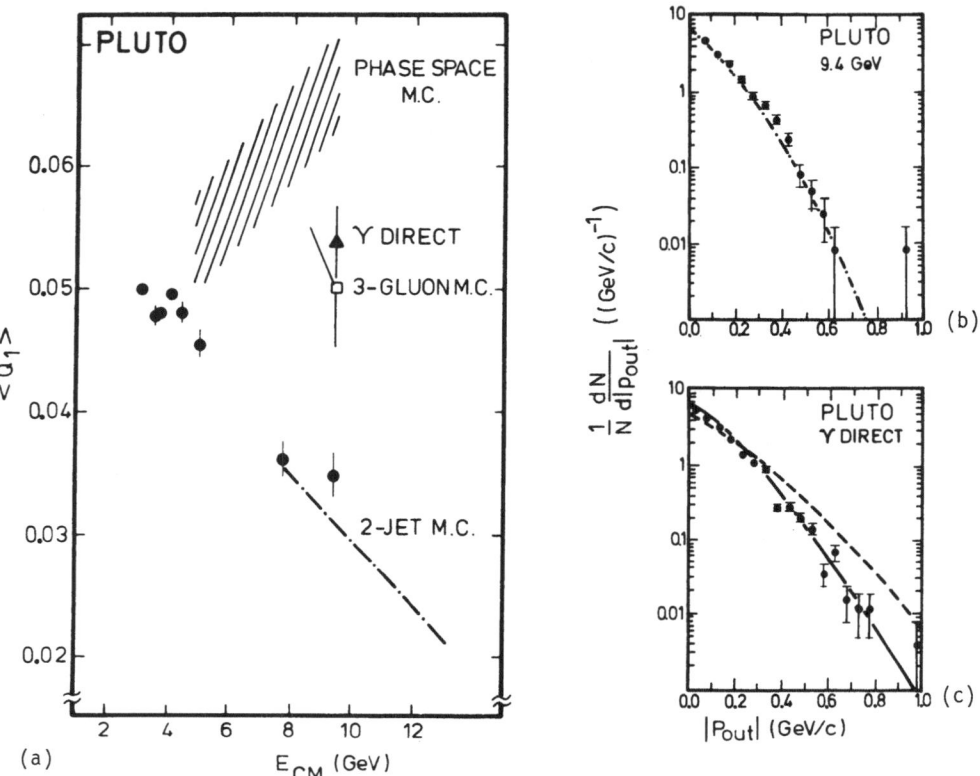

<u>Fig. 15.20.</u> (a) The average observed Q_1 as a function of energy, (b,c) the P_{out} distribution for 9.4 GeV and for Υ direct decay. The dashed-dotted line in (b) represents the two-jet model. The dashed and solid lines in (c) represent the phase space and the three-gluon decay, respectively /234/

Figure 15.20a shows the average Q_1 plotted as a function of E_{cm}. It drops with increasing energy in much the same way as, e.g., the average sphericity and is consistent with the two-jet model. The Υ decays yield a much larger $<Q_1>$ value in agreement with the three-gluon model. Figures 15.20b and c show the p_{out} distributions at 9.4 GeV and from the Υ decay where p_{out} is the momentum component perpendicular to the plane associated with Q_1. The p_{out} distribution is wider for the Υ; it agrees well with the distribution predicted for a three-gluon decay.

The PLUTO group performed a second test devised by DE RUJULA et al. /231/ to measure the flatness. The test requires finding for each event that plane with respect to which the sum of the momentum components perpendicular to it is a minimum

$$A = 4 \min_i \; [(\sum_i |p_{out \; i}|)/(\sum_i |p_i|)]^2, \qquad\qquad (15.19)$$

where A is called the acoplanarity. The average acoplanarity values are given in Table 15.1 together with those measured for sphericity, thrust, etc., and with the theoretical predictions. The average acoplanarity observed for the τ is larger than for the continuum and is consistent with the three-gluon model.

Table 15.1. The observed mean values for sphericity S, Q_1, p_{out}, thrust T, and acoplanarity A. The data are compared with three models computed via the Monte Carlo method to simulate the experimental conditions. The errors given for the values of the phase space and three-gluon decay models include systematic uncertainties. Taken from /234/

	M.C. Two-jet	Data 9.4 GeV	τ Direct	M.C. Three-gluon	Phase space
S	0.22	0.27 ±0.01	0.38 ±0.01	0.35 ±0.03	0.46 ±0.03
Q_1	0.030	0.035 ±0.002	0.054 ±0.003	0.050 ±0.005	0.067 ±0.005
p_{out}	0.115	0.122 ±0.003	0.132 ±0.003	0.140 ±0.006	0.177 ±0.006
T	0.84	0.82 ±0.01	0.76 ±0.01	0.76 ±0.01	0.73 ±0.01
A	0.084	0.096 ±0.005	0.14 ±0.01	0.14 ±0.01	0.16 ±0.01

The DASP2 group also investigated the spatial configuration of events at 9.4 GeV and from τ decay /204, 236/. Charged particles were detected in the nonmagnetic inner detector of the apparatus. Since particle momenta could not be determined, the particle directions were studied in terms of variables related to sphericity, thrust, sphericity, and acoplanarity. Qualitatively, the same behavior was found as observed by the PLUTO group.

The data presented in this section can be summarized as follows. The hadronic final states produced by nonresonant e^+e^- annihilation shows a two-jet structure which becomes more pronounced as the c.m. energy increases. The origin of the jet structure may well be the production of a pair of quark plus antiquark which fragment into hadrons. One of the pieces yet missing in this puzzle is the observation of correlations between the two opposite jets. The distribution of the hadrons within a jet are remarkably similar to those observed from jets produced by hadron-hadron collisions.

The hadrons emitted in the direct decay of the T show a different spatial structure. They are much less collimated along a common axis. The observed features disagree with the two-jet picture but also with pure phase space. All aspects investigated so far are in accord with the assumption that the T decays via a three-gluon intermediate state. This is in strong support of QCD. However, the basic elements of QCD, such as the vector nature of the gluon, the flavor neutrality, the presence of a three-gluon interaction have not yet been established by the data.

16. The Next Generation of e^+e^- Colliding Rings and the First Results from PETRA

The new e^+e^- colliding ring PETRA at DESY was gradually brought into operation in 1978 and the first physics results were published in early 1979. In addition to PETRA two further e^+e^- colliding rings, CESR at Cornell and PEP at Stanford, are scheduled to start operation in 1979 (Table 16.1).

Table 16.1. The total hadronic cross section

Group	2E = 13 GeV		2E = 17 GeV	
	R	ε	R	ε
MARK J /239/	4.6 ± 0.5	0.79	4.9 ± 0.5	0.79
PLUTO /238/	5.0 ± 0.5	0.82	4.3 ± 0.5	0.82
TASSO /240/	5.6 ± 0.7	0.77	4.0 ± 0.7	0.78

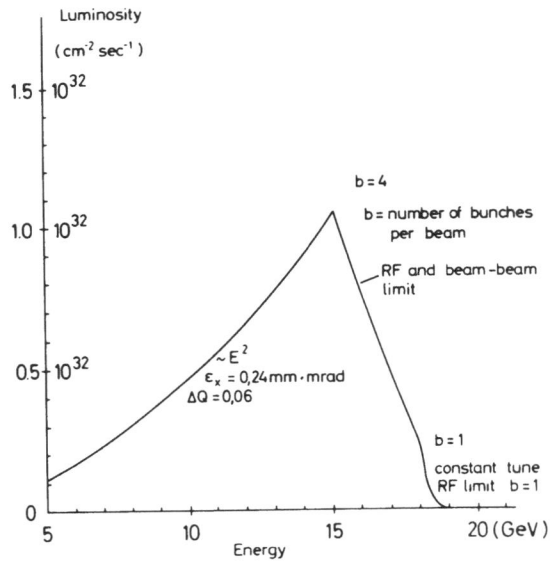

Luminosity vs. beam energy
for 4,5 MW RF – power

Fig. 16.1. Luminosity vs beam energy for 4.5 MW RF-power

In this chapter we will briefly review the first results obtained with PETRA and discuss some of the physics questions which can be investigated using the new accelerators.

The luminosity expected /237/ for PETRA is plotted in Figure 16.1 as a function of beam energy. Note that the luminosity is expected to increase with the square of the beam energy between 5 and 15 GeV. The peak value at 15 GeV is 10^{32} cm^{-2} s^{-1}. Above 15 GeV the luminosity drops rapidly, but it is still around 0.2×10^{32} cm^{-2} s^{-1} at 18 GeV. The predicted luminosity between 5 and 15 GeV corresponds to 860 events/day for the point cross section. Even at 18 GeV the point cross section still yields 50 events/day.

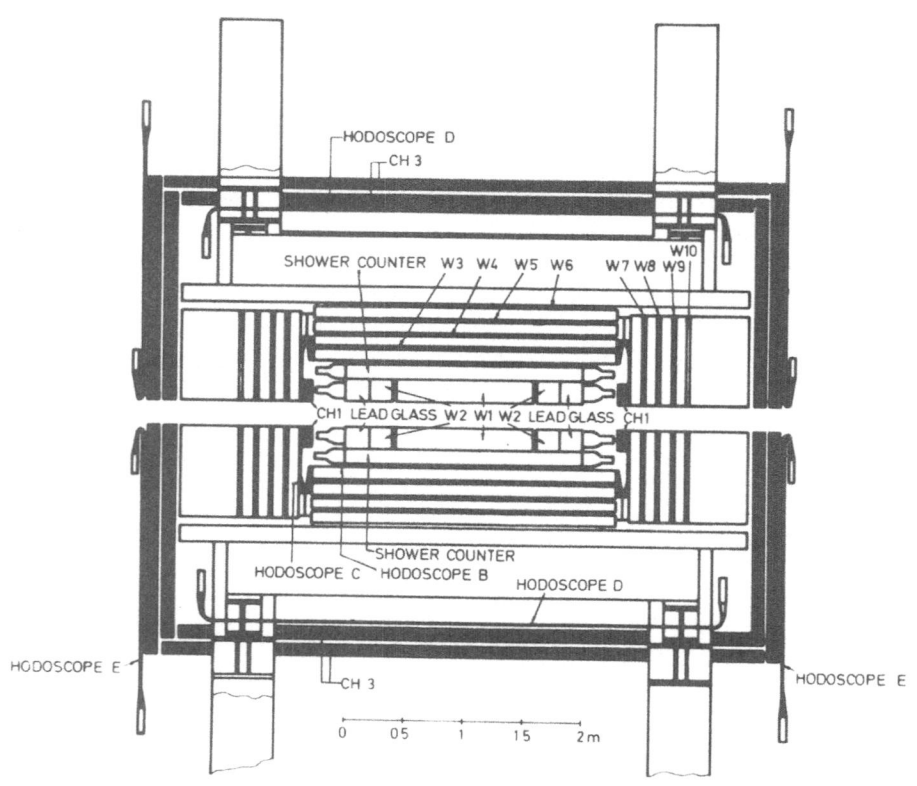

Fig. 16.2. Side view of the MARK J detector

So far, three experiments, PLUTO, MARK J, and TASSO have collected data at 13 and 17 GeV in c.m.s.

The PLUTO detector /238/ as used at DORIS, has been discussed above. For the work at PETRA the muon detection has been improved and the detector has been equipped with forward spectrometers to measure photons and electrons at small angles.

MARK J /239/ is a calorimeter-type detector designed to measure and distinguish hadrons, electrons, neutral particles, and muons. It covers 2π in ϕ and production angles between 9° and 171°. A side view of the detector is shown in Figure 16.2.

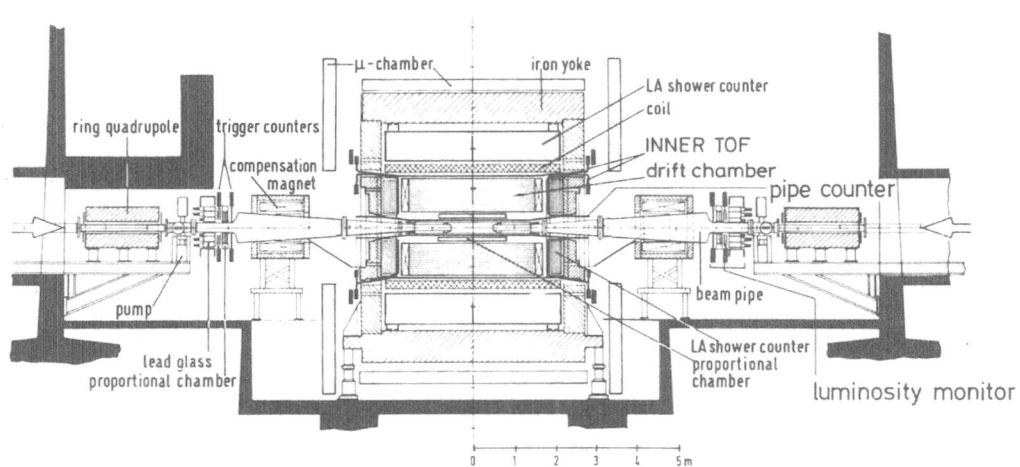

Fig. 16.3. Side view of the TASSO detector

A side view of the TASSO detector /240/ is shown in Figure 16.3. It consists of a large magnetic solenoid, 440 cm long with a radius of 135 cm producing a field of about 0.5 Tesla parallel to the beam axis. The solenoid is filled with tracking chambers and will be surrounded by detector elements to measure the energy and position of photons and to identify charged particles.

All detectors have determined the total annihilation cross section $e^+e^- \rightarrow$ hadrons at 13 and 17 GeV. The values normalized to the point cross section are listed in Table 16.1 together with the estimated detection efficiency for multihadron events. In addition to the statistical error listed in Table 16.1 there are also systematic uncertainties. These are estimated to contribute ±0.7 to R for MARK J and ±20 % for PLUTO and TASSO.

The trigger conditions and the cuts used to evaluate the cross section are very different for the three experiments and the agreement between the data seem to indicate that the corrections which must be applied to the data in order to determine R

252

are well understood. The quark model, including a charge -1/3 b quark, predict R = 3.7. The data at 17 GeV are consistent with this value; however, they are not yet precise enough to exclude either a heavy lepton or a charge 2/3 quark.

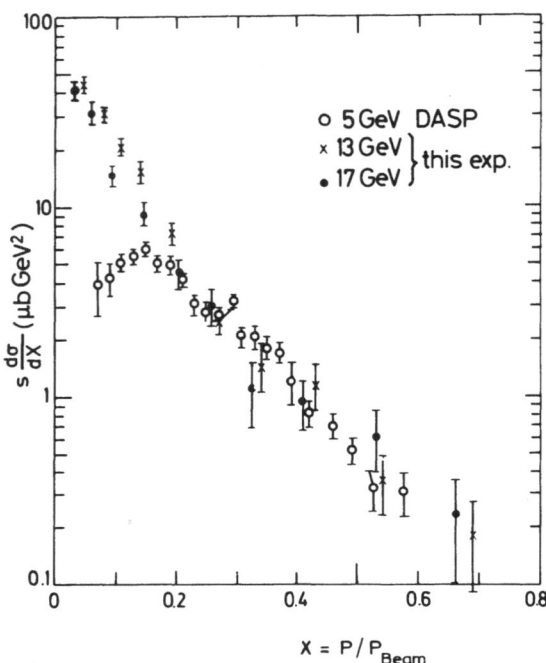

Fig. 16.4. Scaling cross section s dσ/dx for inclusive charged particle production as measured by DASP at 5 GeV and by TASSO /240/ at 13 and 17 GeV

The data indicate that R is larger at 13 GeV than at 17 GeV. To investigate this in more detail, the TASSO group has evaluated /240/ the single-particle inclusive cross section. In the quark model the single-particle inclusive cross section (s/β) $d\sigma/dx_E$ should scale as a function of c.m. energy. Here β is the velocity of the particle, $x_E = 2E_h/W$, and $s = W^2$, the c.m. energy squared. Since the particle mass is not determined the TASSO group used the quantity s dσ/dx with $x = p/p_{beam}$. These cross sections, plotted in Figure 16.4 as a function of x, are consistent with scaling for $x \gtrsim 0.2$ and c.m. energies between 5.0 and 17 GeV, in agreement with the quark model predictions. The cross sections at 13 and 17 GeV are well above the data at 5 GeV for $x < 0.2$. Such a violation of scaling is expected and it gives rise to the increase in multiplicity with s. However, at small x ($x \lesssim 0.2$) the cross section at 13 GeV is about 40 % above the cross section at 17 GeV, which is 2 standard deviations including a systematic uncertainty of 10 % in the relative normalization. This excess at small x is surprising since from the energy dependence one expects an effect in the opposite direction. This and the large R value observed at 13 GeV are reminiscent of the behavior seen above charm threshold in the 4 GeV region, and it might indicate copious $b\bar{b}$ production.

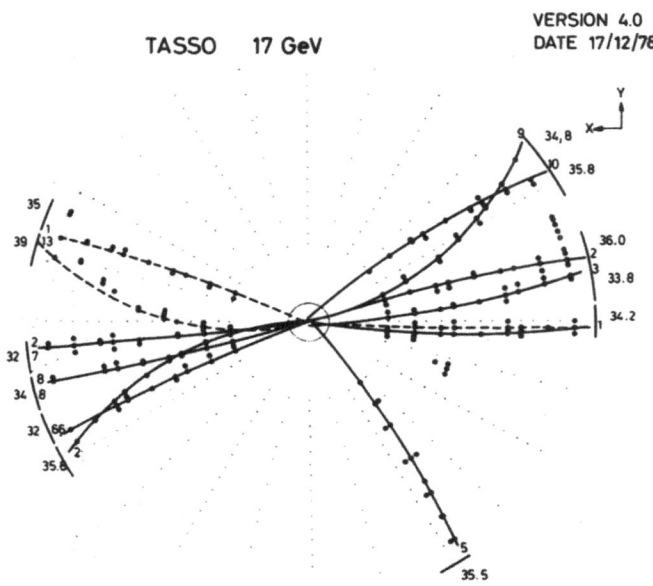

TASSO 17 GeV

VERSION 4.0
DATE 17/12/78

<u>Fig. 16.5.</u> Event of the type $e^+e^- \to$ hadrons as observed by TASSO /240/ at 17 GeV

It has been conjectured that hadron production in e^+e^- annihilation proceeds by quark pair production fragmenting into two roughly collinear jets of hadrons. Figure 16.5 shows a typical event obtained /240/ by the TASSO collaboration. The event is viewed along the beam direction and the jet structure predicted is clearly visible. This impression is borne out by the detailed investigations discussed below.

The normalized sphericity distributions $(1/N)dN/ds$ as measured by PLUTO /238/ and TASSO /240/ at 17 GeV is plotted in Figure 16.6. The distributions peak at low S and shrink with increasing c.m. energy as expected for jetlike events. The mean sphericity (thrust) at 13 and 17 GeV is 0.24 ± 0.02 (0.85 ± 0.01) and 0.19 ± 0.03 (0.87 ± 0.01). These values are in agreement with the values found by PLUTO of 0.26 ± 0.02 (0.82 ± 0.01), respectively, 0.22 ± 0.02 (0.84 ± 0.01) at 13 GeV, respectively, 17 GeV. The values for the spericity decrease with energy (see Figure 15.19) demonstrating that the events indeed become more jetlike with increasing energy.

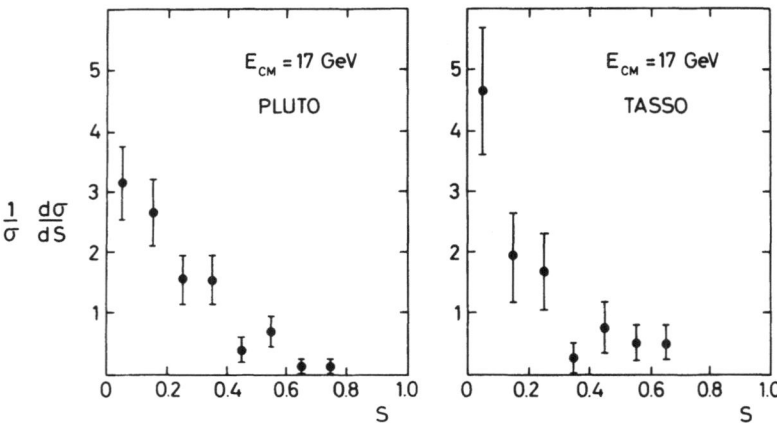

Fig. 16.6. The sphericity distributions as measured by PLUTO /238/ and TASSO /240/ at 17 GeV

References

The following shorthand notations will be used:

"1971 Cornell Conference" for the Proceedings of the 1971 Symposium on Electron and Photon Interactions at High Energies, Cornell, ed. by N. Mistry.

"1975 Stanford Conference" for the Proceedings of the 1975 Symposium on Lepton and Photon Interactions at High Energies, ed. by W.T. Kirk.

"1976 Tbilisi Conference" for the XVIIIth International Conference on High Energy Physics, Tbilisi, USSR (1976).

"1977 Hamburg Conference" for the Proceedings of the 1977 International Symposium on Lepton and Photon Interactions at High Energies, Hamburg, edited by F. Gutbrod.

"1978 Tokyo Conference" for the XIXth International Conference on High Energy Physics, Tokyo (1978).

1   G.K. O'Neill: Bull. Am. Phys. Soc. 3, 158 (1958);
    C. Bernardini, G.F. Corazza, G. Ghigo and B. Touschek: Nuovo Cimento 18, 1293 (1960)
2   W.C. Barber, B. Gittelman, G.K. O'Neill, B. Richter: Phys. Rev. Lett. 16, 1127 (1966)
3   For a review of this work see J. Perez-y-Jorba: Proceedings of the 4th Int. Symposium on Electron and Photon Interactions at High Energies, Liverpool, 1969, ed. by D.W. Braben, 213; and
    J. LeFrancois: Rapporteur talk, 1971 Cornell Conference, 51
4   V. Sidorov: Rapporteur talk, 1971 Cornell Conference, 65
5   B. Bartoli, et al.: Nuovo Cimento 70A, 615 (1970); Phys. Lett. 36B 598 (1971)
    G.G. Barbarino, et al.: Lett. Nuovo Cimento 3, 689 (1972)
    C. Bacci, G. Penso, G. Salvini, B. Stella, R. Baldini-Celio, G. Capon, C. Mencuccini, G.P. Murtas, A. Reale and M. Spinetti: Phys. Lett. 38B, 551 (1972); see also C. Bernardini: Rapporteur talk, 1971 Cornell Conference, 37
6   A. Litke, et al.: Phys. Rev. Lett. 30, 1189 (1973)
    G. Tarnopolsky, et al.: Phys. Rev. Lett. 32, 432 (1974)
7   J.-E. Augustin, et al.: see B. Richter: Rapporteur talk, XVII Int. Conf. on High Energy Physics, London, 1974, IV-37
8   R.F. Schwitters, et al.: Phys. Rev. Lett. 35, 1320 (1975)
9   J.J. Aubert, et al.: Phys. Rev. Lett. 33, 1404 (1974)
10 J.-E. Augustin, et al.: Phys. Rev. Lett. 33, 1406 (1974)
11 G.S. Abrams, et al.: Phys. Rev. Lett. 33, 1453 (1974)
    A.M. Boyarski, et al.: Phys. Rev. Lett. 34, 764 (1975)
12 DASP Collaboration, W. Braunschweig, et al.: Phys. Lett. 57B, 407 (1975); see also B.H. Wiik: Rapporteur talk, 1975 Stanford Conference, 69
13 G.J. Feldman, et al.: Phys. Rev. Lett. 35, 821 (1975)
14 T. Applequist, H.D. Politzer: Phys. Rev. Lett. 34, 43 (1975)
    C.G. Callan, R.L. Kingsley, S.B. Treiman, F. Wilczek and A. Zee: Phys. Rev. Lett. 34, 52 (1975)

256

T. Applequist, A. De Rūjula, H. David Politzer, S.L. Glashow: Phys. Rev. Lett. 34, 365 (1975)
E. Eichten, K. Gottfried, T. Kinoshita, J. Kogut, K.D. Lane and T.-M. Yan: Phys. Rev. Lett. 34, 369 (1975)

15 S.L. Glashow, J. Iliopoulos, L. Maiani: Phys. Rev. D2, 1285 (1970)
16 G. Goldhaber, et al.: Phys. Rev. Lett. 37, 255 (1976)
17 DASP Collaboration, W. Braunschweig, et al.: Phys. Lett. 63B, 471 (1976)
   PLUTO Collaboration, J. Burmester, et al.: Phys. Lett. 64B, 369 (1976)
18 A. Benevenuti, et al.: Phys. Rev. Lett. 35, 1199, 1203 (1975)
19 DASP Collaboration, R. Brandelik, et al.: Phys. Lett. 70B, 132 (1977)
   DASP-Collaboration, R. Brandelik, et al.: Phys. Lett. 80B, 412 (1979)
20 M. Perl, et al.: Phys. Rev. Lett. 35, 1489 (1975)
21 S.W. Herb, et al.: Phys. Rev. Lett. 39, 252 (1977)
   W.R. Innes, et al.: Phys. Rev. Lett. 39, 1240 (1977)
22 PLUTO Collaboration, Ch. Berger, et al.: Phys. Lett. 76B, 243 (1978)
   C.W. Darden, et al.: Phys. Lett. 76B, 246 (1978)
23 J.K. Bienlein, et al.: Phys. Lett. 78B, 360 (1979)
   C.W. Darden, et al.: Phys. Lett. 78B, 364 (1979)
24 For recent reviews of the field see, e.g., the reports by K. Berkelman, G. Feldmann, G.P. Murtas, J. Perez-y-Yorba in the Proceedings of the 1978 Tokyo Conference
25 M. Sands: In Physics with Intersecting Storage Rings, B. Touschek, ed. (Academic Press, New York 1971), 257
   H. Wiedemann: DESY Internal Report 1974
26 A.A. Sokolov, I.M. Ternov: Sov. Phys.-Doklady 8, 1203 (1964);
   V.N. Baier: Sov. Phys. Uspekhi 14, 695 (1972);
   J.D. Jackson: Rev. Mod. Phys. 48, 417 (1976);
   R.F. Schwitters: SLAC-PUB-2258 (1979)
27 V.E. Balakin, A.D. Bukin, E.V. Pakhtusova, V.A. Sidorov, A.G. Khabakhpashev: Phys. Lett. 34B, 663 (1971)
28 C. Bacci, G. Parisi, G. Penso, G. Salvini, B. Stella, R. Baldini-Celio, G. Capon, C. Mencuccini, G.P. Murtas, M. Spinetti, A. Zallo: Frascati report LNF 73/50 (1973)
29 See, e.g., R. Gatto: Proc. Int. Sym. Electron and Photon Interactions at High Energies, Hamburg, 1965. Vol. I, 106
30 H. Salecker: Z. Phys. 160, 385 (1960)
   K. Ringhofer, H. Salecker: Contribution to the 1975 Stanford Conference
31 N.M. Kroll: Nuovo Cimento 45A, 65 (1966)
32 See R. Hofstadter: Rapporteur talk, 1976 Tbilisi Conference, B34
33 W.C. Barber, G.K. O'Neill, B. Gittelman, B. Richter: Phys. Rev. D3, 2796 (1971)
34 J.-E. Augustin, et al.: Phys. Rev. Lett. 34, 233 (1975)
35 B.L. Beron, et al.: Phys. Rev. D17, 2187, 2839 (1978)
36 V. Alles-Borelli, M. Bernardini, D. Bollini, P. Giusti, T. Massam, L. Monari, F. Palmonari, G. Valenti, A. Zichichi: Phys. Lett. 59B, 201 (1975)
37 G. Jarlskog, L. Jönsson, S. Prünster, H.D. Schulz, H.J. Willutzki, G.G. Winter: Phys. Rev. D8, 3813 (1973)
38 F.E. Low: Phys. Rev. 120, 582 (1960)
   F. Calogero, C. Zemach: Phys. Rev. 120, 1860 (1960)
   A. Jaccarini, N. Arteaga-Romero, J. Parisi, P. Kessler: Compt. Rend. 269B, 153, 1129 (1969); Nuovo Cimento 4, 933 (1970)
   N. Arteaga-Romero, A. Jaccarini, P. Kessler, J. Parisi: Phys. Rev. D3, 1569 (1971)
   S.J. Brodsky, T. Kinoshita, H. Terazawa: Phys. Rev. Lett. 25, 972 (1970)
39 S. Orito, M.L. Ferrer, L. Paoluzi, S. Santonico: Phys. Lett. 48B, 380 (1974)
40 R. Feynman: Photon-Hadron Interactions (W.A. Benjamin, New York 1972)
41 See also J.J. Sakurai, D. Schildknecht: Phys. Lett. 40B, 121 (1972); 41B, 489 (1972)
   J.J. Sakurai: Phys. Lett. 46B, 207 (1973)
   A. Bramon, E. Etim, M. Greco: Phys. Lett. 41B, 609 (1972)
42 A. Quenzer: Thesis, ORSAY report LAL 1294 (1977)
   A. Cordier, et al.: Phys. Lett. 81B, 389 (1979)
   J. Perez-y-Yorba: 1978 Tokyo Conference

257

43 V.A. Sidorov: Rapporteur talk, 1976 Tbilisi Conference, N46
44 G.P. Murtas: 1978 Tokyo Conference, B2
45 R.F. Schwitters: Rapporteur talk, 1976 Tbilisi Conference, B34
    G. Hanson: Talk given at the 1976 Tbilisi Conference, SLAC-PUB 1814 (1976)
    V. Lüth: SLAC-PUB 2050 (1977)
46 PLUTO Collaboration, J. Burmester, et al.: Phys. Lett. 66B, 395 (1977)
    See G. Knies: Rapporteur talk at the 1977 Hamburg Conference, 93
    Ch. Berger, et al.: Phys. Lett. 81B, 410 (1979)
47 DASP Collaboration, R. Brandelik, et al.: Phys. Lett. 76B, 361 (1978)
48 MARK-J Collaboration, D. Barber, et al.: MIT-LNS report 100 (1979)
49 TASSO Collaboration, R. Brandelik, et al.: DESY report 79/14 (1979)
50 P. Joos, et al.: Nucl. Phys. B113, 53 (1976)
51 G. Wolf: Rapporteur talk, 1975 Stanford Conference, 795
52 V.L. Auslander, G.I. Budker, Ju.N. Pestov, V.A. Sodorov, A.N. Skrinsky, A.G.
    Khabakpashev: Phys. Lett. 25B, 433 (1967)
    V.E. Balakin, G.I. Budker, E.V. Pakhtusova, V.A. Sidorov, A.N. Skrinsky, G.M.
    Tumaikin, A.G. Kahbakhpashev: Phys. Lett. 34B, 328 (1971)
    J.C. Bizot, et al.: Phys. Lett. 32B, 416 (1970)
    D. Benaksas, G. Cosme, B. Jean-Marie, S. Jullian, F. Laplanche, J. Lefrancois,
    A.D. Liberman, G. Parrour, J.P. Repellin, G. Sauvage: Phys. Lett. 39B, 289
    (1972)
53 G. Wolf: Rapporteur talk, 1971 Cornell Conference, 189
    K.C. Moffeit: Rapporteur talk, 1973 Symposium on Electron and Photon Inter-
    actions at High Energies, Bonn, ed. by H. Rollnik, W. Pfeil, 313
    A. Silverman: Rapporteur talk, 1975 Stanford Conference, 355
54 S. Bartalucci, et al.: DESY reports 76/43 (1976) and 77/59 (1977)
55 M. Ambrosio, et al.: Phys. Lett. 80B, 141 (1978)
    R. Baldini-Celio, et al.: Phys. Lett. 78B, 167 (1978)
    G. Capon: Rapporteur talk, Proceedings of the 1977 European Conference on
    Particle Physics, Budapest, 751
    C. Bemporad: Rapporteur talk, 1977 Hamburg Conference, 165
56 G. Cosme, et al.: Orsay report LAL 78-32 (1978)
    F. Laplanche: Rapporteur talk, 1977 Hamburg Conference, 189
57 J. Ballam, et al.: Nucl. Phys. B76, 375 (1974)
    W. Struczinski, et al.: Nucl. Phys. B108, 45 (1976)
58 SLAC-Berkeley Collaboration, H.H. Bingham, et al.: LBL report 991 (1972);
    Phys. Lett. 41B, 635 (1972)
    M. Davier, I. Derado, D.E.C. Fries, F.F. Liu, R.F. Mozley, A. Odian, J. Park,
    W.P. Swanson, F. Villa, D. Yount: Nucl. Phys. B58, 31 (1973)
59 B. Bartoli, B. Coluzzi, F. Felicetti, V. Silvestrini, G. Goggi, D. Scannicchio,
    G. Marini, F. Massa, F. Vanoli: Nuovo Cimento 70, 615 (1970)
    G. Barbarino, et al.: Nuovo Cimento Lett. 3, 689 (1972)
    F. Ceradini, et al.: Phys. Lett. 43B, 341 (1973)
60 G.J. Gounaris, J.J. Sakurai: Phys. Rev. Lett. 21, 244 (1968)
61 E.B. Dally, et al.: Phys. Rev. Lett. 39, 1176 (1977)
62 G. Bardin, et al.: Nucl. Phys. B120, 45 (1977)
63 DASP Collaboration, W. Braunschweig, et al.: Phys. Lett. 57B, 297 (1975);
    63B, 487 (1976)
64 M. Schliwa: Thesis, University of Hamburg (1979)
65 F. Vannucci, et al.: Phys. Rev. D15, 1814 (1977)
66 C.J. Bebek, C.N. Brown, M. Herzlinger, S.D. Holmes, C.A. Lichtenstein, F.M.
    Pipkin, S. Raither, L.K. Sisterson: Phys. Rev. D13, 25 (1976)
67 B. Hyams, et al.: Nucl. Phys. B100, 205 (1975)
68 R. Anderson, D. Gustavson, J. Johnson, D. Ritson, B.H. Wiik, W.G. Jones, D.
    Kreinick, F. Murphy, R. Weinstein: Phys. Rev. D1, 27 (1970)
69 C. Cordier, et al.: Contribution to the 1977 Hamburg Conference
70 F.M. Renard: Nuovo Cimento 64, 979 (1969)
    G. Kramer, J.L. Uretsky, T.F. Walsh: Phys. Rev. D3, 719 (1971)
    G. Kramer, T.F. Walsh: Z. Phys. 263, 361 (1973)
71 B. Jean-Marie, et al.: SLAC-PUB 1711, LBL-4672 (1976)
72 J.D. Jackson, D.L. Scharre, Nucl. Instrum. Methods 128, 13 (1975)

73   E. Etim, G. Pancheri, B. Touschek: Nuovo Cimento B51, 276 (1967)
     H.D. Schulz: DESY Internal report F39-70/1 (1970)
     M. Greco, G. Pancheri-Srivastava, Y. Srivastava: Nucl. Phys. B101, 234 (1975);
     Phys. Lett. 56B, 367 (1975)
74   D.R. Yennie: Phys. Rev. Lett. 34, 239 (1975)
     M. Greco, A.F. Grillo: Lett. Nuovo Cimento 15, 174 (1976)
75   DASP-Collaboration, W. Braunschweig, et al.: Phys. Lett. 63B, 115 (1976)
76   DASP Collaboration, R. Brandelik, et al.: Submitted to Z. Physik C(1979)
77   PLUTO Collaboration, L. Criegee, et al.: DESY report 75/32 (1975)
     PLUTO Collaboration, J. Burmester, et al.: Contribution to the 1976 Tbilisi
     Conference and DESY report 76/53 (1976)
78   C. Bemporad: 1975 Stanford Conference, 113
79   A.M. Boyarski, et al.: Phys. Rev. Lett. 34, 1357 (1975)
     V. Lüth, Proc. EPS Int. Conf. on High Energy Physics, ed. A. Zichichi (1975),
     54
80   CERN Theory Workshop, CERN-TH 1964 (1974)
81   DASP Collaboration, W. Braunschweig, et al.: Phys. Lett. 57B, 297 (1975);
     Phys. Lett. 63B, 487 (1976)
82   B. Jean-Marie, et al.: Phys. Rev. Lett. 36, 291 (1976)
83   G.J. Feldman, M.L. Perl: Phys. Rep. 33, 285 (1977)
84   G. Goldhaber: LBL-4884 (1976)
85   M. Castellane, G. DiGiugno, J.W. Humphrey, E. Sassi, G. Troise, U. Troya,
     S. Vitale: Data cited by V. Silvestrini, Proc. XVI Int. Conf. on High Energy
     Physics, 1972, Vol. 4, 1
86   M. Conversi, T. Massam, Th. Muller, A. Zichichi: Nuovo Cimento 40A, 690 (1965)
87   D.L. Hartill, B.C. Barish, D.G. Fong, R. Gomez, J. Pine, A.V. Tollestrup, A.W.
     Maschke, T.F. Zipf: Phys. Rev. 184. 1415 (1969)
88   B. Knapp, et al.: Phys. Rev. Lett. 34, 1040 (1975)
89   V. Lüth, et al.: Phys. Rev. Lett. 35, 1124 (1975)
     G.S. Abrams, et al.: Phys. Rev. Lett. 34, 1181 (1975)
     G.S. Abrams: Rapporteur talk, 1975 Stanford Conference, 25
     W. Tannenbaum, et al.: Phys. Rev. Lett. 36, 402 (1976)
90   F.J. Gilman: SLAC-PUB-1720 (1976)
91   W. Bartel, et al.: Phys. Lett. 64B, 483 (1976)
92   G. Goldhaber: LBL report LBL-4224 (1975)
93   H. Kowalski, T.F. Walsh: Phys. Rev. D14, 852 (1976)
94   British-Scandinavian Collaboration, B. Alper, et al.: Nucl. Phys. B87, 19 (1975)
95   See, e.g., J. Engels, K. Schilling, H. Satz: Nuovo Cimento 17A, 535 (1973)
     F. Elvekjaer, F. Steiner: Phys. Lett. 60B, 456 (1976)
     F. Elvekjaer, DESY report 75/53 (1975)
96   DASP Collaboration, R. Brandelik, et al.: Nucl. Phys. B148, 189 (1979)
97   S. Okubo: Phys. Lett. 5, 165 (1963)
     G. Zweig: CERN report TH 401, 412 (1964), unpublished
     J. Iizuka, K. Okada, O. Shito: Progr. Theor. Phys. 35, 1061 (1966)
98   J.S. Whitaker, et al.: Phys. Rev. Lett. 37, 1596 (1976)
99   DASP Collaboration, W. Braunschweig, et al.: Phys. Lett. 53B, 491 (1975)
     G. Wolf: Rapporteur talk, Proc. EPS Int. Conf. on High Energy Physics, ed.
     by Zichichi (1975), 87
100 DASP Collaboration, W. Braunschweig, et al.: Phys. Lett. 67B, 243 (1977);
     67B, 249 (1977)
     S. Yamada: Rapporteur talk, Hamburg Conference 1977, 69
101 W. Bartel, et al.: Phys. Lett. 66B, 489 (1977)
     J. Olsson, Rapporteur talk, 1977 Hamburg Conference, 117
102 PLUTO Collaboration, G. Alexander, et al.: Phys. Lett. 72B, 493 (1978)
     PLUTO Collaboration, J. Burmester, et al.: Phys. Lett. 72B, 135 (1977)
103 DASP Collaboration, R. Brandelik, et al.: Phys. Lett. 74B, 292 (1978)
104 T.F. Walsh: DESY report 76/13 (1976)
     H. Harari: Phys. Lett. 60B, 172 (1976)
     E. Etim, M. Greco: CERN TH-2174 (1976)
     H. Fritzsch, J.D. Jackson: Phys. Lett. 66B, 365 (1977)
     Y.I. Azimov, L.L. Frankfurt, V.A. Khoze: Pisma. Zh. Eksperim. Teor. Fiz 23,
     591 (1976)

105 H. Primakoff, L. Pilachowski, W.A. Simmons, S.F. Tuan: Lett. Nuovo Cimento 15, 583 (1976)
106 T.F. Walsh: Lett. Nuovo Cimento 14, 290 (1975)
    R.N. Cahn, M.S. Chanowith: Phys. Lett. 59B, 277 (1975)
    J. Ellis: Schladming Lectures 1975, Acta Phys. Austriaca Suppl. XVI, 143
    A. Kazi, G. Kramer, D.H. Schiller: Lett. Nuovo Cimento 15, 120 (1976)
107 P. Brauel, Th. Canzler, D. Cords, R. Felst, G. Grindhammer, W.-D. Kollmann, H. Krehbiel, M. Schädlich: Phys. Lett. 61B, 110 (1976)
108 C. Bacci, et al.: Phys. Lett. 64B, 356 (1976)
    G. Barbiellini, et al.: Phys. Lett. 64B, 359 (1976)
    B. Esposito, et al.: Phys. Lett. 64B, 362 (1976)
    See also C. Bemporad: Rapporteur talk, 1975 Stanford Conference, 113
109 A.M. Boyarski, et al.: Phys. Rev. Lett. 34, 762 (1975)
    R.F. Schwitters: Rapporteur talk, 1976 Tbilisi Conference, 34
110 D.C. Holm, et al.: Phys. Rev. Lett. 36, 1236 (1976)
111 Maryland-Princeton-Pavia-San Diego-SLAC Collaboration: Contribution to the 1976 Tbilisi Conference
112 S.L. Glashow, J. Iliopoulos, L. Maiani: Phys. Rev. D2, 1285 (1970)
113 An up-to-date description of the theoretical framework used to describe the new particles with a complete set of references can be found in K. Gottfried: Invited paper at the 1977 Hamburg Conference, 667
    See also:
    J.D. Jackson: Proc. of the 1977 European Conference on Particle Physics, Budapest, 603
    V.A. Novikov, L.B. Okun, M.A. Shifman, A.I. Vainshtein, M.B. Voloshin, V.I. Zakharov: Phys. Rep. 41, 1 (1978)
    M. Krammer, H. Krasemann: Lecture given at the Advanced Summer Institute, Karlsruhe (1978); DESY report 78/66 (1978)
114 T. Appelquist, H.D. Politzer: Phys. Rev. Lett. 34, 43 (1975)
    C.G. Callan, R.L. Kingsley, S.B. Treiman, F. Wilzcek, A. Zee: Phys. Rev. Lett. 34, 52 (1975)
    E. Eichten, K. Gottfried, T. Kinoshita, J. Kogut, K.D. Lane, T.-M. Yan: Phys. Rev. Lett. 34, 369 (1975)
    E. Eichten, K. Gottfried, T. Kinoshita, K.D. Lane, T.-M. Yan: Phys. Rev. Lett. 36, 500 (1976)
115 J. Pumplin, W. Repko, A. Sato: Phys. Rev. Lett. 35, 1538 (1975)
    H.J. Schnitzer: Phys. Rev. Lett. 35, 1540 (1975); Phys. Rev. D13, 74 (1976)
116 P.A. Rapidis, et al.: Phys. Rev. Lett. 39, 526 (1977)
117 DELCO Collaboration, J. Kirkby: Rapporteur talk, 1977, Hamburg Conference, 3
    W. Bacino, et al.: SLAC-PUB-2030 (1977)
118 J. Siegrist, et al.: Phys. Rev. Lett. 36, 700 (1976)
119 PLUTO Collaboration, J. Burmester, et al.: Phys. Lett. 66B, 395 (1977)
    G. Knies: Rapporteur talk, 1977 Hamburg Conference, 93
120 DASP Collaboration, R. Brandelik, et al.: Phys. Lett. 76B, 361 (1978)
    A. Petersen: Thesis, University of Hamburg (1978)
121 J.J. Aubert, et al.: Phys. Rev. Lett. 33, 1404 (1974)
122 J.-E. Augustin, et al.: Phys. Rev. Lett. 33, 1406 (1974)
    G.S. Abrams, et al.: Phys. Rev. Lett. 33, 1453 (1974)
    A.M. Goyarski, et al.: Phys. Rev. Lett. 34, 764 (1975)
123 K. Gottfried: Rapporteur talk, 1977 Hamburg Conference, 667
    M. Caichian, R. Kögerler: Phys. Lett. 80B, 105 (1978)
    A multichannel analysis of the 4 GeV region has been presented by
    D. Horn, D.E. Novoseller: Preprint (1978)
124 R. Barbieri, R. Gatto, R. Kögerler: Phys. Lett. 60B, 183 (1976)
125 DASP Collaboration, W. Braunschweig, et al.: Phys. Lett. 57B, 407 (1975);
    R. Brandelik, et al.: Z. Phys. C1, 233 (1979) and Nucl. Phys., to be published
    See also B.H. Wiik: Rapporteur talk, 1975 Stanford Conference, 69
    S. Yamada: Rapporteur talk, 1977 Hamburg Conference, 69
    E. Gadermann: Thesis, Hamburg University (1978)
126 J.S. Whitaker, et al.: Phys. Rev. Lett. 37, 1596 (1976)
    G.H. Trilling: In Proceeding of the SLAC Summer Institute on Particle Physics, 1976 SLAC report 198

127 C.J. Biddick, et al.: Phys. Rev. 38, 1324 (1977)
    See also H.F.W. Sadrozinski: Rapporteur talk, 1977 Hamburg Conference, 47
128 PLUTO Collaboration, V. Blobel, Proc. of the XIIth Rencontre de Moriond, Flaine, 1977, ed. by Tran Thanh Van, Vol. I., 99
129 J.S. Whitaker, et al.: Phys. Rev. Lett. 37, 1596 (1976)
130 DESY-Heidelberg Collaboration, see J. Olsson: Rapporteur talk, 1977 Hamburg Conference, 117
    W. Bartel, et al.: Phys. Lett. 79B, 492 (1978)
131 F.M. Pierre: LBL report LBL-5324 (1976)
132 G.H. Trilling: Proceedings of the SLAC-Summer Institute on Particle Physics, 1976, SLAC-report 198, 437
133 R. Barbieri, R. Gatto, E. Remiddi: Phys. Lett. 61B, 465 (1976)
134 DASP Collaboration, W. Braunschweig, et al.: Phys. Lett. 67B, 243, 249 (1977)
    S. Yamada: Rapporteur talk, 1977 Hamburg Conference, 69
135 F.A. Berends, R. Gastman: Nucl. Phys. B61, 414 (1973)
136 J. Heintze: Rapporteur talk, 1975 Stanford Conference, 97
137 W.D. Apel, et al.: Phys. Lett. 72B, 500 (1978)
138 F. Buccella: Private Communication
139 Y.I. Azimov, L.L. Frankfurt, V.A. Khoze: Leningrad preprint 239 (1976)
140 A. Bradley, F.D. Gault: Manchester University preprint (1978)
141 H.J. Lipkin, H.R. Rubinstein, N. Isgur: Weizmann Institute preprint WIS-78/23 Ph (1978)
    G. Eilam, B. Margolis, S. Rudaz: Phys. Lett. 80B, 306 (1979)
142 M.S. Chanowitz, F.J. Gilman: Phys. Lett. 63B, 178 (1976)
    H. Harari: Phys. Lett. 64B, 469 (1976)
143 M.K. Gaillard, B.W. Lee, J.L. Rosner: Rev. Mod. Phys. 47, 277 (1975)
144 J. Ellis, M.K. Gaillard, D.V. Nanopoulos: Nucl. Phys. B100, 313 (1975)
145 PLUTO Collaboration, J. Burmester, et al.: Phys. Lett. 67B, 367 (1977)
146 DASP Collaboration, R. Brandelik, et al.: Phys. Lett. 67B, 363 (1977)
147 V. Lüth, et al.: Phys. Lett. 70B, 120 (1977)
148 G. Goldhaber, et al.: Phys. Rev. Lett. 37, 255 (1976)
    I. Peruzzi, et al.: Phys. Rev. Lett. 37, 569 (1976)
149 G.J. Feldman, M.L. Perl: Phys. Rep. 33, 285 (1977)
150 G. Goldhaber: Lectures given at the International School of Physics, Enrico Fermi, Varenna (1977)
151 L.B. Okun, M. Voloshin, Zh. Eksper. Theor. Fiz. 23, 369 (1976)
    C. Rosenzweig: Phys. Rev. Lett. 36, 697 (1976)
    M. Bander, G.L. Shaw, P. Thomas, S. Meshkov: Phys. Rev. Lett. 36, 695 (1976)
    A. De Rujula, et al.: Phys. Rev. Lett. 38, 317 (1977)
152 PLUTO Collaboration, J. Burmester, et al.: Phys. Lett. 68B, 283 (1977)
153 DASP Collaboration, R. Brandelik, et al.: Phys. Lett. 70B, 132 (1977); 80B, 412 (1979)
154 C. Quigg, J. Rosner: Phys. Rev. D17, 239 (1978)
155 DASP Collaboration, R. Brandelik, et al.: Nucl. Phys. B148, 189 (1979)
156 I. Peruzzi, M. Piccolo: Contributed paper to Festschrift for C. Peyrou, LNF-78/12(P), March 1978
157 D. Lüke: Invited talk given at the 1977 Meeting of the Division of Particles and Fields of the A.P.S., Argonne, Illinois, October 1977. SLAC-PUB-2086 (February 1978)
    A. Barbaro-Galtieri: Lectures presented at the XVI International School of Subnuclear Physics, Erice, 1978, LBL report-8537 (1978)
158 D. Fakirov, B. Stech: Nucl. Phys. B133, 315 (1978)
    H. Fritzsch: Phys. Lett. 71B, 429 (1977)
159 I. Hinchliffe, C.H. Llewellyn-Smith: Nucl. Phys. B114, 45 (1976)
    A. Ali, T.C. Yang: Phys. Lett. 65B, 275 (1976)
    F. Bletzacker, H.T. Nieh, A. Soni: Phys. Rev. D16, 732 (1977)
    G.L. Kane: Phys. Lett. 70B, 227 (1977)
160 DASP Collaboration, W. Braunschweig, et al.: Phys. Lett. 63B, 471 (1976)
    DASP Collaboration, R. Brandelik, et al.: Phys. Lett. 70B, 125 (1977); 70B, 387 (1977)
161 DELCO Collaboration, J. Kirkby: Rapporteur talk, 1977 Hamburg Conference, 3
162 J.M. Feller, et al.: Phys. Rev. Lett. 40, 274 (1978)

    A. Barbaro-Galtieri: Rapporteur talk, 1977 Hamburg Conference, 21
163 J.M. Feller, et al.: Phys. Rev. Lett. 40, 1677 (1978)
164 CERN-Dortmund-Heidelberg-Saclay Collaboration,
    F. Dydak: Private communication
165 T.D. Cheng, L.F. Li: Phys. Rev. Lett. 38, 381 (1977)
166 P. Fayet: Nucl. Phys. B78, 14 (1974)
167 M.L. Perl, et al.: Phys. Rev. Lett. 35, 1489 (1975); Phys. Lett. 63B, 366 (1976)
168 M.L. Perl: Proceedings of the XII Rencontre de Moriond, Flaine, Vol. 1, 75 1977;
    1977 Hamburg Conference, 145
    G. Flügge: Z. Phys. C 1, 121 (1979)
169 DASP Collaboration, R. Brandelik, et al.: Phys. Lett. 73B, 109 (1978)
170 H.B. Thacker, J.J. Sakurai: Phys. Lett. 36B, 103 (1971)
    Y.S. Tsai: Phys. Rev. D4, 2821 (1971)
    J.D. Bjorken, C.H. Llewellyn-Smith: Phys. Rev. D7, 887 (1973)
    K. Fujikawa, N. Kawamoto: Phys. Rev. D14, 59 (1976)
    Y.I. Azimov, L.L. Frankfurt, V.A. Khoze: Zh. Eksperim. I Teor. Fiz. 30, 63 (1977)
171 N. Kawamoto, A.I. Sanda: Phys. Lett. 76B, 446 (1978)
    F.J. Gilman, D.H. Miller: Phys. Rev. D17, 1846 (1978)
    Y.S. Tsai: Private communication
172 C.H. Llewellyn-Smith: Proc. Royal Soc. A355, 585 (1977)
173 M. Bernardini, D. Bollini, P.L. Brunini, E. Fiorentino, T. Massam, L. Monari,
    F. Palmonari, F. Rimondi, A. Zichichi: Nuovo Cimento 17, 383 (1973)
174 S. Orito, R. Visentin, F. Ceradini, M. Conversi, S. d'Angelo, L. Paoluzi,
    R. Santonico: Phys. Lett. 48B, 165 (1974)
175 M.L. Perl, et al.: Phys. Lett. 70B, 487 (1977); 63B, 466 (1976)
    G.J. Feldman, et al.: Phys. Rev. Lett. 38, 117 (1977)
    F.B. Heile, et al.: Nucl. Phys. B138, 187 (1978)
176 Contribution to the Hamburg Conference, see H. Sadrozinski: Rapporteur talk,
    1977 Hamburg Conference, 47
177 M. Cavalli-Sforza, et al.: Phys. Rev. Lett. 36, 558 (1976)
178 PLUTO Collaboration, J. Burmester, et al.: Phys. Lett. 68B, 301 (1977)
179 PLUTO Collaboration, J. Burmester, et al.: Phys. Lett. 68B, 297 (1977)
180 A. Barbaro-Galtieri, et al.: Phys. Rev. Lett. 39, 1058 (1977)
181 DASP Collaboration, S. Yamada: 1977 Hamburg Conference, 69
182 J.G. Smith, et al.: Phys. Rev. D18, 1 (1979)
183 W. Bartel, et al.: Phys. Lett. 77B, 331 (1978)
184 W. Bacino, et al.: Phys. Rev. Lett. 41, 13 (1978)
    J, Kirz: Data reported at the 1978 Tokyo Conference
185 G.J. Feldman, et al.: Phys. Rev. Lett. 38, 117 (1977)
186 F. Gutbrod, Z. Rek: Z. Phys. C 1, 171 (1979)
187 PLUTO Collaboration, G. Alexander, et al.: Phys. Lett. 81B, 84 (1979)
188 W. Alles, Ch. Boyer, A.J. Buras: CERN-TH-220 (1977)
189 W. Bacino, et al.: SLAC-PUB-2249 (1979)
190 T. Kinoshita, A. Sirlin: Phys. Rev. 108, 844 (1957)
191 G.J. Feldman: International Conference on Neutrino Physics, Purdue University
    1978; also SLAC-PUB-2138
192 PLUTO Collaboration, G. Alexander, et al.: Phys. Lett. 78B, 162 (1978)
193 D. Hitlin: Talk at the 1978 Tokyo Conference
194 G.J. Feldman: Rapporteur talk, 1978 Tokyo Conference
195 PLUTO Collaboration, G. Alexander, et al.: Phys. Lett. 73B, 99 (1978)
196 W. Wagner: Thesis; Technische Hochschule Aachen report PITHA 7801 (1978)
    G. Knies: Proc. of the Sixth Trieste Conference on Particle Physics, 1978,
    ICTP report IC/78/76, 142
197 M.L. Perl: Rapporteur talk, 1977 Hamburg Conference, 145
198 R.B. Palmer: Results presented at the 13th Rencontre de Moriond (1978),
    Vol. II, Gauge Theories and Leptons, ed. by J. Tran Thanh Van, 361
199 S.W. Herb, et al.: Phys. Rev. Lett. 39, 252 (1977)
    W.R. Innes, et al.: Phys. Rev. Lett. 39, 1240 (1977)
    L.M. Lederman: Rapporteur talk, 1978 Tokyo Conference
200 PLUTO Collaboration, Ch. Berger, et al.: Phys. Lett. 76B, 243 (1978)

262

201 C.W. Darden, et al.: Phys. Lett. 76B, 246 (1978)
202 J.K. Bienlein, et al.: Phys. Lett. 78B, 360 (1978)
203 PLUTO Collaboration, Ch. Berger, et al.: DESY report 79/19 (1979)
204 C.W. Darden, et al.: Phys. Lett. 78B, 364 (1978); 80B, 419 (1979)
205 G. Heinzelmann: Data presented at the 1978 Tokyo Conference
206 G. Flügge: Rapporteur talk, 1978 Tokyo Conference
     H. Spitzer: Talk given at the 1978 Kyoto Conference
207 J.L. Rosner, C. Quigg, H.B. Thacker: Phys. Lett. 74B, 350 (1978)
208 S.D. Drell, D.J. Levy, T.M. Yan: Phys. Rev. 187, 2159 (1969); D1, 1035, 1617,
     2402 (1970)
209 R. Gatto, P. Menotti, I. Vendramin: Nuovo Cimento Lett. 4, 79 (1972)
     R. Gatto, G. Preparata: Nucl. Phys. B47, 313 (1972)
210 V.N. Gribov, L.N. Lipatov: Phys. Lett. 37B, 78 (1971)
211 D.J. Bjorken, S.J. Brodsky: Phys. Rev. D1, 1416 (1970)
212 J. Engels, H. Satz, K. Schilling: Nuovo Cimento 17A, 535 (1973)
213 Taken from A.J. Sadoff, et al.: Phys. Rev. Lett. 32, 955 (1974)
214 G. Hanson: Data presented at the 1976 Tbilisi Conference, 81
215 U. Timm: Proc. of the 1977 European Conference on Particle Physics, Budapest, 727
     A. Bäcker: Thesis, Gesamthochschule Siegen (1977); Internal report DESY F33-77/03
216 DASP Collaboration, R. Brandelik, et al.: Phys. Lett. 67B, 358 (1977); Nucl.
     Phys. B148, 189 (1979)
217 British Scandinavian Collaboration, B. Alper, et al.: Nucl. Phys. B78, 19 (1975)
218 E.M. Riordan, et al.: SLAC-PUB-1634 (1975)
219 G. Schierholz, M.G. Schmidt: Nucl. Phys. B101, 429 (1975)
220 PLUTO Collaboration, G. Knies: Rapporteur talk, 1977 Hamburg Conference, 93
221 P.A. Rapidis, et al.: SLAC-PUB 2184 (1978)
222 A. Barbaro-Galtieri: Lectures presented at the XVI International School of Sub-
     nuclear Physics, Erice 1978, LBL report 8537 (1978)
223 R.F. Schwitters: Rapporteur talk, 1975 Stanford Conference, 5
     R.F. Schwitters, et al.: Phys. Rev. Lett. 35, 1320 (1975)
     G. Hanson, et al.: Phys. Rev. Lett. 35, 1609 (1975)
224 R.F. Schwitters: Rapporteur talk, 1976 Tbilisi Conference, B34
     G. Hanson: Talk, 1976 Tbilisi Conference; SLAC-PUB-1814 (1976)
225 G. Kramer, T.F. Walsh: Z. Phys. 263, 316 (1973)
226 G.G. Hanson: Results presented at the 13th Rencontre de Moriond (1978) Vol. II,
     Gauge Theories and Leptons, ed. by J. Tran Thanh Van; SLAC-PUB-2118
227 J.D. Bjorken, S.J. Brodsky: Phys. Rev. D1, 1416 (1970)
228 K. Koller, T.F. Walsh: Phys. Lett. B72, 227 (1977); B73, 504 (1978); Nucl. Phys.
     B140, 449 (1978)
     T.A. De Grand, et al.: Phys. Rev. D16, 3251 (1977)
     S.J. Brodsky, et al.: Phys. Lett. B73, 203 (1978)
229 See, e.g., H. Fritzsch: Schladming Lectures 1978, Acta Phys. Austriaca, Supnl.
     XIX, 249
     H. Fritzsch, K.H. Streng: Phys. Lett. B74, 90 (1978)
230 K. Hagiwara: Nucl. Phys. B137, 164 (1978)
     A. de Rujula, et al.: Nucl. Phys. B138, 387 (1978)
231 S. Brandt, et al.: Phys. Lett. 12, 57 (1964)
     E. Fahri: Phys. Rev. Lett. 39, 1587 (1977)
232 PLUTO Collaboration, Ch. Berger, et al.: Phys. Lett. B78, 176 (1978)
233 H. Georgi, M. Machacek: Phys. Rev. Lett. 39, 1237 (1977)
234 PLUTO Collaboration, Ch. Berger, et al.: DESY report 78/71 (1978), to be
     published in Phys. Lett.
235 R.D. Field, R.P. Feynman: Nucl. Phys. B136, 1 (1978)
236 C.W. Darden, et al.: Paper contributed to the 1978 Tokyo Conference; Internal
     report DESY F15-78/1 (1978)
237 PETRA Proposal, DESY Hamburg (1976)
238 PLUTO Collaboration, Ch. Berger, et al.: Phys. Lett. 81B, 410 (1979)
239 MARK-J Collaboration, D. Barber, et al.: MIT-LNS report 100 (1979)
240 TASSO Collaboration, R. Brandelik, et al.: Phys. Lett. 83B, 261 (1979)

E. Amaldi, S. Fubini, G. Furlan

Pion-Electroproduction

Electroproduction at Low Energy and
Hadron Form Factors

1979. 47 figures, 13 tables. VIII, 162 pages
(Springer Tracts in Modern Physics,
Volume 83)
ISBN 3-540-08998-5

Contents:
Introduction. – Quantities of Physical
Interest. – Theoretical Approaches. –
Main Features of the Experiments, Pre-
liminary Tests and Measurements. –
Hadron Form Factors from Electropro-
duction. – Other Developments. –
Appendices. – References.

Modern Three-Hadron Physics

Editor: A. W. Thomas

1977. 30 figures. XI, 250 pages
(Topics in Current Physics, Volume 2)
ISBN 3-540-07950-5

Contents:
I. R. Afnan, A. W. Thomas: Fundamentals
of Three-Body Scattering Theory. –
L. R. Dodd: Analytic Structure of On-
Shell Three-Body Amplitudes. – *R. D.
Amado:* Theory of Three-Body Final
States. – *D. D. Brayshaw:* The Boundary
Condition Method. – *R. Aaron:* A Rela-
tivistic Three-Body Theory. – *E. F. Redish:*
Applications of Three-Body Methods to
Many-Body Hadronic Systems.

Collective Ion Acceleration

1979. 63 figures, 9 tables. VII, 231 pages
(Springer Tracts in Modern Physics,
Volume 84)
ISBN 3-540-09066-5

Contents:
C. L. Olson: Collective Ion Acceleration
with Linear Electron Beams. –
U. Schumacher: Collective Ion Acceleration with Electron Rings.

Elementary Particle Physics

1976. 37 figures. VI, 145 pages
(Springer Tracts in Modern Physics,
Volume 79)
ISBN 3-540-07778-2

Contents:
H. Rollnik, P. Stichel: Compton Scattering. – *E. Paul:* Status of Interference Experiments with Neutral Kaons.

Springer-Verlag
Berlin
Heidelberg
New York

Zeitschrift für Physik C

Particles and Fields

Europhysics Journal

In cooperation with the Deutsche Physikalische Gesellschaft

Editors in Chief:
G. Kramer, II. Institut für Theoretische Physik, Luruper
Chaussee 149, 2000 Hamburg 52, FRG
H. Satz, Fakultät für Physik, Universität Bielefeld,
Postfach 86 40, 4800 Bielefeld 1, FRG

Editors: **K. Fujikawa,** Tokyo; **K. Gottfried,** Cornell; **K. Kajantie,**
Helsinki; **A. Krzywicki,** Orsay; **P. V. Landshoff,** Cambridge;
J. J. Sakurai, UCLA; **P. Söding,** DESY; **B. Stech,** Heidelberg;
J. Steinberger, CERN

Aims and Scope:
Zeitschrift für Physik C, Particles and Fields is devoted to
the experimental and theoretical investigation of elementary
particles. In view of the steadily growing interplay of theory
and experiment in this field, particular emphasis is given to a
clear and complete presentation of research.

The topics covered include: strong, electromagnetic, and weak
interactions of elementary particles, the constituent structure
of elementary particles, interactions and classification of con-
stituents, and symmetry and unification schemes of different
interactions.

Zeitschrift für Physik appears in three Parts: **A:** Atoms and
Nuclei, **B:** Condensed Matter and Quanta, **C:** Particles and
Fields. Each part may be ordered separately.

Coodinating editor for Zeitschrift für Physik, parts A, B and C:
O. Haxel, Heidelberg.

Springer-Verlag
Berlin
Heidelberg
New York

Please ask for your sample copy.

Send your order or request to your bookseller or
Springer-Verlag
Journal Promotion Department, P.O. Box 105 280
D-6900 Heidelberg, FRG